MCAS BIOLOGY

Michael F. Renna
Principal
Preparatory Academy for Writers MS/HS
Queens, New York
Former Assistant Principal, Supervision Science
Hillcrest High School, Queens, New York

Carl M. Raab
Science Consultant
NYC Department of Education
Former Assistant Principal, Supervision Science
Fort Hamilton High School, Brooklyn, New York

Amsco School Publications, Inc.
315 Hudson Street, New York, N.Y. 10013

The publisher would like to thank the following people who reviewed the manuscript:

Wendell Cerne
Director, K-12 Science & Health
Waltham Public Schools
Waltham, Mass.

Susan P. Mooney
Biology Teacher
Haverhill High School
Haverhill, Mass.

Laurie Pancoast
Biology Teacher
Amesbury High School
Amesbury, Mass.

Cover Photo: Bullfrog (*Rana catesbeiana*), © Stephen Maka / Photex / Corbis
Cover Design: Meghan J. Shupe
Text Design: Howard Petlack /A Good Thing, Inc.
Production Manager: Joe Campanella
Editor: Carol Davidson Hagarty
Composition: Nesbitt Graphics, Inc.
Artwork: Hadel Studio

Please visit our Web site at: ***www.amscopub.com***

When ordering this book, please specify:
either **R 100 W** *or* **MCAS BIOLOGY**

ISBN: 978-1-56765-941-2

Printed in the United States of America

1 2 3 4 5 6 7 8 9 10   15 14 13 12 11 10 09

Contents

To The Student

About This Book

All Massachusetts high school students enrolled in biology must take the MCAS Biology test. The purpose of this book is to help you prepare for that examination. The book provides a comprehensive review of the main content areas that are covered in a biology course and which are tested on the biology exam.

The nine chapters of this book are organized into the following content areas, based on the learning standards in the Biology content strand of the Massachusetts *Science and Technology/Engineering Curriculum Framework* (2006): cell biology and biochemistry; life processes; human anatomy and physiology; reproduction and development; genetics and heredity; evolution and biodiversity; ecology and environment; and scientific inquiry and laboratory skills. Each chapter opens with the MCAS Biology Standards and Broad Concepts that are covered in that chapter. Numerous question sets that contain MCAS-type multiple-choice and open-response questions are placed after major topic sections in each chapter. These question sets give you the opportunity to assess and reinforce your understanding of the biological facts and concepts covered.

A Diagnostic Test and Diagnostic Checklist follow this introduction. The Diagnostic Test is modeled on the actual MCAS Biology test. It has the same number and types of multiple-choice and open-response questions as are on the MCAS test. The questions were developed to be similar to those you will see on the actual exam. The Diagnostic Checklist will help you determine your areas of strength and areas of weakness in the biology content and skills areas. This will help you focus your attention on those topics that need more review.

Numerous figures and tables are presented throughout the text to help illustrate important concepts. In addition, there are two complete Practice Tests at the back of the book. Again, the number, type, and format of the Practice Test questions are the same as those you will encounter on the MCAS Biology test. You will take these tests after you have completed this book, to prepare for the MCAS exam.

Throughout the book, important science terms are defined in text and printed in **boldface** type; these words are also presented with full definitions in the Glossary. Many other scientific terms that are less important to memorize are printed in *italics* type for emphasis; these words do not appear in the Glossary, but are listed in the Index.

About The MCAS Biology Test

Later this year, you will take the Massachusetts Comprehensive Assessment System (MCAS) exam in Biology. According to a statement issued by the Massachusetts DOE, the "high school MCAS Biology test [is] based on learning standards in the Biology content strand of the Massachusetts *Science and Technology/Engineering Curriculum Framework* (2006)."

The test is given in two parts (Session 1 and Session 2), which are administered on consecutive days. The test generally consists of 45 questions: 40 multiple-choice and 5 open-response questions. Each session contains both multiple-choice and open-response question types. Each multiple-choice question has four answer choices: A, B, C, or D. You will be asked to choose the **best** or **most likely** choice for your answer.

The open-response questions usually have more than one part (labeled a, b, and c). There is typically a paragraph to read, or a series of diagrams to analyze, and then two or three related questions (i.e., parts a, b, and c) on the material. You will be provided space in your Student Answer Booklet for your answers. The percent of each topic covered in the exam is broken down as follows (approximately): Biochemistry & Cell Biology, 25%; Genetics, 20%; Anatomy & Physiology, 15%; Ecology, 20%; and Evolution & Biodiversity, 20%.

The MCAS Biology test was designed so that it could be taken without the aid of a calculator; however, students are allowed to have calculators with them during the test. More information about the MCAS Biology test and curriculum framework is available at: www.doe.mass.edu/frameworks/current.html.

Diagnostic Test

SESSION 1

DIRECTIONS

This session contains twenty-one multiple-choice questions and two open-response questions. You may work out solutions to multiple-choice questions in your notebook.

1. A characteristic of a DNA molecule that is **not** a characteristic of a protein molecule is that the DNA molecule
   A. can replicate itself.
   B. can be very large.
   C. is found in nuclei.
   D. is composed of subunits.

2. Two closely related species of birds live in the same tree. Species *A* feeds on ants and termites, whereas species *B* feeds on caterpillars. Which of the following provides the **best** explanation for why the two species can coexist successfully?
   A. Each occupies a different niche.
   B. Species *A* and *B* can interbreed.
   C. They reproduce by different methods.
   D. All birds compete for food.

3. The production of energy-rich ATP molecules in a cell is the direct result of
   A. recycling light energy to be used in the process of photosynthesis
   B. releasing the stored energy of organic compounds through respiration
   C. breaking down starch molecules by the process of digestion
   D. copying information during the process of protein synthesis

4. In the following diagram, what does *X* most likely represent?

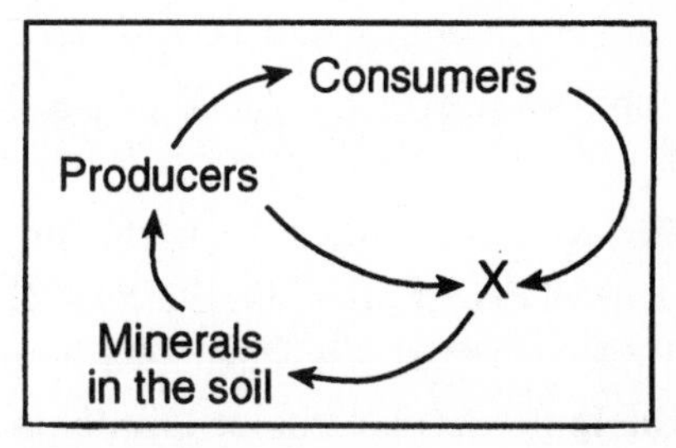

   A. autotrophs
   B. decomposers
   C. herbivores
   D. carnivores

5. Ten breeding pairs of rabbits are introduced onto an island having no natural predators and a good supply of food and freshwater. The rabbit population will **most likely**
   A. remain constant due to equal birth and death rates.
   B. die out due to an increase in the mutation rate.
   C. increase until it exceeds carrying capacity.
   D. decrease and then increase indefinitely.

6. Which statement **best** expresses the relationship between the three structures represented in the diagram below?

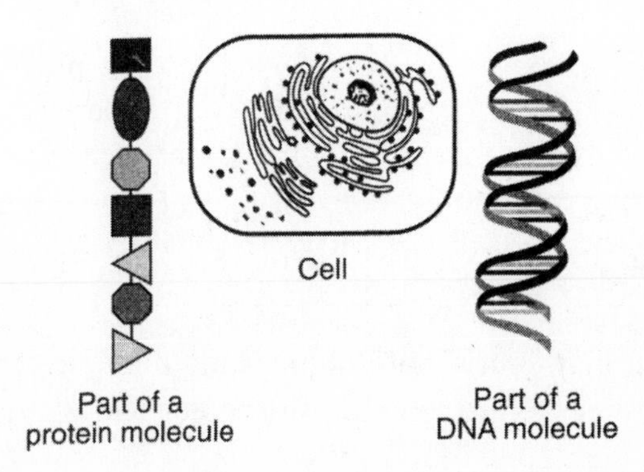

A. DNA is produced from protein absorbed by the cell.
B. Protein is composed of DNA that is produced in the cell.
C. DNA controls the production of protein in the cell.
D. Cells make DNA by digesting protein.

7 All the cells of an organism are constantly engaged in different chemical reactions. This fact is supported by the presence in each cell of thousands of different types of

A. enzymes
B. chloroplasts
C. nuclei
D. organelles

The following section focuses on the presence of similar types of cytochrome c in the cells of different organisms. Read the information below and use it to answer the four multiple-choice questions and one open-response question that follow.

Cytochrome c is an enzyme located in the mitochondria of many types of cells. The number of differences in the amino acid sequences of cytochrome c from different species is compared to human cytochrome c in the data table below.

Differences in Amino Acid Sequences

Organism	*Number of Differences in Cytochrome c Compared to Humans*
Tuna	21
Mold	48
Moth	31
Dog	11
Horse	12
Chicken	13
Monkey	1

You may work out solutions to multiple-choice questions 8 through 11 in your notebook.

8 Of the organisms listed below, which one has a DNA code for cytochrome c that is **most** similar to that of a human?

A. tuna
B. chicken
C. moth
D. dog

9 The fact that all of these organisms contain cytochrome c could lead to which inference?

A. Cytochrome c is essential for the reproduction of all organisms.
B. These organisms all evolved from an ancestor that produced cytochrome c.
C. Mutations in genes that code for cytochrome c always occur during DNA replication.
D. Only heterotrophic organisms can make cytochrome c.

10 Cytochrome c is **most likely** a

A. type of protein molecule.
B. material that contains genes.
C. carbohydrate that is absorbed by cells.
D. component of the plasma membrane.

11 The **best** explanation for the fact that cytochrome c is found in the mitochondria of many different organisms is that this enzyme is involved in the process of cellular

A. digestion
B. division
C. respiration
D. excretion

Question 12 is an open-response question.

- **BE SURE TO ANSWER AND LABEL ALL PARTS OF THE QUESTION.**
- **Show all your work (diagrams, tables, or computations).**
- **If you do the work in your head, explain in writing how you did the work.**

12 Biochemical similarities often reflect evolutionary relationships between organisms.

a. According to the table, the monkey has only one difference from humans in the amino acid sequence of its cytochrome c. Based on your knowledge of biology, give the **most likely** explanation for this finding.

b. Construct a bar graph that shows the differences in cytochrome c compared to humans, going from the least to the most differences. What correlation between the numbers and the relationship to humans is evident from your bar graph?

You may work out solutions to multiple-choice questions 13 through 22 in your notebook.

Base your answer to question 13 on the diagram below, which represents nutritional relationships among organisms.

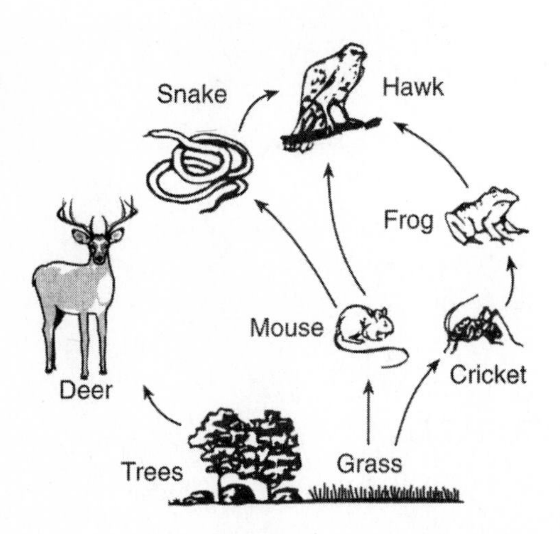

13 Which of the following is most likely to lead to an **increase** in the mouse population over time?

A. an increase in the frog and hawk populations

B. a decrease in the snake and hawk populations

C. an increase in the number of deer and snakes

D. a decrease in the amount of available sunlight

14 Which process uses carbon dioxide molecules?

A. cellular respiration

B. asexual reproduction

C. active transport

D. autotrophic nutrition

15 A mutation occurs in the liver cells of a certain field mouse. Which statement concerning the spread of this mutation throughout the mouse population is correct?

A. It will spread because it is a beneficial trait.

B. It will spread because it is a dominant gene.

C. It will not spread because it is a recessive gene.

D. It will not spread because it is not in a gamete.

16 The diagram below illustrates the movement of substances in a process that manufactures energy for many organisms. This process occurs within a plant cell's

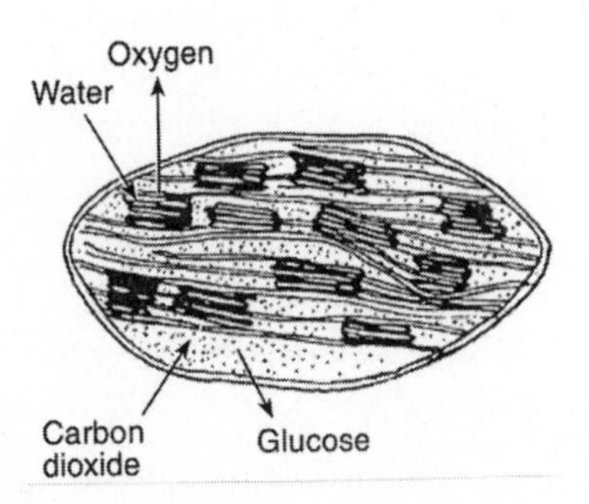

A. chloroplasts

B. ribosomes

C. mitochondria

D. vacuoles

17 When a planarian (worm) is cut in half, each half usually grows back into a complete worm over time. This situation most closely resembles

A. asexual reproduction in which a mutation has occurred
B. sexual reproduction in which each half represents one parent
C. asexual reproduction of one single-celled organism
D. sexual reproduction of two single-celled organisms

18 The aquatic plant represented in the diagram below was exposed to light for several hours. Which gas would **most likely** be found in the greatest amount in the bubbles?

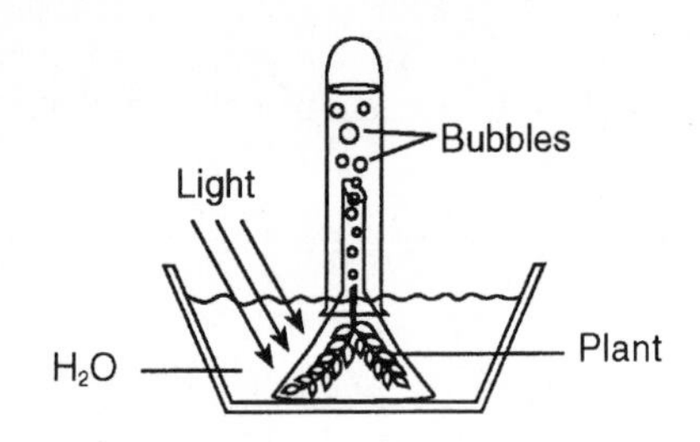

A. oxygen
B. ozone
C. nitrogen
D. carbon dioxide

19 In one variety of corn, the kernels turn red when exposed to sunlight. In the absence of sunlight, the kernels remain yellow. Based on this information, which of the following **best** explains what determines the color of these corn kernels?

A. the process of selective breeding
B. the corn's rate of photosynthesis
C. the effect of environment on gene expression
D. the chemical composition of the soil

20 Which sequence correctly shows the increasingly complex levels of organization in multicellular organisms?

A. organelle → cell → tissue → organ → organ system → organism
B. cell → organelle → tissue → organ → organ system → organism
C. organelle → tissue → cell → organ → organ system → organism
D. cell → organism → organ system → organ → tissue → organelle

21 The evolutionary pathways of seven living species are shown in the diagram below. Which two species are likely to have the **most** similar DNA base sequences?

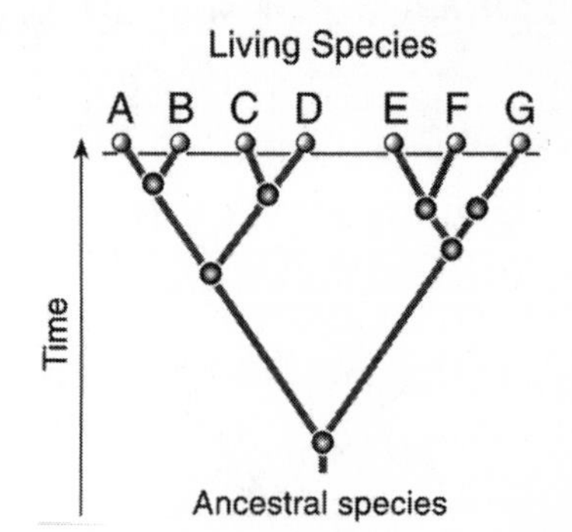

A. *D* and *E*
B. *B* and *C*
C. *E* and *G*
D. *C* and *D*

22 Which statement is true of **both** mitosis and meiosis?

A. Both are involved in asexual reproduction.
B. Both occur only in the reproductive cells.
C. The final number of chromosomes is reduced by half.
D. DNA replication occurs before division of the nucleus.

Question 23 is an open-response question.

- **BE SURE TO ANSWER AND LABEL ALL PARTS OF THE QUESTION.**
- **Show all your work (diagrams, tables, or computations) .**
- **If you do the work in your head, explain in writing how you did the work.**

23 The diagram below illustrates some steps in genetic engineering.

a. State **one** way that enzymes are used in step 2.

b. Describe the result of step 3 (in terms of DNA).

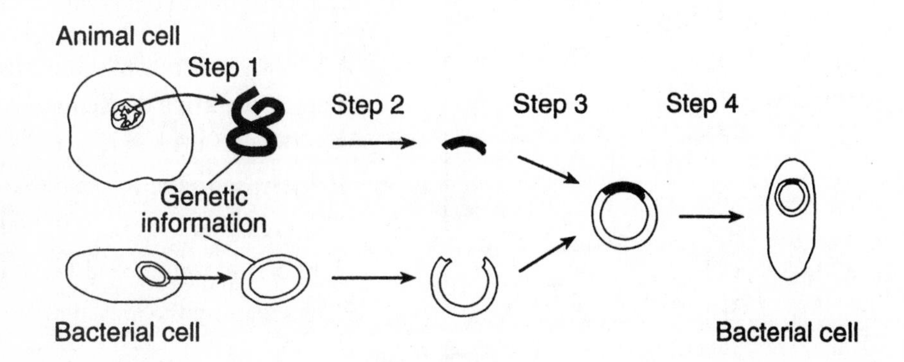

SESSION 2

DIRECTIONS

This session contains nineteen multiple-choice questions and three open-response questions. You may work out solutions to multiple-choice questions in your notebook.

24 The diagram below represents a portion of an organic molecule. This molecule controls cellular activity by directing the synthesis of

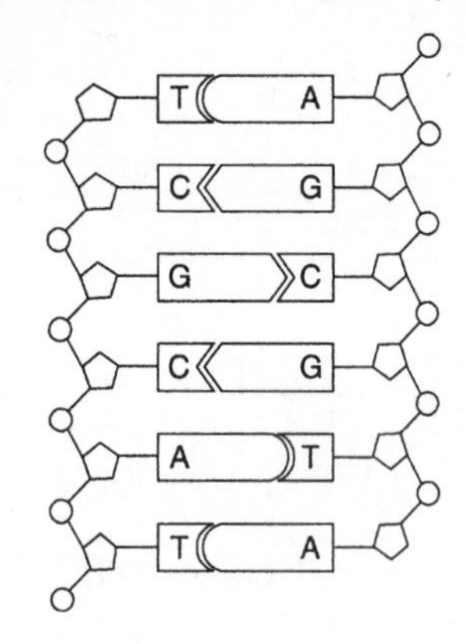

A. carbohydrates
B. lipids
C. minerals
D. proteins

25 Meiosis and fertilization are important for the survival of many species. Which of the following statements describes the **result** of these two processes that makes them so important?

A. a large number of gametes
B. increasingly complex organisms
C. genetic variability among offspring
D. the cloning of superior offspring

26 The diagram below, which represents a series of reactions that can occur in an organism, illustrates the relationship between

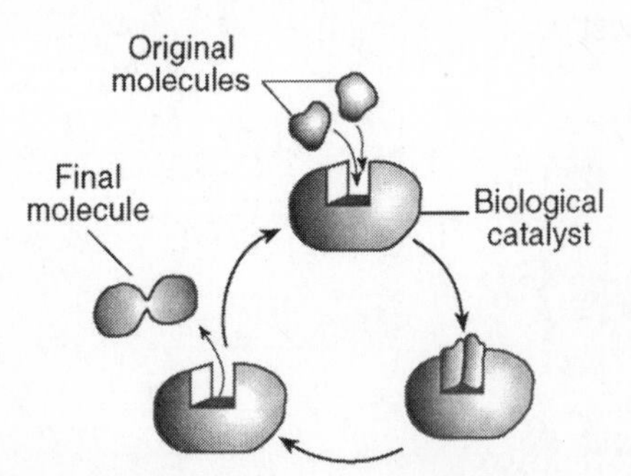

A. enzymes and synthesis
B. amino acids and glucose
C. antigens and immunity
D. ribosomes and sugars

27 A forest is cut down and replaced by a cornfield. A **negative** consequence of this practice would be a (an)

A. increase in the oxygen that is released
B. increase in the size of local predators
C. decrease in the biodiversity of the area
D. decrease in the amount of soil washed away

28 An investigation was carried out and the results are shown below. Substance *X* resulted from a metabolic process that produces ATP in yeast (a single-celled fungus). Which statement correctly identifies substance *X*?

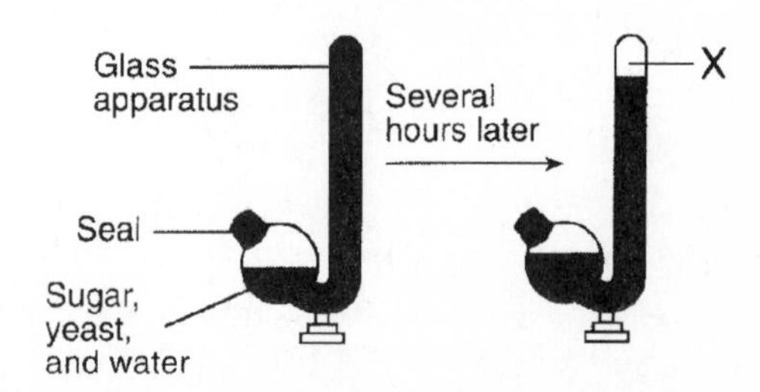

A. It is oxygen released by protein synthesis.
B. It is glucose that was produced in photosynthesis.
C. It is starch that was produced during digestion.
D. It is carbon dioxide released by respiration.

29 Which statement **best** describes the relationship between cells, DNA, and proteins?

A. Cells contain DNA, which controls the production of proteins.
B. DNA is composed of proteins, which carry coded information about cell functions.

C. Proteins are used to produce cells, which link amino acids together into DNA.
D. Cells are linked together by proteins to make different kinds of DNA molecules.

30 The theory of biological evolution by natural selection includes which concept?
A. Species of organisms found on Earth today have adaptations that are not always found in earlier species.
B. Fossils are the remains of present-day species and were all formed at the same time.
C. Individuals may acquire physical characteristics after birth and pass these acquired characteristics on to their offspring.
D. The smallest organisms are always eliminated by the larger organisms within the ecosystem.

Base your answer to questions 31 on the information below and on your knowledge of biology.

Analysis of a sample taken from a pond showed variety in both the number and types of organisms present. The data collected are in the table below.

Types of Organisms	*Number Present in Sample*
Bass (fish)	Two
Frogs (amphibian)	Forty
Insect larvae	Hundreds
Phytoplankton	Thousands

31 If the frogs feed on insect larvae, what is the role of the frogs in this pond ecosystem?
A. herbivore
B. consumer
C. parasite
D. host

Question 32 is an open-response question.

- **BE SURE TO ANSWER AND LABEL ALL PARTS OF THE QUESTION.**
- **Show all your work (diagrams, tables, or computations).**
- **If you do the work in your head, explain in writing how you did the work.**

32 A population of gray squirrels lived in the trees surrounding four houses in a city. The houses and trees were removed, and a tall office building was constructed in their place. Some of the squirrels were able to survive by relocating to the trees in a park nearby.
a. State **one** specific way the relocated squirrels would most likely interact with a gray squirrel population that has lived in the park for many years.
b. State **one** specific way the relocated squirrels will change an abiotic factor in the park ecosystem.
c. State **one** specific natural factor in the park ecosystem that will limit the growth of the squirrel population; explain your answer.

You may work out solutions to multiple-choice questions 33 through 43 in your notebook.

33 The graph below shows the number of birds in a population over time. Which statement best explains section *X* of the graph?

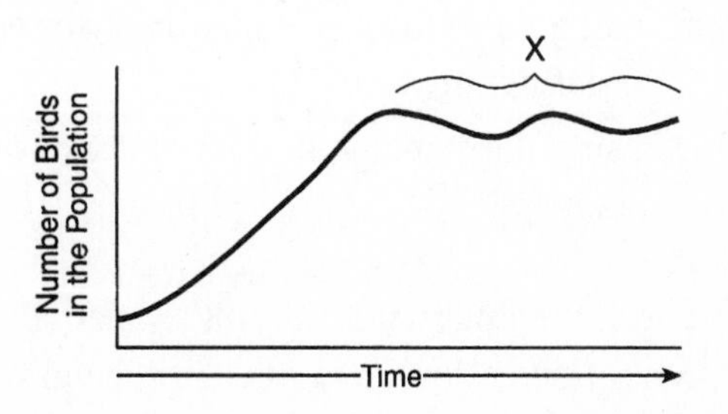

A. Interbreeding between members of the population increased their mutation rate.
B. An increase in the bird population caused a decrease in the predator population.
C. The population reached the environment's carrying capacity and its growth rate stabilized.
D. Another species came to the area and provided food for the birds.

Base your answer to question 34 on the following information and diagram.

An experiment was carried out to determine which mouthwash was most effective against bacteria commonly found in the mouth. Four paper discs were each dipped into a different brand of mouthwash. The discs were then placed onto the surface of a culture plate that contained food, moisture, and bacteria commonly found in the mouth. The diagram below shows the growth of bacteria on the plate after 24 hours.

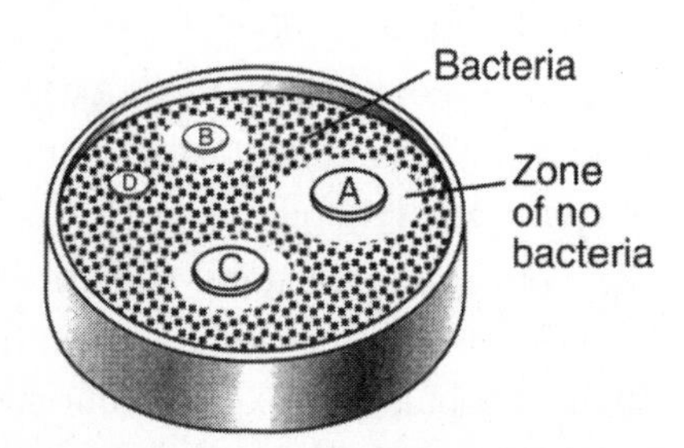

34 Which change in procedure would have improved the experiment?

A. using a smaller plate with less food and moisture
B. using bacteria from many habitats other than the mouth
C. using the same size paper discs for each mouthwash
D. using the same type of mouthwash on each disc

35 The following graph shows the relative concentrations of different ions inside and outside of an animal cell. Which process is directly responsible for the net movement of K^+ and Mg^{++} **into** the animal cell?

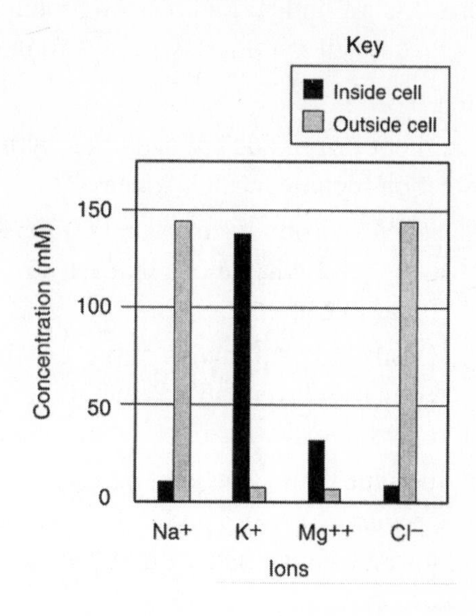

A. electrophoresis
B. active transport
C. diffusion
D. circulation

36 The information that controls the production of proteins in a cell must pass from the nucleus to the

A. cell membrane
B. mitochondria
C. chloroplasts
D. ribosomes

37 When a white insect lands on the bark of a white birch tree, its light color has a high adaptive value. If the birch trees become covered with black soot, then the white color of this type of insect would most likely

A. retain its adaptive value
B. increase in adaptive value
C. mutate to a darker color
D. decrease in adaptive value

38 What is the volume of the liquid in the graduated cylinder shown on page 9?

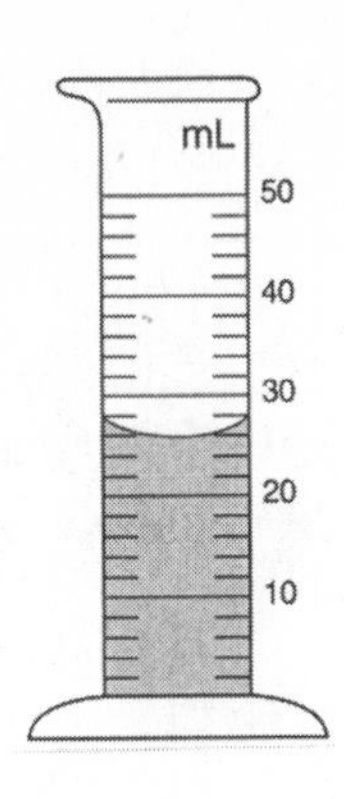

A. 22 mL
B. 24 mL
C. 26 mL
D. 28 mL

39 Strawberries can reproduce by means of runners, which are stems that grow horizontally above the ground. At the point where a runner touches the ground, a new plant develops. Which of the following statements **best** explains why the new plant is genetically identical to the parent plant?

A. The new plant was produced sexually.
B. Nuclei traveled through the runner to fertilize it.
C. The new plant was produced asexually.
D. There were no other strawberry plants to fertilize it.

40 The graph below shows the effect of temperature on an enzyme's action on a protein. Which change would **not** affect the enzyme's rate of action?

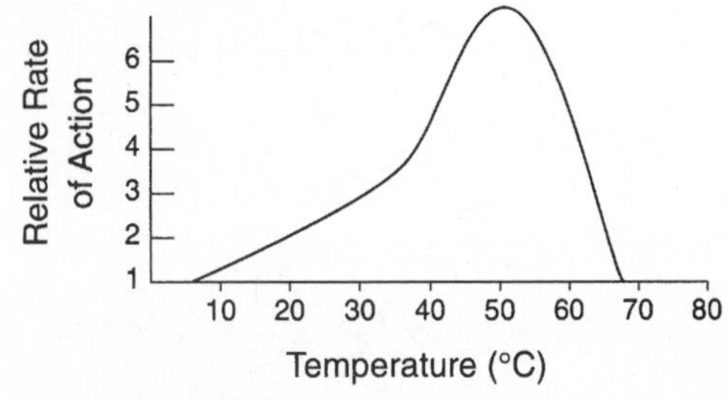

A. the addition of cold water when the reaction is at 50°C
B. an increase in temperature from 70°C to 80°C
C. the removal of the protein when the reaction is at 30°C
D. a decrease in temperature from 40°C to 10°C

41 A scientist is planning an experiment to test the effect of heat on the function of a certain enzyme. Which would **not** be an appropriate first step?

A. doing research about enzymes in a library
B. having discussions with other scientists
C. completing a data table of expected results
D. using what is already known about the enzyme

42 The diagram below represents an energy pyramid. At each successive level, going from *A* to *D*, what happens to the amount of available energy?

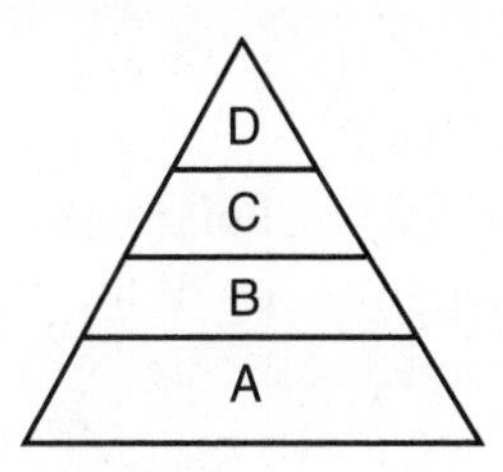

A. it increases, only
B. it decreases, only
C. it increases, then decreases
D. it remains the same

43 Leaves contain openings known as stomata, which are opened and closed by specialized cells to allow gas exchange with the outside environment. Which phrase **best** represents the net flow of gases involved in photosynthesis that would go through a leaf's stomata on a sunny day?

A. carbon dioxide moves in; oxygen moves out
B. oxygen moves in; nitrogen moves out
C. carbon dioxide moves in; ozone moves out
D. ozone moves in; carbon dioxide moves out

Questions 44 and 45 are open-response questions.

- **BE SURE TO ANSWER AND LABEL ALL PARTS OF EACH QUESTION.**
- **Show all your work (diagrams, tables, or computations).**
- **If you do the work in your head, explain in writing how you did the work.**

Base your answer to question 44 on the following chemical equation and on your knowledge of biology.

$$CO_2 + H_2O \xrightarrow[\text{CHLOROPHYLL}]{\text{LIGHT ENERGY}} C_6H_{12}O_6 + O_2$$

44 An important life process, carried out by producers, is described by this equation.

a. Identify the process described and the **two** vital products of this reaction.

b. Explain why cellular respiration is basically the **opposite** of the process shown in the equation.

c. Identify the **two** waste products of cellular (aerobic) respiration.

45 The illustration below shows a stable ecosystem that has a variety of producers and consumers.

a. Identify the **two** autotrophic organisms shown in this ecosystem. What role do they fill in the food chain? What is their source of energy?

b. Identify at least **two** heterotrophic organisms in the illustration. Would they have more available energy or less available energy than the autotrophic organisms shown? Support your answer by drawing an energy pyramid that includes all the organisms shown. You may sketch your diagram in your Student Answer Booklet.

c. What important group of organisms is **not** represented in this illustration?

Diagnostic Checklist

Use this checklist to evaluate your Diagnostic Test answers. The checklist is designed so that you can determine which skills you need to improve in your preparation for the MCAS Biology exam. Answer columns are provided so that you can compare your answers to those given by your teacher. If you miss an answer, check the box for that item next to the Correct Answer column. Once you have checked all your answers, note which sections you need to review by referring to the sections listed in the Reporting Category column. Page numbers in the second column provide the location of those sections in this book.

Item No.	Page Nos.	Reporting Category	Standard	Your Answer	Correct Answer*	Check if missed
1.	126	Genetics	3			[ ]
2.	165	Ecology	6			[ ]
3.	41	Biochemistry and Cell Biology	1, 2			[ ]
4.	161	Ecology	6			[ ]
5.	165	Ecology	6			[ ]
6.	126	Genetics	3			[ ]
7.	24	Biochemistry and Cell Biology	1, 2			[ ]
8.	138	Evolution and Biodiversity	5			[ ]
9.	138	Evolution and Biodiversity	5			[ ]
10.	22	Biochemistry and Cell Biology	1, 2			[ ]
11.	22, 24	Biochemistry and Cell Biology	1, 2			[ ]
12.	138	Evolution and Biodiversity	5			
13.	161	Ecology	6			[ ]
14.	31	Biochemistry and Cell Biology	1, 2			[ ]
15.	92-93	Anatomy and Physiology	4			[ ]
16.	30-31	Biochemistry and Cell Biology	1, 2			[ ]
17.	90	Anatomy and Physiology	4			[ ]
18.	31	Biochemistry and Cell Biology	1, 2			[ ]
19.	109	Genetics	3			[ ]
20.	50	Anatomy and Physiology	4			[ ]
21.	138	Evolution and Biodiversity	5			[ ]
22.	94	Anatomy and Physiology	4			[ ]
23.	131	Genetics	3			
24.	125	Genetics	3			[ ]
25.	93	Anatomy and Physiology	4			[ ]
26.	24	Biochemistry and Cell Biology	1, 2			[ ]
27.	170	Ecology	6			[ ]
28.	41	Biochemistry and Cell Biology	1, 2			[ ]
29.	22	Biochemistry and Cell Biology	1, 2			[ ]
30.	142	Evolution and Biodiversity	5			[ ]
31.	160	Ecology	6			[ ]
32.	165	Ecology	6			
33.	165	Ecology	6			[ ]
34.	177	Scientific Inquiry and Lab Skills	SIS			[ ]
35.	37	Biochemistry and Cell Biology	1, 2			[ ]
36.	128	Genetics	3			[ ]

(*continued*)

Item No.	Page Nos.	Reporting Category	Standard	Your Answer	Correct Answer*	Check if missed
37.	145	Evolution and Biodiversity	5			[ ]
38.	185	Scientific Inquiry and Lab Skills	SIS			[ ]
39.	91	Anatomy and Physiology	4			[ ]
40.	177	Scientific Inquiry and Lab Skills	SIS			[ ]
41.	176	Scientific Inquiry and Lab Skills	SIS			[ ]
42.	161	Ecology	6			[ ]
43.	33	Biochemistry and Cell Biology	1, 2			[ ]
44.	31	Biochemistry and Cell Biology	1, 2			
45.	161	Ecology	6			

*Shaded cells represent open-response items.

CHAPTER 1 Cell Biology and Biochemistry

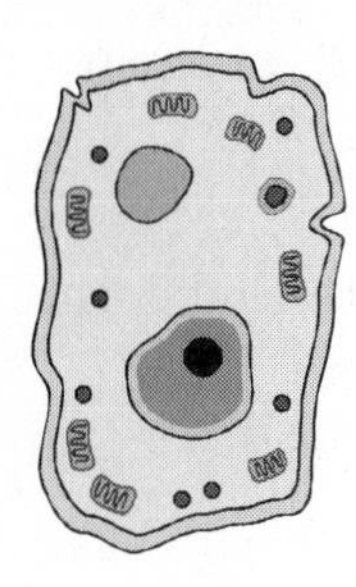

Standards 1.1, 1.2, 1.3 (The Chemistry of Life)

BROAD CONCEPT: Chemical elements form organic molecules that interact to perform the basic functions of life.

Standards 2.1, 2.2, 2.4, 2.5, 2.8 (Cell Biology)

BROAD CONCEPT: Cells have specific structures and functions that make them distinctive. Processes in a cell can be classified broadly as growth, maintenance, and reproduction.

CONCEPT OF LIFE

Scientists have not yet agreed on a single definition of life. Thus, life is often defined in terms of certain activities, or **life functions**, that are performed by all living things.

Life Functions

All living things, or **organisms**, carry on several basic life functions. *Regulation* involves the control and **coordination** of the life functions. The process of **nutrition** provides all the substances that are used by an organism for the growth, repair, and functioning of its **tissues**. Nutrition among heterotrophs includes the activities involved in *ingestion* (obtaining food from the environment), **digestion** (processing food for use by the organism), and *egestion* (removal of solid wastes). *Transport* includes the **circulation**, or distribution, of materials to all the cells of the organism, and the movement of materials across the cell membranes. After the materials are delivered to the cells, the process of **cellular respiration** can occur. Cellular respiration includes the chemical activities that release energy from organic molecules for use by the cells. During respiration, the chemical bonds of **glucose** are broken down, and the energy released is stored in the compound **ATP (adenosine triphosphate)**. An organism uses the energy in ATP to perform its life functions. ATP functions much like a rechargeable battery—when it gets "run down," it is recharged by the breakdown of glucose.

Other chemical reactions are involved in building up **molecules**, rather than breaking them down. During **synthesis** reactions, small molecules combine to form larger ones. These products of synthesis are the raw materials that are used for growth. *Growth* is an increase in size brought about by an increase in cell size and cell number. **Excretion** includes all the activities that are involved in the removal of cellular waste products from an organism. These wastes include **carbon dioxide**, water, salts, and nitrogen-containing compounds. **Reproduction** results in the production of new individuals. However, since each organism has a limited life span, reproduction is necessary more for the survival of a **species** (a group of related organisms) than for the individual organism itself.

Metabolism. All of the chemical activities that an organism must carry on to sustain life are known as its **metabolism**, or *metabolic* activities. Breaking apart glucose molecules to release their energy, and the growth and repair of tissues to maintain a functioning body, are both examples of metabolic activities.

Homeostasis. The maintenance of a stable internal environment, despite changes, or **deviations**, in the external environment is known as **homeostasis**. An example of homeostasis is the maintenance of a constant body temperature in spite of temperature fluctuations in the external environment. The ability to maintain homeostasis is critical to survival. If maintenance of homeostasis fails, the organism becomes ill and, in some cases, may die.

QUESTIONS

MULTIPLE CHOICE

1. The tendency of an organism to maintain a stable internal environment is called A. homeostasis B. nutrition C. reproduction D. synthesis
2. The energy available for use by the cell is obtained from the life function of A. reproduction B. cellular respiration C. transport D. synthesis
3. The chemical process by which complex molecules such as proteins are made from simple molecules is called A. regulation B. respiration C. synthesis D. excretion
4. Which life function includes the absorption and circulation of essential substances throughout an organism? A. transport B. excretion C. ingestion D. nutrition
5. Which term includes all the chemical activities carried on by an organism? A. regulation B. metabolism C. digestion D. respiration
6. Which life activity is *not* required for the survival of an individual organism? A. nutrition B. respiration C. reproduction D. synthesis
7. In a single-celled organism, such as an amoeba, materials are taken from the environment and then moved throughout its cytoplasm. These processes are known as A. absorption and circulation B. food processing and energy release C. energy release and synthesis D. coordination and regulation
8. In an organism, the coordination of the activities that maintain homeostasis in a constantly changing environment is a process known as A. digestion B. regulation C. synthesis D. respiration

OPEN RESPONSE

9. Identify and briefly describe the life function that provides the substances an organism uses for growth and repair of its tissues.
10. Why are living things as different as single-celled amoebas and multicelled humans both considered to be organisms?
11. You are working as a biologist in a laboratory. An unknown specimen is brought in for analysis. Describe the steps you would take to determine if the specimen is a living organism or simply a collection of nonliving molecules.

CELLULAR STRUCTURE OF LIVING THINGS

All living things are composed of **cells**. Some organisms consist of only one cell, while others may consist of billions of cells. Processes that are essential for the survival of an organism are performed by its cells. Most of the cells in a multicellular organism are capable of performing all the life functions independently, as well as with the other cells of the body.

Cells can be classified on the basis of whether they contain certain organelles, and on the types and structures of those organelles. **Prokaryotic cells** are characterized by the lack of a membrane-bound nucleus and organelles. Their genetic material is suspended within the cytoplasm. However, prokaryotic cells do contain ribosomes. Bacteria are examples of prokaryotic cells. **Eukaryotic cells** have a membrane-bound nucleus and most or all of the organelles that will be discussed in the section on cell structure. Aerobic cellular respiration occurs within the mitochondria of eukaryotes. The ribosomes of eukaryotes differ in structure from the ribosomes of prokaryotes. The cells of animals, plants, protists, and fungi are all eukaryotic.

The Cell Theory

The **cell theory,** which is one of the major theories of biology, can be stated as follows: (a) Every organism is made up of one or more **cells**; (b) the cell is the basic unit of structure and function in all living things (for example, cells make, or *synthesize*, proteins and release energy); and (c) all cells come only from preexisting cells (that is, new cells are formed when previously existing cells divide).

Development of the Cell Theory. During the last four centuries, development of the microscope and other technologies have made it possible for biologists to observe and study cells. The cell theory was developed from the work of a number of scientists. First, *Anton van Leeuwenhoek* (1632–1723) made powerful, but fairly simple, microscopes that he used to study living cells; he was the first person to observe sperm cells, bacteria, and protozoa. Then, *Robert Hooke* (1635–1703) made compound microscopes (microscopes with two or more lenses) that he used to observe thin slices of cork; he used the term "cells" to describe the small compartments that made up the cork tissue. In 1831, *Robert Brown* concluded from his studies that all plant cells contain a nucleus. Later, in 1838, *Matthias Schleiden* concluded that all plants are made up of cells, and in 1839, *Theodor Schwann* concluded that all animals are made up of cells. Finally, in 1855, *Rudolph Virchow* concluded that all cells arise only from preexisting cells. All of these scientific ideas and observations formed the basis of the cell theory.

Exceptions to the Cell Theory. Recent discoveries have led scientists to identify several exceptions to the cell theory. For example, mitochondria and chloroplasts, which are cell organelles, contain genetic material (DNA) and can duplicate themselves within living cells. Another exception is the **virus**, which is not a living cell. It consists of an outer coat of protein surrounding a core of DNA or RNA. A virus can reproduce while it is inside a living host cell, but outside the host organism, it shows no signs of life. As such, viruses are not included in any of the six kingdoms of living things. Recently, scientists have discovered tiny particles known as *viroids* and *prions*, which also are not living cells. However, like viruses, these infectious particles have the ability to reproduce themselves and cause diseases when they enter a living host organism.

There are exceptions to the cell theory among multicellular organisms as well. For example, some tissues in multicellular plants and animals do not appear to be made up of clearly identifiable cells. In humans, skeletal muscle tissue does not show distinct boundaries between the cells. In plants, some tissues found in seeds are also not clearly cellular. A group of protists known as *slime molds* have tissues that do not appear to be made up of individual cells, either. Finally, the very first living cells on Earth most likely developed from non-cellular matter (that is, not from preexisting cells).

Cell Structure. Cells contain a variety of small structures, called **organelles**, which perform specific functions (Figure 1-1).

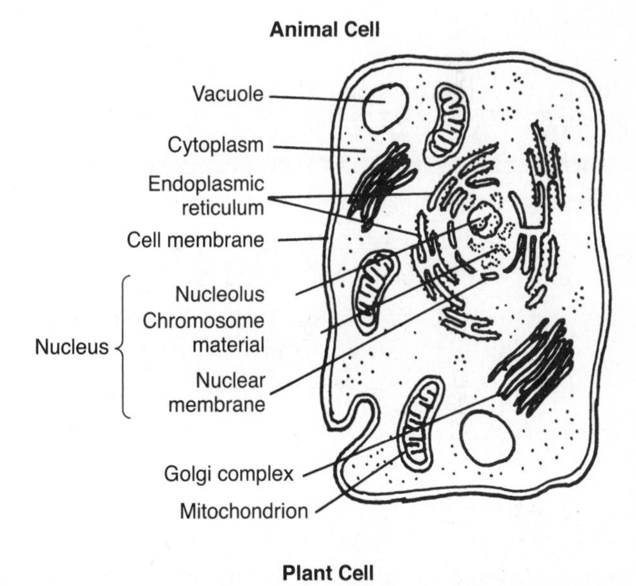

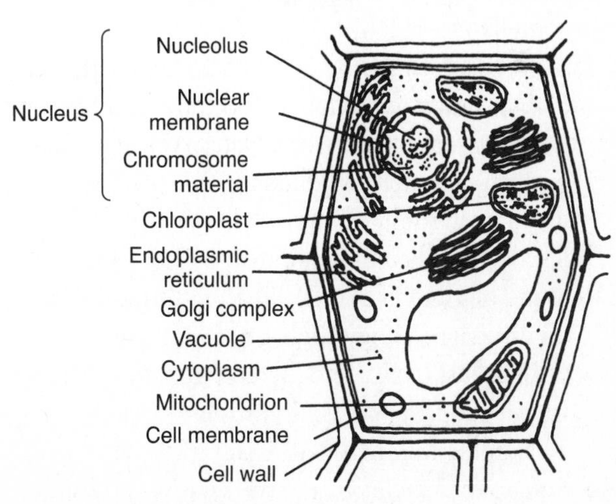

Figure 1-1. Typical cell organelles.

The **cell**, or **plasma**, **membrane** surrounds the cell and separates its contents from the environment. The membrane consists of a double lipid layer (phospholipid bilayer) in which large protein molecules are embedded. The cell membrane is *selectively permeable*, which means that some substances can pass through it, while others cannot. In this way, the membrane regulates the passage of materials into and out of the cell, thereby controlling the cell's chemical makeup.

There are various ways that materials may move across the cell membrane. For example, *diffusion* between the cell and its environment moves molecules from areas of greater concentration to areas of lesser concentration. (Other processes are discussed in Chapter 2.)

The **cytoplasm** is the fluid that fills the space between the cell membrane and the nucleus. Many metabolic reactions occur in the cytoplasm, which consists mainly of water. The organelles are suspended within the cytoplasm.

The **nucleus**, which is surrounded by a nuclear membrane, contains the genetic material found in the chromosomes. The **chromosomes** are made up of DNA and protein (called *histones*). The *nucleolus* is where the ribosomes are produced. Information that controls the production of proteins must pass from the nucleus to the ribosomes in the cytoplasm.

Ribosomes are tiny organelles that are attached to the cytoskeleton or to the membranes of the *endoplasmic reticulum* (organelle where lipid parts of the cell membrane are made). Proteins are made at the ribosomes.

Mitochondria are the sites of most aerobic cellular respiration reactions, the energy-releasing process that uses oxygen to produce ATP from nutrient molecules (such as glucose). Most of the ATP produced by aerobic respiration is synthesized in the mitochondria. The mitochondria contain their own DNA and are capable of duplicating themselves.

Vacuoles are fluid-filled organelles surrounded by membranes. In single-celled organisms, digestion occurs in food vacuoles. Plant cells contain very large vacuoles that may fill much of the cell's interior. In animal cells, there are relatively few vacuoles, and they are small.

Chloroplasts are small organelles that contain photosynthetic pigments; they are found in the cytoplasm of plants, algae, and some protists (which are mostly single-celled, eukaryotic organisms). Photosynthesis takes place in the chloroplasts. Like mitochondria, the chloroplasts also contain their own DNA and can duplicate.

The *cell wall* is a nonliving structure found outside the cell membranes of plants, algae, and fungi. It provides strength and rigidity, but does not interfere with the passage of materials into or out of the cell. The *Golgi apparatus* is similar in structure to the endoplasmic reticulum and is involved in the secretion of cell products. *Lysosomes* are organelles that contain digestive enzymes and which assist in breaking down large molecules to smaller molecules. Enzymes in the lysosomes also destroy foreign microbes such as bacteria that may invade a cell.

The *cytoskeleton* is a structure made up of protein microtubules and microfilaments to which many organelles are attached. It assists in maintaining the cell's shape. *Centrioles* are composed of microtubules and are involved in cell division. *Cilia* and *flagella* are hairlike structures found in some cells; they are involved in locomotion. Generally, cilia are short and more numerous than the long whiplike flagella.

QUESTIONS

MULTIPLE CHOICE

12. The unit of structure and function of all living things is a (an) A. organ B. atom C. cell D. nucleolus

13. According to the cell theory, which statement is correct? A. Viruses are true living cells. B. All cells are basically different in structure. C. Mitochondria are found only in plant cells. D. All cells come from preexisting cells.

14. Chloroplasts and mitochondria are examples of A. cells B. tissue C. organelles D. organs

15. Which statement best describes the term *theory* as used, for example, in the "cell theory"? A. A theory is never revised as new scientific evidence is presented. B. A theory is an assumption made by scientists and implies a lack of certainty. C. A theory refers to a scientific explanation that is strongly supported by a variety of experimental data. D. A theory is a hypothesis that has been supported by one experiment performed by two or more scientists.

16. The term "selectively permeable" describes which of these structures? A. nucleus B. cell wall C. cytoplasm D. cell membrane

17. The structure that separates the cell from the environment is the A. nucleus B. cell membrane C. mitochondrion D. vacuole

18. Plant cell organelles that contain photosynthetic pigments are A. chloroplasts B. ribosomes C. chromosomes D. cell walls

19. Unlike those of an animal, the cells of a plant would most likely have which structure? A. cell membrane B. nucleus C. vacuole D. cell wall

20. The sites of protein synthesis in the cytoplasm are the A. ribosomes B. chromosomes C. nuclei D. vacuoles

21. The fluid-filled area in which most life activities of a cell take place is the A. cell membrane B. chloroplast C. cytoplasm D. vacuole

22. Transport of materials into and out of a cell is most closely associated with the A. nucleus B. cell wall C. ribosome D. cell membrane

23. Which organelle contains the genetic material that controls most cell activities? A. nucleus B. cell membrane C. vacuole D. endoplasmic reticulum

24. The organelles that are the sites of aerobic cellular respiration in both plant and animal cells are the A. mitochondria B. vacuoles C. chloroplasts D. nuclei

25. An increase in the concentration of ATP in a muscle cell is a direct result of which life function? A. cellular respiration B. reproduction C. digestion D. excretion

26. A nonliving cell structure is a A. cell membrane B. nucleus C. cell wall D. mitochondrion

27. In a cell, information that controls the production of proteins must pass from the nucleus to the A. cell membrane B. chloroplasts C. mitochondria D. ribosomes

28. The arrows in the diagram below indicate the movement of materials into and out of a single-celled organism. The movements indicated by these arrows are directly involved in A. the maintenance of homeostasis B. photosynthesis only C. excretion only D. the digestion of minerals

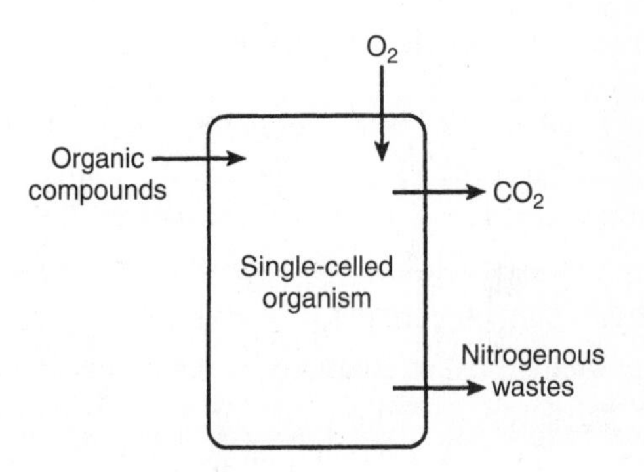

OPEN RESPONSE

29. In the diagram of a single-celled organism shown below, the arrows indicate various activities taking place. Select one of these activities and identify the life process in humans that carries out a similar function.

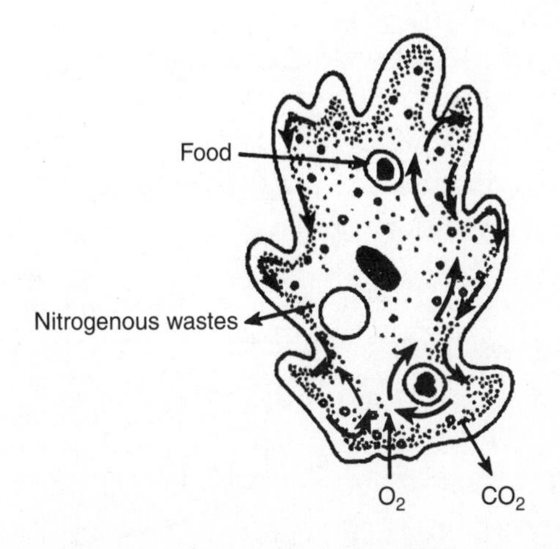

30. Explain why scientists have difficulty including viruses in any category of living organisms.

TOOLS AND METHODS OF CELL STUDY

There are various scientific tools and methods that enable the up-close study of cell structures and functions. These different techniques and types of equipment are used to study cells and cell parts at varying levels of magnification and in different conditions. For example, some tools are used for the study of living cells, while others can be used only for the examination of preserved (dead) cells. Some of these tools and techniques are described here.

Compound Light Microscope

A microscope that uses two lenses to form an enlarged image is called a *compound light microscope.* Light passes through the specimen, the objective lens, and the ocular lens, or *eyepiece*, before reaching the eye. The objective lens produces a magnified image that is further enlarged by the ocular lens. The main parts of a compound light microscope are shown in Figure 1-2. The functions of these parts are listed in Table 1-1.The amount of enlargement of an image produced by the lenses of a microscope is its *magnifying power.* For a compound microscope, magnifying power is found by multiplying the magnifying power of the objective lens by the magnifying power of the ocular lens. For example, if the magnifying power of the objective is 40× (40 times) and that of the ocular is 10× (10 times), the total magnification is $40 \times 10 = 400\times$ (400 times). The greater the magnification of a specimen, the smaller the *field of view*, or observable area. The *resolution*, or resolving power, is the capacity of the microscope to show as separate two points that are close together.

Other Types of Microscopes

A microscope that has an ocular lens and an objective lens for each eye is called a binocular microscope. The *dissecting microscope* is a type of binocular microscope. It produces a three-dimensional image, has relatively low magnifying power, and is used for viewing fairly large, opaque specimens. A *phase-contrast microscope*, which has more magnifying power, is used to observe specimens that cannot be seen with an ordinary light microscope. The *electron microscope* can magnify an object more than 400,000×. Unlike other microscopes, it uses an electron beam focused by electromagnets, instead of light and lenses. One disadvantage of the electron microscope is that only dead specimens can be viewed with it.

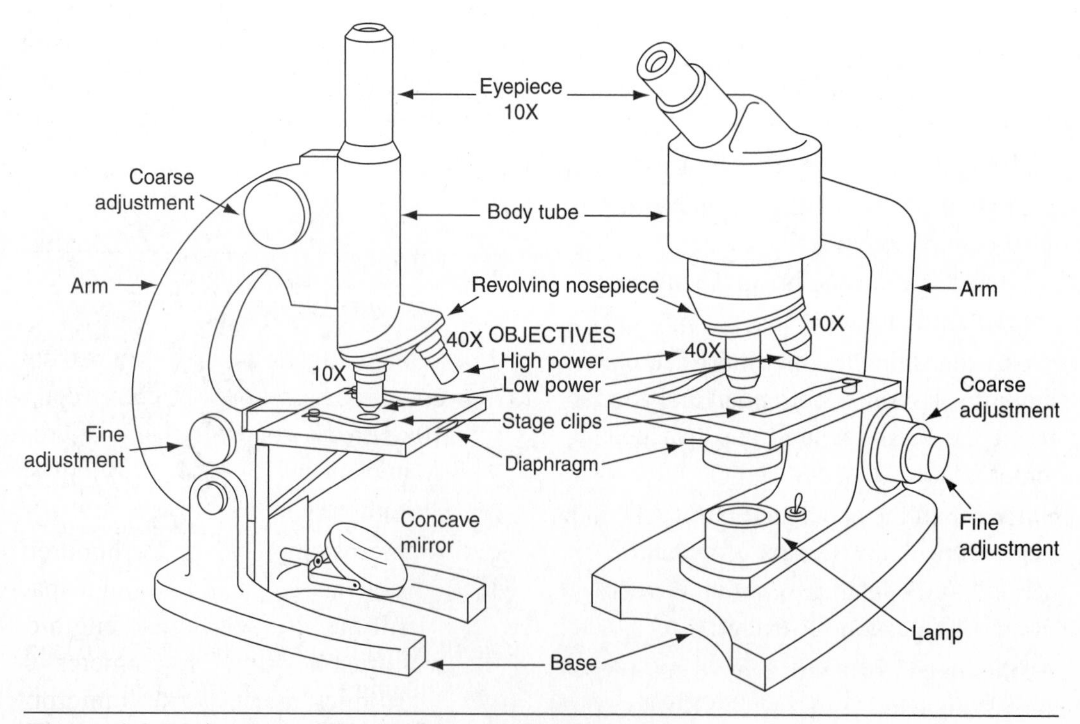

Figure 1-2. Main parts of the compound light microscope.

Table 1-1. **Parts of the Compound Light Microscope and Their Functions**

Part	*Function*
Base	Supports the microscope
Arm	Used to carry microscope; attaches to the base, stage, and body tube
Body tube	Holds the objective lens and eyepiece
Stage	Platform on which the glass slide with the specimen is placed (over the opening in the stage through which light passes)
Clips	Hold the slide in position on the stage
Nosepiece	Holds the objective lenses; rotates so that the different objective lenses can be moved in line with the specimen and eyepiece
Coarse adjustment	Larger knob used for rough-focusing with the low-power objective
Fine adjustment	Smaller knob used for focusing with the high-power objective and for final focusing with the low-power objective
Mirror or Lamp	Directs light to the specimen (on the stage)
Diaphragm	Controls the amount of light reaching the specimen
Objective lenses	Lenses for magnification mounted on the nosepiece
Ocular lens	Lens for magnification at top of the body tube; commonly called the *eyepiece*

Techniques of Cell Study

Tiny instruments that can be used, with the aid of a microscope, to remove or transfer the parts of a cell are *microdissection* instruments. For example, with the use of microdissection instruments, a nucleus can be transferred from one cell to another (as is in done in the process of cloning). The laboratory instrument that is used to separate small particles or materials on the basis of density is the *ultracentrifuge.* In fact, various cell organelles can be isolated by the process of ultracentrifugation. The ultracentrifuge spins the sample in a test tube at very high speeds so that particles of different densities settle to the bottom of the test tube in layers. In addition, cell structures can be made clearly visible by the use of various *staining* techniques. Depending on its specific chemical makeup, a particular stain will be absorbed only by certain parts of the cell. For example, methylene blue and iodine are stains that are absorbed by the nucleus. Other parts of the cell can be made visible with other stains.

The unit used in measuring structures that can be viewed with a compound light microscope is the *micrometer* (μm). One micrometer equals 0.001 millimeter (mm); 1000 micrometers equal 1 millimeter. The diameter of the low-power field of view of a compound light microscope is commonly about 1500 μm. A paramecium is about 250 μm (0.25 mm) long. (Measurement with a microscope is discussed in greater detail in Chapter 9.)

QUESTIONS

MULTIPLE CHOICE

31. Which of the following plant cell structures could most likely *not* be seen when using the 10× objective of a compound microscope?
A. nucleus B. cell wall C. cytoplasm
D. endoplasmic reticulum

32. A microscope reveals one hundred similar cells arranged end-to-end in a space of 1 millimeter. The average length of each cell must be A. 0.1 micrometer
B. 10 micrometers C. 100 micrometers
D. 1000 micrometers

33. Which instrument would provide the most detailed information about the internal structure of a chloroplast? A. a compound light microscope B. a phase-contrast microscope C. an electron microscope D. an ultracentrifuge

34. If the low-power objective and the eyepiece both have a magnifying power of 10×, the total magnifying power of the microscope is A. 10× B. 100× C. 1× D. 20×

35. To separate the parts of a cell by differences in density, a biologist would probably use A. a microdissection instrument B. an ultracentrifuge C. a phase-contrast microscope D. an electron microscope

36. Which microscope magnification should be used to observe the largest field of view of an insect wing?
A. 20× B. 100× C. 400× D. 900×

37. The diameter of the field of view of a compound light microscope is 1.5 millimeters. This may also be expressed as A. 15 micrometers B. 150 micrometers C. 1500 micrometers D. 15,000 micrometers

38. To transplant a nucleus from one cell to another cell, a scientist would use A. an electron microscope B. an ultracentrifuge C. microdissection instruments D. staining techniques

39. A student used a compound microscope to measure the diameters of several red blood cells and found that the average length was 0.008 millimeter. What was the average length of a single red blood cell in micrometers? A. 0.8 B. 8 C. 80 D. 800

40. A student using a compound microscope estimated the diameter of a cheek cell to be about 50 micrometers. What is the diameter of this cheek cell in millimeters? A. 0.050 B. 0.500 C. 5.00 D. 50.0

41. A student has a microscope with a 10 eyepiece and 10× and 40× objectives. She observed 40 onion skin cells across the diameter of the low-power field. How many cells would she observe under high power?
A. 1 B. 40 C. 10 D. 4

42. After examining cells from an onion root tip under high power, a student switches to the low-power objective without moving the slide. He would most likely see A. more cells and less detail B. more cells and more detail C. fewer cells and less detail D. fewer cells and more detail

43. The diagram below represents the field of view of a microscope. What is the approximate diameter in micrometers of the cell shown in the field? A. 50 B. 500 C. 1000 D. 2000

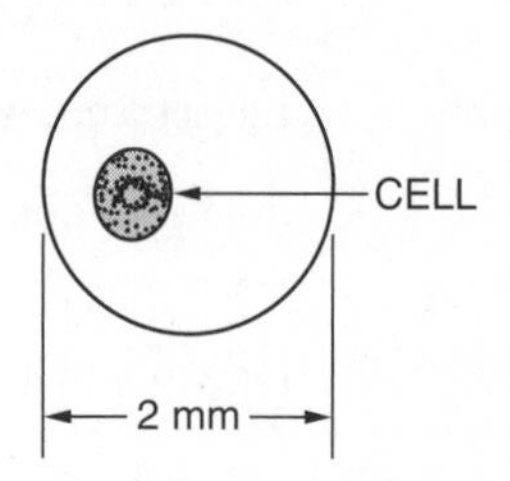

OPEN RESPONSE

44. Select any three parts that are labeled in the diagram below and, for each part selected:

- identify the part; and
- state the function of that part.

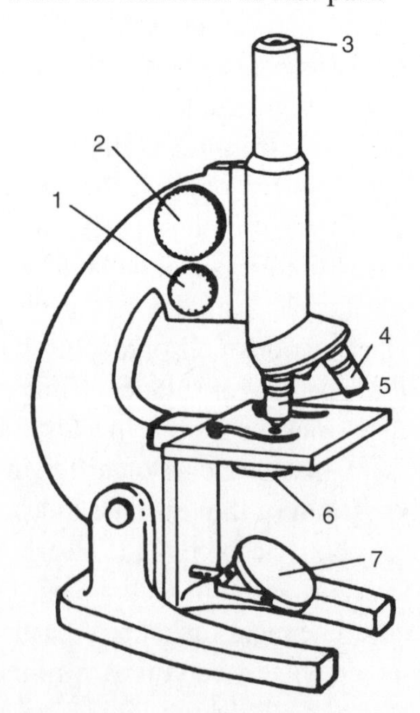

45. Explain how a biology student can calculate the magnification of a specimen when the powers of the eyepiece lens and the objective lens are known.

Answer question 46 based on the following information and data table.

46. A lab microscope has two interchangeable eyepieces and four objective lenses. The table below shows various combinations of the eyepiece and objective lenses and the apparent magnification of the specimen image produced. Use the information provided in the table to complete the missing data.

Eyepiece Lens	*Objective Lens*	*Magnification of Image*
10×		100×
	40×	400×
15×	90×	
10×		150×
		900×

47. Briefly state how you think the development of the compound microscope and other magnifying instruments greatly advanced the science of biology.

BIOCHEMISTRY

The chemical reactions necessary to sustain life take place in the cells. The study of the chemical reactions of living things is called *biochemistry*.

Elements

A substance that cannot be broken down into simpler substances is called an **element**. Examples of elements include hydrogen, oxygen, sodium, and potassium. The most abundant elements in living things are **carbon, hydrogen, oxygen**, and **nitrogen**. Elements found in lesser amounts in living things include sulfur, sodium, phosphorus, magnesium, iodine, iron, calcium, chlorine, potassium, and others. Sodium, for example, is an element that is involved in a number of important cell functions, such as transmission of nerve impulses. Iron plays an important role in the composition of blood in humans and many other animals, where it is involved in oxygen transport.

Atoms

All elements are made up of particles called **atoms**, which are the smallest particles that have the characteristics of their particular element. Each element has a different kind of atom. The atoms of different elements differ in the numbers of protons, neutrons, and electrons they contain. A *compound* is formed when two or more elements combine chemically. For example, water (H_2O) is formed by the chemical bonding (combination) of two hydrogen atoms and one oxygen atom.

Chemical Bonding

The formation of compounds involves either the transfer or the sharing of electrons between atoms, resulting in the formation of chemical bonds. When atoms lose or gain electrons, they become electrically charged particles called ions, and an *ionic bond* is formed. When atoms share electrons, a *covalent bond* is formed. When a compound forms, it has properties that are different from those of the elements that make it up.

Inorganic and Organic Compounds

There are two basic classes of chemical compounds found in living things: inorganic compounds and organic compounds.

Compounds that do not contain both carbon and hydrogen atoms are **inorganic** compounds. Inorganic compounds found in cells include water, salts, carbon dioxide, and inorganic acids, such as hydrochloric acid (HCl).

Compounds that contain both carbon and hydrogen atoms are **organic** compounds. Because carbon atoms can form four covalent bonds with other atoms, organic compounds are often large and complex. The major categories of organic compounds are carbohydrates, proteins, lipids, and nucleic acids.

Carbohydrates

Carbohydrates include sugars and starches, which are used primarily as sources of energy and as food-storage compounds. Other carbohydrates function as structural compounds. In plants, for example, cellulose gives structure and rigidity to the cell walls.

In bacteria, special types of carbohydrates also make up the cell wall. Carbohydrates are made up of carbon, hydrogen, and oxygen, and the ratio of hydrogen to oxygen is always 2 to 1, the same ratio that occurs in water molecules. The simplest carbohydrates are the *monosaccharides*, or **simple sugars**. Glucose, galactose, and fructose, each with the formula $C_6H_{12}O_6$, are simple sugars.

Some carbohydrates, such as maltose and sucrose (both $C_{12}H_{22}O_{11}$) are *disaccharides*, sugars whose molecules are made up of two monosaccharide molecules bonded together. For example, maltose is a disaccharide molecule that is formed from two glucose molecules that are bonded together.

Complex carbohydrates that are made up of chains of monosaccharides are called *polysaccharides*. Starch, cellulose, and glycogen are polysaccharides that are made up of chains of glucose molecules. In plants, **starch** is a food storage compound, and cellulose makes up the cell walls. In animals, glycogen is the food-storage compound, found primarily in the liver.

Proteins

Enzymes, many hormones, and various structural parts of organisms are **proteins**. Proteins are very large molecules, or *polymers*; proteins are made up of smaller *subunits* called **amino acids.**

Structure of Amino Acids. Amino acids contain the elements carbon, hydrogen, oxygen, and nitrogen. Some also contain sulfur. Figure 1-3 shows the generalized structure of an amino acid.

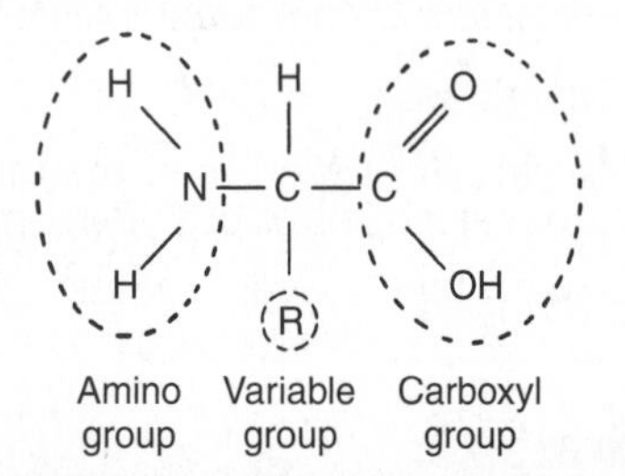

Figure 1-3. Generalized structure of an amino acid.

The $-NH_2$ is an amino group; the $-COOH$ is a carboxyl, or acid, group; and the *R* represents a *variable* group. The *R* group is the part of the amino acid structure that differs from one amino acid to another. Twenty different amino acids are found in the cells of living things.

As in carbohydrates, amino acids combine chemically to form more complex molecules. When two amino acids combine, they form a *dipeptide*. The bond that holds the amino acids together is called a *peptide bond*. More amino acids may combine with a dipeptide to form a *polypeptide*. A protein is made up of one or more polypeptide chains. There are a great many types of protein molecules in living things. These molecules differ in the number, kinds, and sequences of amino acids they contain.

Lipids

Fats, oils, and waxes belong to the group of organic compounds called **lipids**. They function mainly as sources of energy and as components of structures such as cell membranes. In many organisms, lipids can be stored to provide energy for times when food is scarce. Lipids that are solid at room temperature are *fats*, while those that are liquid are *oils*. Another type of lipids, called *steroids*, includes cortisone, vitamin D, and the sex hormones. Lipids contain carbon, hydrogen, and oxygen. The ratio of hydrogen atoms to oxygen atoms is greater than 2 to 1 and varies from one lipid to another. The building blocks of lipids are fatty acids and glycerol.

Nucleic Acids

DNA and RNA are both **nucleic acids**, the very important organic molecules involved in the transfer of genetic information from one generation to the next. **DNA (deoxyribonucleic acid)** is located in the nucleus, mitochondria, and chloroplasts of cells, where it stores information for the cell. DNA molecules are capable of duplicating, or *replicating*, themselves. In addition to being the cell's genetic material, DNA codes for the production of proteins. **RNA (ribonucleic acid)** is similar to DNA in that both molecules are *polymers*; this means they are composed of thousands of smaller chemical units (*monomers*) that are repeated over and over again, thus forming giant molecules. RNA assists in carrying out the vital information (such as the instructions for protein synthesis) stored in the DNA molecules. In some viruses, RNA, rather than DNA, is the genetic material that allows the virus to function. For example, HIV—the human immunodeficiency virus that causes AIDS—is a type of RNA virus. (Nucleic acids are discussed further in Chapter 6.)

QUESTIONS

MULTIPLE CHOICE

48. What is the principal inorganic solvent in cells? A. salt B. water C. alcohol D. carbon dioxide

49. Fats that are stored in human tissue contain molecules of A. glycerol and fatty acids B. amino acids C. monosaccharides D. disaccharides

50. One of the carbon compounds found in a cell has twice as many hydrogen atoms as oxygen atoms. This compound most likely belongs to the group of substances known as A. nucleic acids B. lipids C. proteins D. carbohydrates

51. Which formula represents an organic compound? A. NH_3 B. H_2O C. NaCl D. $C_{12}H_{22}O_{11}$

52. Starch is classified as a A. disaccharide B. polypeptide C. nucleotide D. polysaccharide

53. Which organic compound is correctly matched with the subunit that composes it? A. maltose—amino acid B. starch—glucose C. protein—fatty acid D. lipid—sucrose

OPEN RESPONSE

54. There are only 20 different amino acids found in living things, yet there are thousands of different proteins. Explain why this is possible.

55. Explain why protein molecules are considered to be polymers.

56. Examine each of the following four molecular structures. Identify each molecule as organic or inorganic and explain why it is classified as such.

```
  H     H
   \   /
     O                 O=C=O
    (1)                 (2)

     H                    H
     |     O              |
H — C — C//          H — C — H
     |     \              |
     R      OH            H
    (3)                  (4)
```

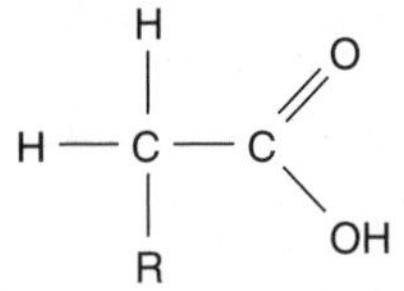

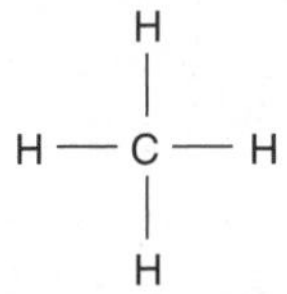

Base your answer to question 57 on the following information and data table.

57. A lab was set up for students to analyze three unknown samples of organic molecules—a lipid, a carbohydrate, and a protein. The results of their lab tests are shown in the table below. Based on these results, identify each sample as a protein, carbohydrate, or lipid, and then state the reason for your identification of each molecule.

Unknown Sample	*Elements Contained*	*Molecular Characteristics*
A	C, H, O, and N	Polymer, high molecular mass
B	C, H, and O	Very little oxygen, much hydrogen
C	C, H, and O	Twice as much hydrogen as oxygen

58. There are four major types of organic molecules that are important in living things: carbohydrates, lipids, proteins, and nucleic acids. Select any two molecules and, for each one chosen:

- describe the structure of the molecule;
- state *two* ways that the molecule is useful to living organisms.

ENZYMES

The Role of Enzymes

Chemical reactions occur continuously in living things. All of these reactions require the presence of special proteins called **enzymes**, which regulate the reaction rates. In general, enzymes speed up the rate of a reaction (by lowering the activation energy). This is important because most of the reactions needed within cells to sustain life normally occur at a relatively slow rate. Enzymes are **catalysts**, substances that change the rate of a chemical reaction but are themselves unchanged by the reaction.

Enzymes are named after their *substrates*, the substances they act on. The name of an enzyme generally ends in *ase*. For example, a lipase acts on lipids, a protease acts on proteins, and maltase acts on the sugar maltose.

Enzyme Structure

An enzyme is a large, complex protein that consists of one or more polypeptide chains. The polypeptide chains that make up an enzyme are folded in a highly specific way, forming pockets on the enzyme surface into which the substrate molecule or molecules fit. The specific part of the enzyme where the substrate fits is called the *active site*.

In addition to the protein, some enzymes contain a nonprotein component called a *coenzyme*. If the coenzyme part is missing, the enzyme will not function. *Vitamins* often function as coenzymes. Although most enzymes are proteins, some enzymes are composed of another type of organic molecule, that is, RNA. This type of enzyme is sometimes called a *ribozyme*.

For an enzyme to affect the rate of a chemical reaction, the substrate must become attached to the active site of the enzyme, forming an *enzyme-substrate complex*. The enzyme's action occurs while the enzyme and substrate are bound together. At this time, bonds of the substrate may be weakened, causing it to break apart, or bonds may form between substrate molecules, joining them together. After the reaction is complete, the enzyme and product(s) separate, and the enzyme molecule becomes available to act on other substrate molecules. Although it is important for the substrate to bind to the enzyme, this must be temporary so that the enzyme can act again on other substrates. In fact, some poisons are deadly because they form permanent bonds with enzyme active sites, thus preventing the enzymes from functioning again.

Models of Enzyme Action

Different models can be used to describe the mechanism of enzyme action. According to the *lock-and-key model*, the active site on an enzyme has a unique three-dimensional shape that can form a complex with only one type of substrate. The substrate fits an active site just as a key fits a lock (Figure 1-4). However, it is important to note that the enzyme molecule is not rigid. Rather, according to the *induced fit model* of enzyme action, when it binds to a substrate at the active site, the enzyme molecule bends somewhat to cause, or induce, a closer fit between itself and the substrate. This enhanced fit allows the enzyme to function in a more effective manner.

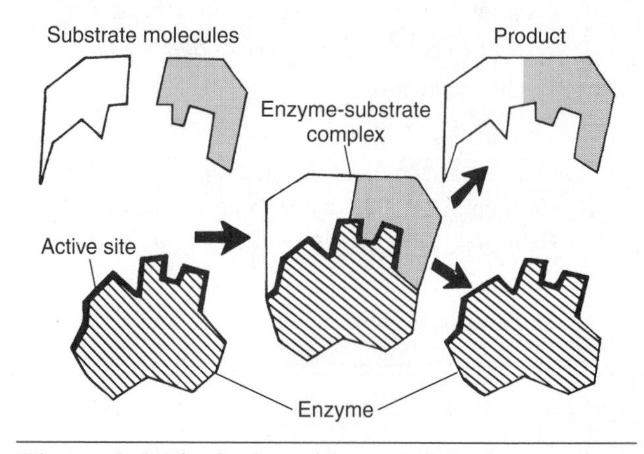

Figure 1-4. The lock-and-key model of enzyme action.

Factors That Influence Enzyme Action

The rate of enzyme action is affected by temperature, concentrations of enzyme and substrate, and pH.

Temperature. The rate of enzyme action varies with temperature. Up to a point, the rate increases with increasing temperature (Figure 1-5). The temperature at which the enzyme functions most efficiently is called the *optimum temperature*. If the temperature rises above the optimum, the rate of enzyme action begins to decrease. This decrease in enzyme action occurs because the higher temperature destroys the shape of the enzyme. This process, known as *denaturation*, changes the shape of the enzyme's active site so that it no longer fits the substrate. In humans, the normal body temper-

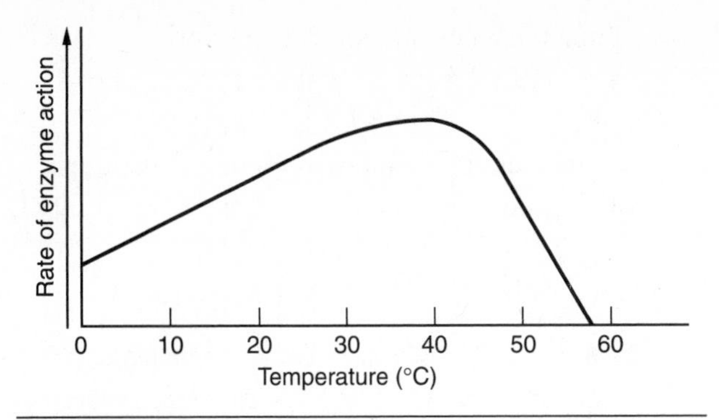

Figure 1-5. The effect of changing temperature on the rate of enzyme action.

ature of about 37°C is also the optimum temperature for most human enzymes. Denaturation of these enzymes begins at about 40°C, upsetting the body's homeostasis.

Enzyme and Substrate Concentrations. The rate of enzyme action varies with the amount of available substrate. The rate of enzyme action increases as the substrate concentration increases (Figure 1-6). At the point where all enzyme molecules are reacting, the rate levels off, and addition of more substrate has no further effect.

pH. Enzyme action varies with the pH of the environment. The **pH** scale is a measure of the hydrogen ion (H^+) concentration of a solution. Solutions with a pH of 7 are neutral. Those with a pH below 7 are acids, while those with a pH above 7 are bases (Figure 1-7 at the bottom of this page).

Each enzyme has a particular pH at which it functions most efficiently. For example, most enzymes in human blood function best in neutral solutions. Pepsin, an enzyme found in the stomach (an acidic environment), works best at a pH of 2; and trypsin, an enzyme in the small intestine (an alkaline environment), works best at a pH of 8 (Figure 1-8). Changes in pH can alter the shape of an enzyme's active site, causing the enzyme to be nonfunctional.

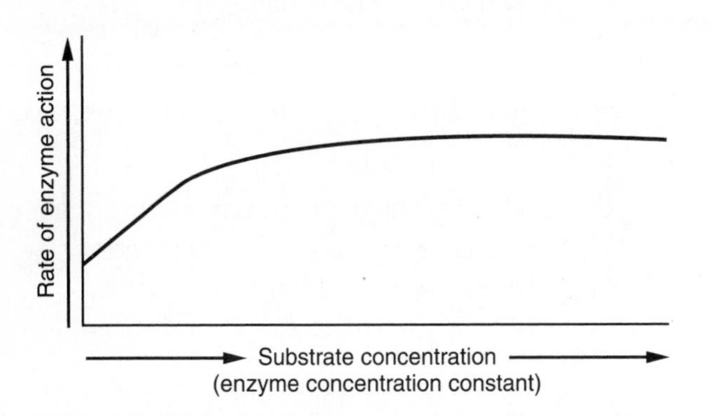

Figure 1-6. The effect of changing substrate concentration on the rate of enzyme action.

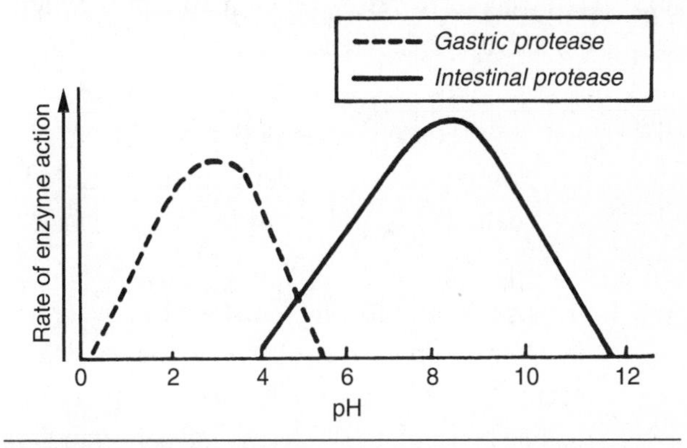

Figure 1-8. The effect of pH on the rate of enzyme action.

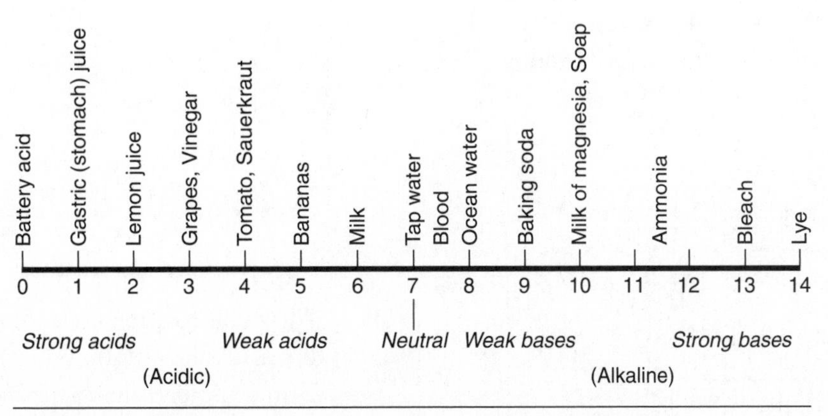

Figure 1-7. The pH scale ranges from acidic to basic.

QUESTIONS

MULTIPLE CHOICE

59. Which of the following is characteristic of an enzyme? A. It is a carbohydrate. B. It is destroyed after each chemical reaction. C. It provides energy for any chemical reaction. D. It regulates the rate of a specific chemical reaction.

60. The "lock-and-key" model of enzyme action illustrates that a particular enzyme molecule will A. form a permanent enzyme-substrate complex B. be destroyed and resynthesized several times C. interact with a specific type of substrate molecule D. react at identical rates under all conditions

61. An enzyme-substrate complex may result from the interaction of molecules of A. glucose and lipase B. fat and trypsin C. sucrose and maltase D. protein and protease

62. The part of the enzyme molecule into which the substrate fits is called the A. active site B. coenzyme C. polypeptide D. protease

63. A nonprotein molecule necessary for the functioning of a particular enzyme is called a A. catalyst B. polypeptide C. coenzyme D. substrate

64. Which of the following variables has the *least* direct effect on the rate of an enzyme regulated reaction? A. temperature B. pH C. carbon dioxide concentration D. enzyme concentration

65. The diagram below represents a beaker containing a solution of various molecules involved in digestion. Which structures represent products of digestion? A. *A* and *D* B. *B* and *C* C. *B* and *E* D. *D* and *E*

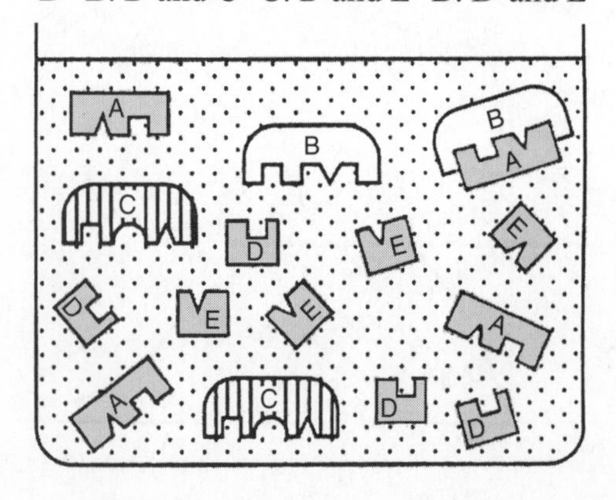

66. Enzymes have an optimum temperature at which they work best. Temperatures above and below this temperature will decrease enzyme activity. Which graph best illustrates the effect of temperature on enzyme activity?

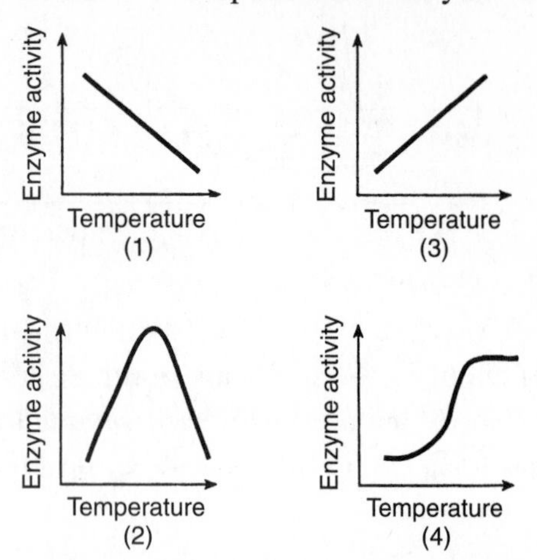

67. A word equation is shown below. This reaction is most directly involved in the process of A. reproduction B. protein synthesis C. replication D. digestion

starch molecules $\xrightarrow{\textbf{biological catalyst}}$ **simple sugars**

68. The change in shape of enzyme molecules that occurs at high temperatures is known as A. synthesis B. specificity C. replication D. denaturation

Base your answers to questions 69 through 71 on the following graph and on your knowledge of biology. The graph represents the rate of enzyme action when different concentrations of enzyme are added to a system with a fixed amount of substrate.

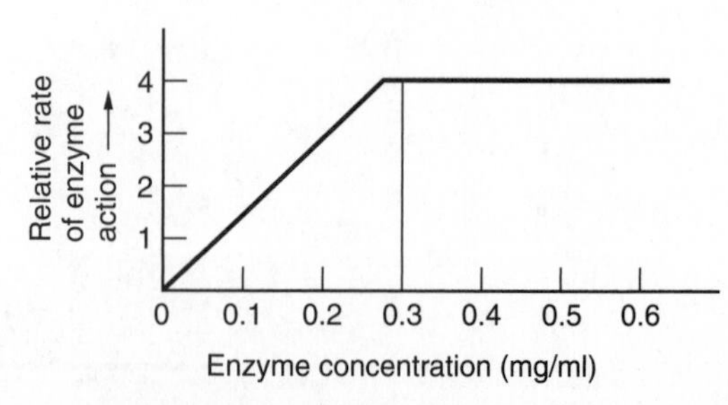

69. At which enzyme concentration does all of the available substrate react with the enzyme? A. 0.1 mg/mL B. 0.2 mg/mL C. 0.3 mg/mL D. 0.5 mg/mL

70. When the enzyme concentration is increased from 0.5 mg/mL to 0.6 mg/mL, the rate of enzyme action A. decreases B. increases C. remains the same

71. If more substrate is added to the system at an enzyme concentration of 0.4 mg/mL, the rate of the reaction would most likely A. decrease B. increase C. remain the same

Base your answers to questions 72 and 73 on the following graphs. Graph I shows the relationship between temperature and the relative rates of activity of enzymes A *and* B. *Graph II shows the relationship between pH and the relative rates of activity of enzymes* A *and* B.

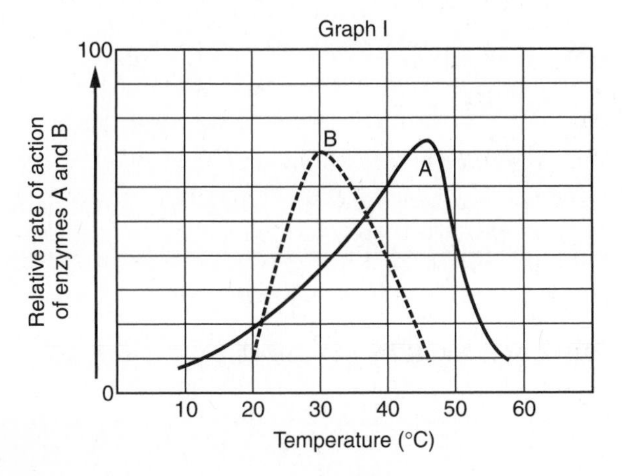

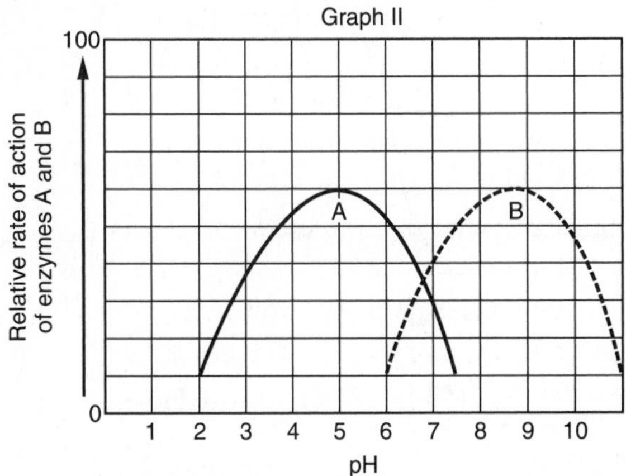

72. Under which conditions is enzyme *A* most effective? A. at 40°C and a pH of 5 B. at 45°C and a pH of 5 C. at 45°C and a pH of 9 D. at 50°C and a pH of 9

73. The optimum environment for enzyme *B* is A. a basic medium B. an acidic medium C. either an acidic or a basic medium D. a neutral medium

OPEN RESPONSE

Use your knowledge of enzymes and biology to answer questions 74 and 75.

74. Fresh pineapple contains an enzyme that digests proteins. Adding fresh pineapple to gelatin (a protein) prevents it from setting or gelling. Adding cooked or canned pineapple does not have this effect and the gelatin can set normally. Explain why these different effects occur.

75. When an apple is cut open, the inside soon turns brown. This is because enzymes that are released from the cut cells react with certain molecules in the apple. Rubbing lemon juice (which contains citric acid) on the cut apple prevents it from browning. Explain why this is so.

76. The human stomach contains an enzyme called pepsin, which actively breaks down (digests) protein molecules found in food. Based on your knowledge of biology, answer the following:

- At what temperature would you expect pepsin to work best? Give a reason for your answer.
- Why would drinking very cold beverages have a negative effect on digestion of food in the stomach?

77. Draw a diagram in which you show how the enzyme maltase combines with two glucose molecules to form maltose. Label the enzyme, substrate, active site, enzyme-substrate complex, and end product.

78. An incomplete graph is shown below. What label could appropriately be used to replace letter *Z* on the horizontal axis?

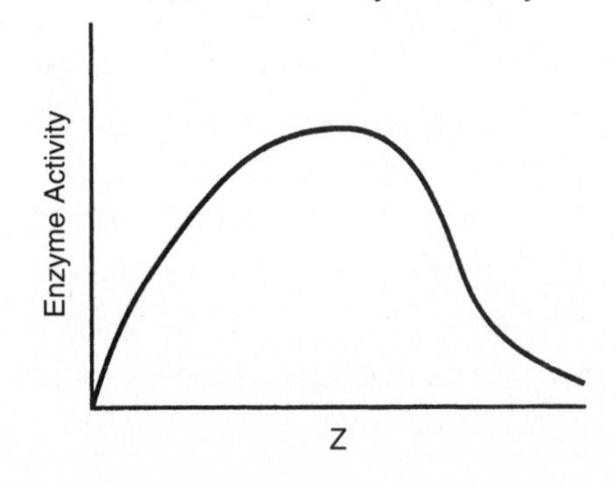

Base your answers to questions 79 through 81 on the two different cells shown below. Only cell A *produces substance* X. *Both cells* A *and* B *use substance* X.

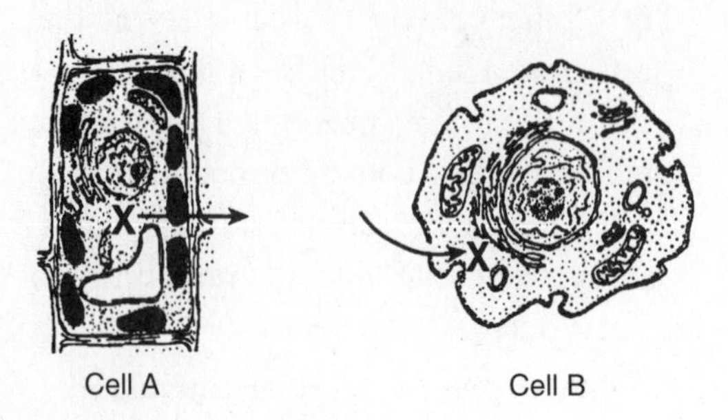

79. Identify substance *X*.

80. Identify the type of organelle in cell *A* that produces substance *X*.

81. Identify the type of organelle found in both cell *A* and cell *B* that uses substance *X*.

82. The enzyme catalase is found in almost all living tissues. This enzyme catalyzes the breakdown of harmful hydrogen peroxide in the body. Liver tissue is particularly rich in catalase content. Like all enzymes, catalase is affected by temperature fluctuations. Design and describe an experiment in which a person can study the activity of catalase over a range of temperatures, from 0°C to 80°C. Be sure to include an appropriate control and a data table in your experimental design.

READING COMPREHENSION

Base your answers to questions 83 through 86 on the information below and on your knowledge of biology. Source: *Living Environment Regents Exam*—June 2004, p. 7. (Adapted from: Paul Recer, *The Daily Gazette*, April 26, 1988.)

Researchers Find New Means to Disrupt Attack by Microbes

Some of the most common and deadly bacteria do their mischief by forming a sticky scum called biofilm. Individually, the microbes are easy to control, but when they organize themselves into biofilms they can become deadly, said Dr. Barbara Iglewski of the University of Rochester.

Biofilms are actually intricately organized colonies of billions of microbes, all working in a coordinated way to defend against attack and to pump out a toxin that can be deadly.

Once they are organized, the bacteria are highly resistant to antibiotics and even strong detergents often cannot wash them away or kill them.

Iglewski and colleagues from Montana State University and the University of Iowa report in *Science* that they have discovered how the microbes in the colonies communicate and found that once this conversation is interrupted, the deadly bugs can be easily washed away.

Using *Pseudomonas aeruginosa*, a common bacteria that is a major infection hazard in hospitals and among cystic fibrosis patients, the researchers isolated a gene that the bacteria uses to make a communications molecule. The molecule helps the microbes organize themselves into a biofilm—a complex structure that includes tubes to carry in nutrients and carry out wastes, including deadly toxins.

In their study, the researchers showed that if the gene that makes the communications molecule was blocked, the *Pseudomonas aerginosa* could form only wimpy [weak], unorganized colonies that could be washed away with just a soap that has no effect on a healthy colony.

83. What is one characteristic of a biofilm? A. presence of tubes to transport material into and out of the colony B. presence of a nervous system for communication within the colony C. ease with which healthy colonies can be broken down by detergents D. lack of resistance of the bacterial colony to antibiotics

84. Which statement best describes *Pseudomonas aeruginosa* bacteria? A. They cause mutations in humans. B. They are easy to control. C. They cause major infection problems in hospitals. D. They are deadly only to people with cystic fibrosis.

85. The tubes in biofilms function much like the human A. muscular and nervous systems B. circulatory and excretory systems C. digestive and endocrine systems D. reproductive and respiratory system

86. Bacteria that form biofilms may be controlled most effectively by A. antibiotics B. detergents C. cutting the tubes through which the bacteria communicate D. blocking the expression of a gene that helps the colonies to organize

CHAPTER 2

Life Processes in Living Things

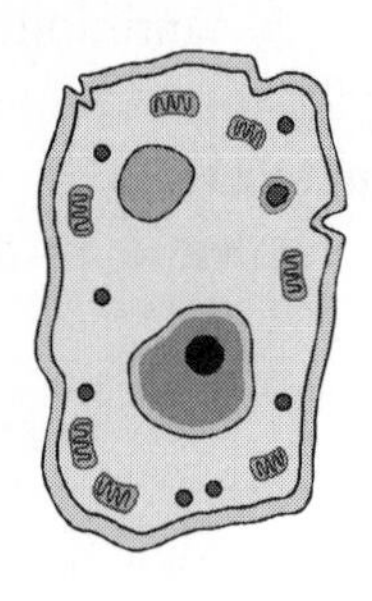

Standards 2.3, 2.4 (Cell Biology)

BROAD CONCEPT: Cells have specific structures and functions that make them distinctive. Processes in a cell can be classified broadly as growth, maintenance, and reproduction.

Standards 4.1, 4.2, 4.3, 4.4, 4.5, 4.7 (Anatomy and Physiology)

BROAD CONCEPT: There is a relationship between the organization of cells into tissues, and tissues into organs. The structure and function of organs determine their relationship within body systems of an organism. Homeostasis allows the body to perform its normal functions.

Most living organisms perform all the same **life processes**, or life functions. These include such life processes as *nutrition*, the ability to obtain and process food; *transport*, the distribution of nutrients and essential materials to the cells; *respiration*, the release in cells of the chemical energy stored in glucose; *excretion*, the removal of metabolic wastes; and *regulation*—the control and coordination of the other life functions.

Different kinds of organisms have specific structures and behavioral patterns that enable them to perform the life functions efficiently within their physical surroundings, or *environment*. These structures and behavioral patterns are called **adaptations**.

NUTRITION

Nutrition includes those activities by which organisms obtain and process food for use by the cells. The cells use **nutrients** from foods for energy, growth, repair, and regulation. Nutrition may be autotrophic or heterotrophic. In **autotrophic** nutrition, the organism can synthesize organic molecules from inorganic substances obtained from the environment. In **heterotrophic** nutrition, the organism must ingest needed organic molecules from other organisms in the environment.

Photosynthesis: Autotrophic Nutrition

The most common type of autotrophic nutrition is **photosynthesis**. Organisms that carry out photosynthesis are called *autotrophs*, a group that includes all plants, all **algae**, and some **bacteria**. Photosynthetic organisms use carbon dioxide and water taken from the environment, along with energy from sunlight, to synthesize the organic compound **glucose**. Most of the chemical energy available to living organisms comes either directly or indirectly from photosynthesis. Also, most of the oxygen in the atmosphere is a by-product of photosynthesis.

Photosynthetic Pigments. Photosynthesis requires the presence of certain colored molecules called *pigments*, which absorb light energy and convert it to a form of chemical energy that can be used by living things. *Chlorophylls* are the green pigments found in photosynthetic organisms. In most of these organisms, the chlorophyll is found in organelles called *chloroplasts*. Within the chloroplasts, the pigments are located inside sacs called *thylakoids*, which form many stacks, or *grana* (Figure 2-1).

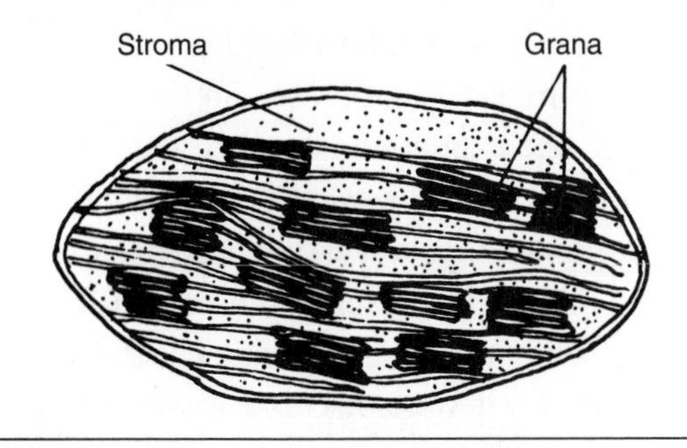

Figure 2-1. Structure of a chloroplast.

In addition to the chlorophylls, chloroplasts may contain a variety of other pigments. These additional pigments, called *accessory pigments*, absorb those wavelengths of light that chlorophyll cannot absorb. Chlorophyll is unable to absorb the green wavelengths of light. That is why plants appear green; chlorophyll reflects the green light. The orange pigment, *carotene*, can absorb green light, and thus the energy of green wavelengths of light are made available for photosynthesis.

Chemistry of Photosynthesis. The process of photosynthesis is a complex biochemical pathway, involving a series of chemical reactions. However, it can be summarized by the following equation:

$$\text{carbon dioxide} + \text{water} \xrightarrow{\text{light energy}} \text{glucose} + \text{oxygen} + \text{water}$$

$$6CO_2 + 12H_2O \rightarrow C_6H_{12}O_6 + 6O_2 + 6H_2O$$

The *reactants*, carbon dioxide and water, are the raw materials of photosynthesis. Light energy absorbed by chlorophyll is converted to chemical energy; this is used to synthesize the *product*, glucose, from the carbon dioxide and water. Oxygen and water are released as waste products or, more appropriately, *by-products* of photosynthesis.

Photosynthesis occurs in two distinct phases, one that requires light directly (the light-dependent reactions) and one that does not directly depend on light energy (the light-independent reactions). In the first phase, the energy of light is converted into the form of chemical energy called *ATP*. It is also during this phase that photosynthetic organisms release oxygen into the atmosphere as a result of splitting the water molecules that enter as a raw material. During the second phase of photosynthesis, carbon dioxide from the atmosphere is combined with the hydrogen left over from the splitting of water to form the organic molecule glucose. Energy to drive this reaction comes from the ATP produced in the light-dependent reactions. The two phases of photosynthesis are, therefore, interdependent.

The glucose produced by photosynthesis is used, when needed, as an energy source in cellular respiration for the production of ATP. It can also be converted to starch, an insoluble food storage compound. Before the starch can be used in any cellular process, it must be broken down to glucose by enzymes within the cell. Glucose also can be used in the synthesis of other organic compounds, such as lipids and proteins.

In photosynthetic organisms, the processes of photosynthesis and cellular respiration are interrelated. Photosynthesis produces glucose from the raw materials carbon dioxide and water. The glucose, in the presence of oxygen, is then used in cellular respiration as an energy source. The waste products of cellular respiration are carbon dioxide and water. Thus, the raw materials of photosynthesis (carbon dioxide and water) are the by-products of cellular respiration; and the raw materials of cellular respiration (glucose, oxygen, and water) are the products and by-products of photosynthesis. The interrelationship of these two processes is an excellent example of the recycling of substances within organisms and between organisms and their environment. This relationship can be summarized in the following chemical equation:

$$6CO_2 + 12H_2O \leftrightarrow C_6H_{12}O_6 + 6O_2 + 6H_2O$$

Reading the equation from left to right illustrates photosynthesis; reading the equation from right to left illustrates cellular respiration.

QUESTIONS

MULTIPLE CHOICE

1. By which process are carbon dioxide and water converted to carbohydrates such as glucose? A. transpiration B. respiration C. fermentation D. photosynthesis

2. The conversion of light energy into chemical bond energy occurs within the cells of A. molds B. yeasts C. algae D. grass-hoppers
3. Glucose molecules may be stored in plants in the form of A. oxygen B. starch C. nucleic acids D. amino acids
4. Organisms capable of manufacturing organic molecules from inorganic raw materials are classified as A. autotrophs B. heterotrophs C. aerobes D. anaerobes
5. The basic raw materials for photosynthesis are A. water and carbon dioxide B. oxygen and water C. sugar and carbon dioxide D. carbon dioxide and oxygen
6. Which word equation represents the process of photosynthesis?

   A. carbon dioxide + water → glucose + oxygen + water

   B. glucose → alcohol + carbon dioxide

   C. maltose + water → glucose + glucose

   D. glucose + oxygen → carbon dioxide + water
7. Autotrophic activity in plant cells is most closely associated with the organelles called A. mitochondria B. ribosomes C. vacuoles D. chloroplasts
8. In terms of nutrition, the main difference between animals and plants is that green plants are able to A. synthesize glucose B. break down carbohydrates C. carry on aerobic respiration D. form ATP molecules

OPEN RESPONSE

Base your answers to questions 9 and 10 on the following statement and on your knowledge of biology.

Carbon exists in a simple organic molecule in a leaf and in an inorganic molecule in the air that animals exhale.

9. Identify the simple organic molecule formed in the leaf and the process that produces it.
10. Identify the inorganic molecule that animals exhale, and state why it is important to plants.
11. State one function of each of the following in the process of photosynthesis: light; chlorophyll; carbon dioxide; and water.
12. Why is photosynthesis called one of the most important processes on Earth? Give at least one example to support your answer.
13. Bromthymol blue is a chemical that turns to bromthymol yellow in the presence of carbon dioxide. When the carbon dioxide is removed, the solution returns to a blue color. Two green water plants were placed in separate test tubes, each containing water and bromthymol yellow. Both test tubes were corked. One tube was placed in the light, the other in the dark. After several days, the liquid in the tube exposed to the light turned blue. Based on these results, answer the following:
    - Why did the bromthymol solution turn blue in the tube exposed to light?
    - What does it illustrate about the activity of plants during photosynthesis?
    - What do you think happened in the tube that was placed in the dark?

Use the information below and your knowledge of biology to answer the following question.

14. A mixture of chloroplasts and water was kept under a bright light at a temperature of 25°C. Another batch of this mixture was kept in a dark corner of the same room. A small pipette was attached to each container of mixture to measure the amount of oxygen released by the chloroplasts during photosynthesis. The data table shows the volume of oxygen produced by each mixture over a 24-hour period.

Total Volume of Oxygen Released by Chloroplasts (mL)

Time (hours)	*Incubated in Light*	*Incubated in Dark*
0	0.00	0.00
6	0.42	0.01
12	0.96	0.01
18	1.78	0.01
24	2.36	0.01

- Describe the difference recorded in the amount of oxygen produced by the two chloroplast mixtures.
- Explain why there was a difference in the volume of oxygen produced by the two chloroplast mixtures.
- Give a scientific reason for why the mixture incubated in the dark produced only 0.01 mL of oxygen.
- Make a line graph showing the results of the experiment. Use different colors to plot the data for each of the two mixtures.
- State one way the researcher could modify the experiment to show that the results are reliable.

Adaptations for Photosynthesis

Algae and green plants are autotrophic organisms that carry on photosynthesis. A large percentage of Earth's photosynthesis occurs in unicellular (single-celled) algae present in the oceans. The raw materials necessary for photosynthesis are absorbed directly from the water into the cells of the algae. Most of the photosynthesis in terrestrial (land-dwelling) plants takes place in their leaves.

Structure of Leaves. Most leaves are thin and flat, providing a large surface area for the absorption of light. The outermost cell layer of the leaf, the *epidermis*, protects the internal tissues from water loss, mechanical injury, and attack by fungi (Figure 2-2). In some plants, the epidermis is covered by a waxy coating, called the *cuticle*, which provides additional protection against water loss and infection.

There are many tiny openings in the epidermis and cuticle, usually on the undersurface of the leaf. These openings, called *stomata*, allow the exchange of carbon dioxide, oxygen, and water vapor between the environment and the moist, inner tissues of the leaf. Each stomate is surrounded by a pair of chloroplast-containing guard cells. By changing shape, the guard cells open or close the stomate opening.

Beneath the upper epidermis is the *palisade layer*, which is made up of tall, tightly packed cells filled with chloroplasts. Most of the photosynthetic activity of the leaf occurs in this layer. The cells of the upper epidermis are clear, so that light striking the leaf passes through to the chloroplasts in the palisade layer.

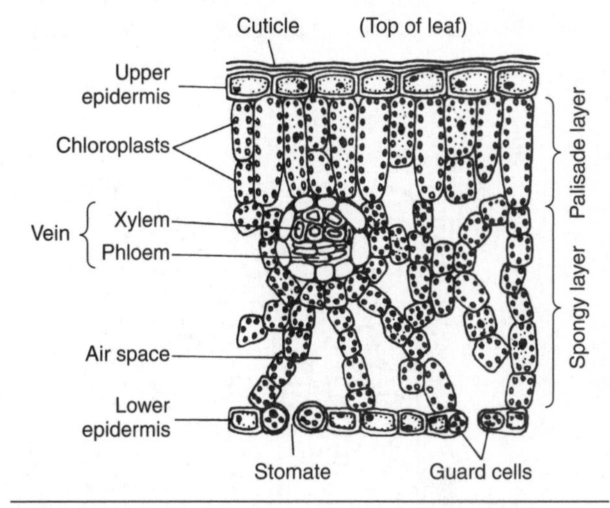

Figure 2-2. Cross section of a typical leaf.

Between the palisade layer and the lower epidermis of the leaf is the *spongy layer*, which is made up of loosely arranged cells separated by interconnecting air spaces. The air spaces are continuous with the stomata. Gases from the environment enter the leaf through the stomata and diffuse from the air spaces into the cells. Other gases diffuse out of the cells into the air spaces and then out of the leaf through the stomata. The cells of the spongy layer contain chloroplasts and carry on some photosynthesis.

Specialized *conducting tissues* carry out the transport of materials in plants. The conducting tissues of the leaf are found in bundles called *veins*. These tissues carry water and dissolved minerals from the roots through the stems to the leaves, and they carry food from the leaves to the rest of the plant.

QUESTIONS

MULTIPLE CHOICE

15. Water is lost from the leaves of a plant through A. spongy cells B. root hairs C. veins D. stomata

16. The waxy covering over the surface of a leaf is the A. cuticle B. epidermis C. palisade layer D. spongy layer

Base your answers to questions 17 through 20 on the following diagram, which shows a leaf cross section, and on your knowledge of biology.

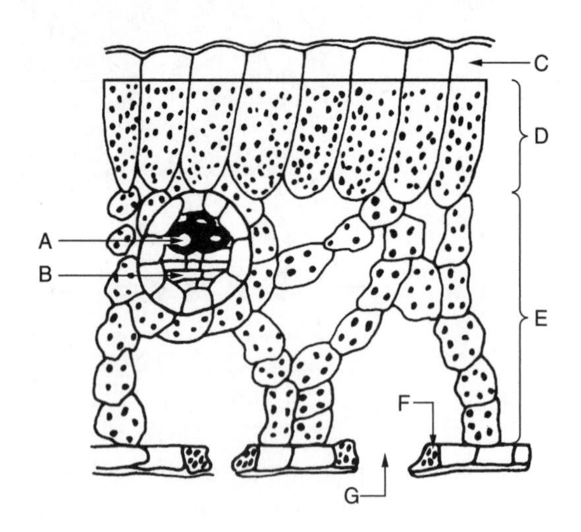

17. Which letter indicates the principal region of food manufacture? A. *E* B. *B* C. *C* D. *D*

18. Which letter indicates the area where carbon dioxide passes out of the leaf? A. *A* B. *G* C. *C* D. *D*

19. Which letter indicates a structure that regulates the size of a stomate? A. *A* B. *B* C. *F* D. *G*

20. Water and dissolved nutrients are carried by the tissues labeled A. *D* and *E* B. *C* and *D* C. *A* and *B* D. *E* and *F*

OPEN RESPONSE

Refer to the following diagrams of three different leaf types to answer question 21.

21. How does the structure of each leaf enable it to carry out photosynthesis in the particular habitat in which the plant lives? (See diagrams for typical habitat of each leaf type; for example, *alpine* refers to high mountain regions.)

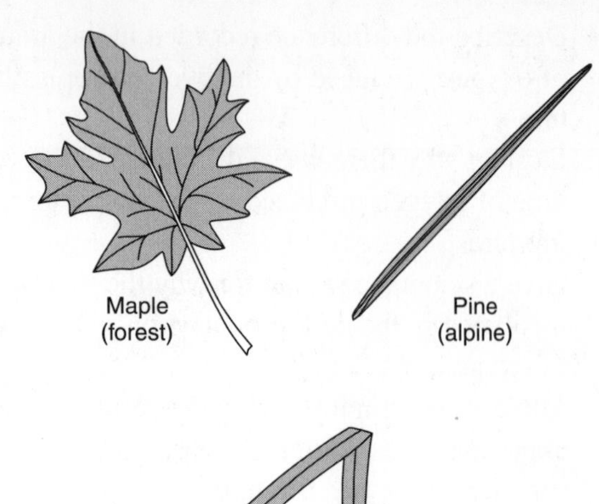

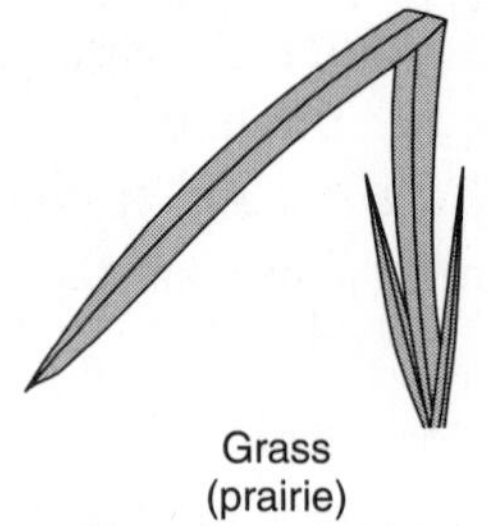

22. A student placed one of her tropical houseplants outside on her porch during the summer to receive some natural sunshine. A few days later, a rubbish fire broke out in a nearby vacant lot, spreading soot all over the neighborhood. Within two weeks, the plant's leaves started to turn yellow and drop off the stem. Give two scientific explanations of how the soot may have caused the leaves to turn yellow and drop off.

23. Explain how a cactus's needle-shaped leaves help the cactus plant survive in the desert.

24. Explain how the following structures help a typical leaf carry out photosynthesis: the cuticle; chloroplasts; stomata; and guard cells.

HETEROTROPHIC NUTRITION

Organisms that cannot synthesize organic molecules from inorganic raw materials are *heterotrophs*; they must obtain preformed organic molecules from the environment for nutrition. Heterotrophic organisms include most bacteria, some protists, and all **fungi** and animals. Heterotrophic nutrition involves the processes of ingestion and digestion, usually beginning with the mechanical breakdown of food. Large pieces of food are broken down into smaller pieces by cutting, grinding, and tearing; the smaller pieces then provide greater surface area for the action of enzymes during chemical digestion.

Digestion. In some heterotrophs, chemical digestion is *intracellular*—it occurs within the cell (or cells) of the organism. In most heterotrophs, however, digestion is *extracellular*—it occurs in a sac or a tube outside the cells. The end products of digestion are then absorbed into the cells.

Adaptations for Heterotrophic Nutrition

Heterotrophs obtain nutrients in a variety of ways.

Protists. In protists, single-celled organisms such as the amoeba and paramecium, digestion is intracellular. In the amoeba, food particles are surrounded and engulfed by extensions of the cell called *pseudopods*, in a process known as *phagocytosis*. Within the cell, food particles are enclosed in a food vacuole. In the paramecium, food particles are ingested through a fixed opening called the *oral groove*. They are moved into this opening by the beating of tiny cytoplasmic "hairs" called *cilia*. The food particles are then enclosed in a food vacuole, which circulates in the cytoplasm (Figure 2-3).

In both the amoeba and paramecium, the food vacuole merges with a *lysosome*, an organelle that contains digestive enzymes. Food within the vacuole is digested by these enzymes, and the end products of digestion are then absorbed into the cytoplasm. In the amoeba, wastes are expelled from the cell through the cell membrane. In the paramecium, wastes are expelled through a fixed opening called the *anal pore*.

Humans. The human digestive system is essentially like that of most other **multicellular** (many-celled) animals. Food moves in one direction through a tube, and specialized organs carry out its mechanical breakdown and chemical digestion.

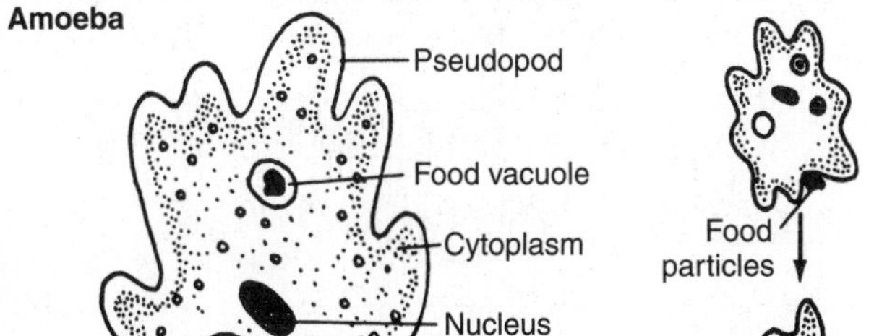

Figure 2-3. Nutrition in the amoeba and paramecium.

QUESTIONS

MULTIPLE CHOICE

25. Based on their type of nutrition, all animals are classified as A. autotrophic B. heterotrophic C. photosynthetic D. phagocytic

26. Digestion that occurs in a sac or a tube is referred to as A. phagocytic B. intracellular C. extracellular D. heterotrophic

27. A fruit fly is classified as a heterotroph, rather than as an autotroph, because it is unable to A. transport needed materials throughout its body B. release energy from organic molecules C. manufacture its own food D. divide its cells by mitosis

28. The principal function of mechanical digestion is the A. storage of food molecules in the liver B. production of more surface area for enzyme action C. synthesis of enzymes necessary for food absorption D. breakdown of large molecules to smaller ones by the addition of water

29. In the paramecium, most intracellular digestion occurs within structures known as A. ribosomes B. endoplasmic reticula C. mitochondria D. food vacuoles

30. Which organism ingests food by engulfing it with pseudopods? A. grasshopper B. paramecium C. amoeba D. earthworm

OPEN RESPONSE

31. How does mechanical digestion aid the process of chemical digestion?

32. Briefly compare intracellular digestion and extracellular digestion.

Base your answer to the following question on the information below and your knowledge of biology and experimental procedures.

33. A biology student performed an experiment to determine the rate of digestion by protease (a protein-digesting enzyme) on cooked egg white. He set up three sets of six test tubes each. Into the first set, he placed the same amount of water and protease plus two grams of cooked egg white into each test tube. The egg white was left in one piece in each tube. To the second set, he added the same amounts of water, protease, and egg white, but this time he cut the two grams of egg white into eight small pieces before placing it into each test tube. The third set of test tubes also received the same amounts of water, protease, and egg white, but the egg white was finely chopped up before being placed into each test tube.

- What hypothesis was the student most likely trying to test?
- Predict what should occur in each setup and give a scientific explanation for your prediction.
- The student omitted a control in his experiment. Describe an appropriate control group that could be used in this investigation.

TRANSPORT IN CELLS

Transport involves the absorption of materials through an organism's cell membranes and into its body fluids, and the circulation of materials throughout its body.

The Cell Membrane

The cell membrane surrounds the cell and regulates the passage of materials into and out of the cell.

Structure of the Cell Membrane. The currently accepted model of the structure of the cell membrane is called the *fluid mosaic model.* According to this model, the cell membrane consists of a double layer of lipids within which large protein molecules are located (Figure 2-4).

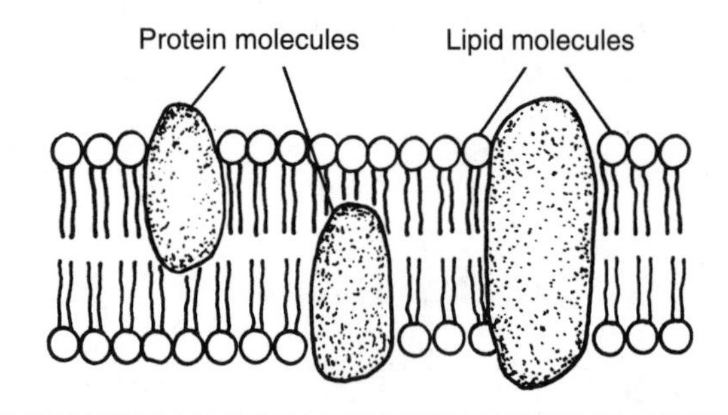

Figure 2-4. The fluid mosaic model of cell membrane structure.

Function of the Cell Membrane. The cell membrane selectively regulates the passage of substances into and out of the cell. Small molecules, including water, carbon dioxide, oxygen, and some end products of digestion, pass easily through the cell membrane. Most larger molecules, such as proteins and starch, cannot pass through the cell membrane. However, molecular size is not the only factor that affects the movement of substances into or out of a cell. The cell membrane may contain a number of special *receptor sites* to which molecules bind as they enter or leave the cell. Binding to the receptor site is often the only way these molecules can enter or leave a cell.

The shape of these receptor site molecules is highly specific to the shape of the molecule being transported. For example, a common receptor site on many cell membranes regulates the passage of sodium ions and potassium ions. Maintenance of homeostasis is largely dependent on the proper functioning of receptor sites on cells; and the failure of a receptor site can lead to a serious illness. Cystic fibrosis, for example, is linked to the failure of chloride ion receptor sites on the cell membranes, resulting in the accumulation of mucus in a person's lungs.

Diffusion and Passive Transport

All ions and molecules are in constant, random motion. When such particles collide, they bounce off each other and travel in new directions. As a result of their motion and collisions, the particles tend to spread out from an area where they are in higher concentration to an area where they are in lower concentration, a process known as **diffusion**. The difference in molecule or ion concentration between two such areas is known as the *concentration gradient*.

Molecules and ions that can pass through a cell membrane tend to move into or out of the cell by diffusion. The direction of diffusion depends on the relative concentration of the substance inside and outside the cell and usually results in a balance, or **equilibrium**, in the substance's concentration. Diffusion is a type of **passive transport**; it occurs because of the energy of the molecules and ions, and it does not require the input of additional energy by the cell. The diffusion of water through a cell membrane is called *osmosis*. Water molecules move easily from a region of higher concentration to a region of lower concentration until they reach an equilibrium. In *facilitated diffusion*, special receptor molecules in the cell membrane bond with, and move molecules and/or ions across, the cell membrane.

Active Transport

The process of **active transport** requires an input of energy by the cell to move molecules across its membrane because, in active transport, the substances are being moved *against* their concentration gradient. In active transport, protein molecules embedded in the cell membrane act as *carriers* that aid in the transport of materials across the membrane.

Pinocytosis and Phagocytosis

Large, dissolved molecules can pass through a cell membrane by the process of *pinocytosis* (Figure 2-5). In pinocytosis, the cell membrane folds inward. The outer surface of the cell membrane then closes over, and the large molecule is enclosed in a vacuole inside the cell. In contrast, *phagocytosis* is the process by which a cell engulfs large, undissolved particles by flowing around them and enclosing them in a vacuole. For example, amoeba use their pseudopods to engulf food particles during phagocytosis (refer to Figure 2-3).

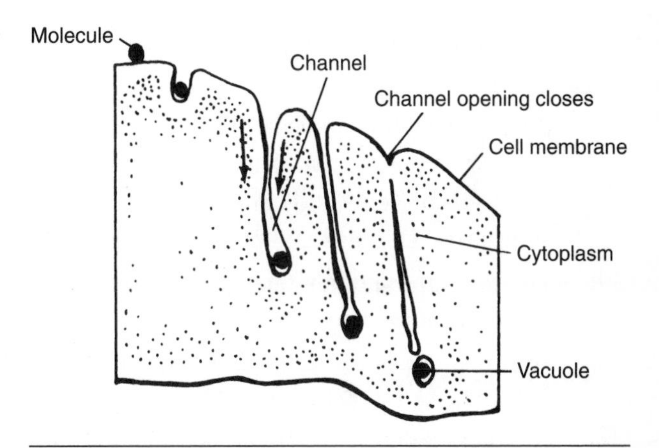

Figure 2-5. Pinocytosis.

Circulation

Circulation involves the movement of materials both within cells and throughout multicellular organisms. The movement of materials within a cell, *intracellular circulation*, takes place by diffusion and by *cyclosis*. Cyclosis is the natural streaming of cytoplasm that occurs within all cells. Intracellular circulation may also involve the movement of materials through the channels of the endoplasmic reticulum. The transport of materials throughout multicellular organisms is called *intercellular circulation*. Depending on the complexity of the organism, intercellular circulation may occur by diffusion or it may involve a specialized circulatory system with conducting, or *vascular*, tissues. Such a system, in which blood flow is restricted to vessels, is called a *closed circulatory system*.

QUESTIONS

MULTIPLE CHOICE

34. Which process would describe the movement of molecules through a membrane, going from a region of higher concentration to a region of lower concentration?
A. osmosis B. cyclosis C. passive transport D. active transport

35. In the human body, the potassium ion can pass easily through cell membranes, yet the potassium ion concentration is higher inside many cells than it is outside these cells. This condition is mainly the result of A. passive transport B. active transport C. osmosis D. pinocytosis

36. Chemical analysis indicates that the cell membrane is composed mainly of A. proteins and starch B. proteins and cellulose C. lipids and starch D. lipids and proteins

37. The flow of materials through the membrane of a cell against the concentration gradient is known as A. passive transport B. active transport C. osmosis D. pinocytosis

38. A biologist observed a plant cell in a drop of water and illustrated it as in diagram *A*. He added a 10 percent salt solution to the slide, observed the cell, and illustrated it as in diagram *B*. The change in appearance of the cell resulted from more A. salt flowing out of the cell than into the cell B. salt flowing into the cell than out of the cell C. water flowing into the cell than out of the cell D. water flowing out of the cell than into the cell

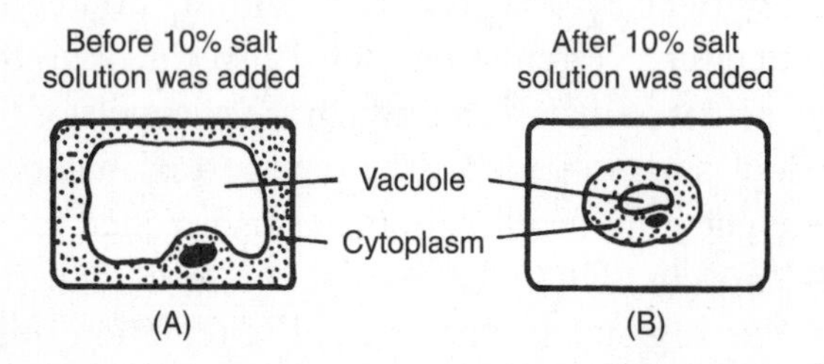

39. The natural streaming of the cytoplasm that occurs within all cells is called
A. pinocytosis B. phagocytosis C. osmosis D. cyclosis

40. When a cell uses energy to move materials across its membrane, the process is known as A. osmosis B. active transport C. diffusion D. passive transport

41. The diffusion of water molecules into and out of cells is called A. cyclosis B. pinocytosis C. osmosis D. active transport

42. The movement of molecules into and out of cells is most dependent on the A. selectivity of the cell membrane B. selectivity of the cell wall C. number of vacuoles D. number of chromosomes

43. The process by which amoeba ingest food particles is called A. pinocytosis B. osmosis C. phagocytosis D. cyclosis

Base your answers to questions 44 and 45 on the following information and diagram, and on your knowledge of biology.

An investigation was set up to study the movement of water through a membrane. The results are shown in the diagram below.

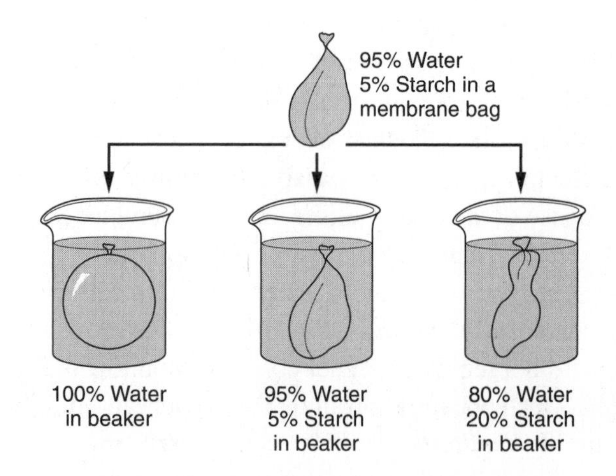

44. Based on these results, which statement correctly predicts what will happen to red blood cells when they are placed in a beaker containing a water solution in which the salt concentration is much higher than the salt concentration in the red blood cells? A. The red blood cells will absorb water and increase in size. B. The red blood cells will lose water and decrease in size. C. The red blood cells will first absorb water, then lose water and maintain their normal size. D. The red blood cells will first lose water, then absorb water, and finally double in size.

45. A red blood cell placed in distilled water will swell and burst due to the diffusion of A. salt from the red blood cell into the water B. water into the red blood cell C. water from the blood cell into its environment D. salt from the water into the red blood cell

Base your answers to questions 46 and 47 on the diagram below, which illustrates a process by which protein molecules may enter a cell, and on your knowledge of biology.

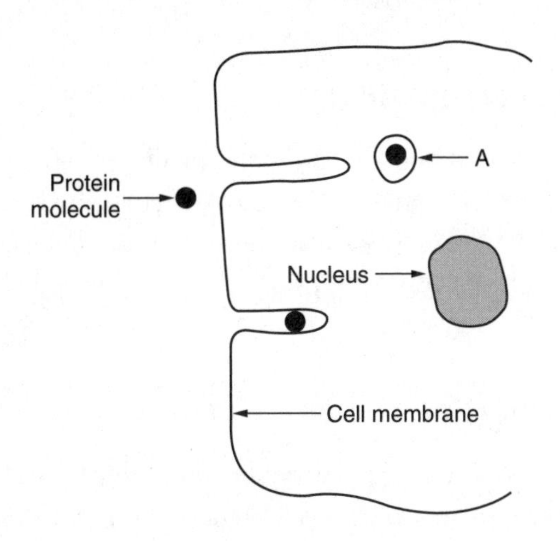

46. Which process is illustrated in this diagram? A. pinocytosis B. osmosis C. diffusion D. passive transport

47. Structure *A* is most likely a A. ribosome B. mitochondrion C. nucleolus D. vacuole

OPEN RESPONSE

48. Describe the differences between active transport and passive transport. Give one example of each type of transport.

49. Freshwater protists live in an environment that is close to 100 percent water. The inside of their cell (cytoplasm) is about 90 percent water. Explain the problem these protists face with respect to maintaining homeostasis in their environment. How might they be able to overcome this problem?

Use the information in the paragraph and table below to answer the following questions.

50. A biology student was attempting to determine the percent of water present in the cells of *Elodea* (an aquatic plant). She placed leaves of *Elodea* in varying concentrations of saltwater solutions and observed when plasmolysis (cell shrinking) occurred. The table summarizes the results of her experiment.

Solution Concentration	*Observed Plasmolysis*
0.5% NaCl	None
1.0% NaCl	None
1.5% NaCl	None
2.0% NaCl	Very slight
2.5% NaCl	Pronounced
3.0% NaCl	Pronounced

- According to the data in the table, what percent of *Elodea* cells is water? Explain how you arrived at this conclusion.
- Give a scientific explanation for what caused the cells to shrink at a certain concentration of salt water. What process causes plasmolysis of the cells?

TRANSPORT IN ORGANISMS

Transport in Plants

The transport of materials in plants involves cyclosis, osmosis, diffusion, and active transport. Some plants contain specialized transport, or *vascular*, tissues while others do not.

Roots. Roots are structures that are specialized for the absorption of water and minerals from the soil and the transport of these materials to the stem. Roots also anchor the plant in the soil and may contain stored nutrients in the form of starch.

The surface area of the root is increased (for greater absorption) by the presence of *root hairs* just behind the growing tip (Figure 2-6). Water and minerals from the soil are absorbed through the membranes of the root hairs by osmosis, diffusion, and active transport. Materials are transported throughout the plant by two kinds of vascular tissues, *xylem* (for water) and *phloem* (for food).

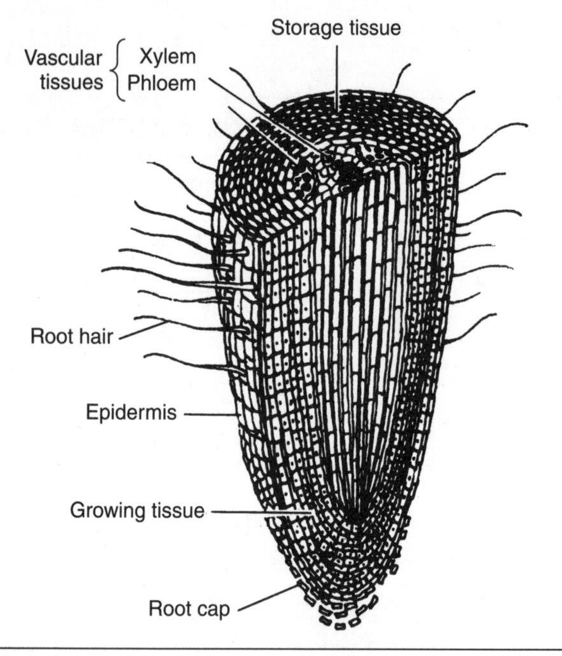

Figure 2-6. Structure of a root tip.

Stems. Although the structure of stems is more complex than that of roots, the xylem and phloem of the stem are continuous with the xylem and phloem of the roots.

Leaves. The xylem and phloem of the leaves, which are in bundles called *veins*, are also continuous with the xylem and phloem of the roots and stem.

Transport in Protists

Protists and other unicellular organisms have no specialized transport system. Materials enter and leave the cell by diffusion and active transport, and are circulated within the cell by diffusion and cyclosis (Figure 2-7).

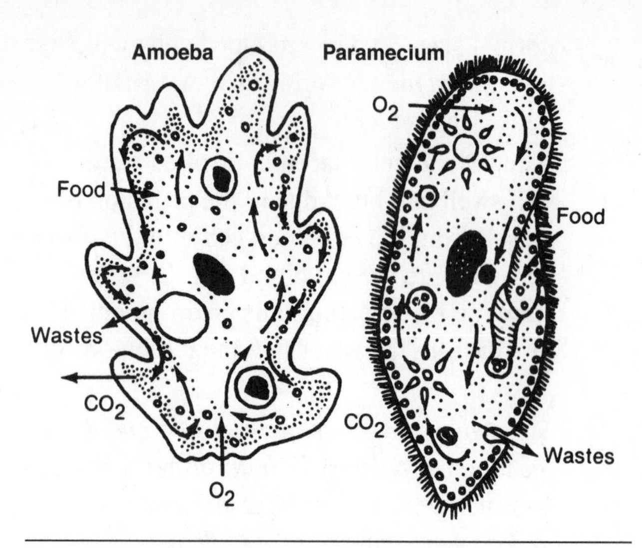

Figure 2-7. Transport in amoeba and paramecium.

Transport in Animals

Simple multicellular animals (such as sponges), whose cells are in direct contact with the surrounding water, have no specialized transport system. All other (that is, more complex) multicellular animals do have specialized systems for the transport of materials.

Transport in Humans

The human circulatory system is a closed system. This means that blood is contained within a system of vessels as it moves through the body by the pumping action of the heart (Figure 2-8). Human blood contains the pigment *hemoglobin*, which carries oxygen to the body tissues.

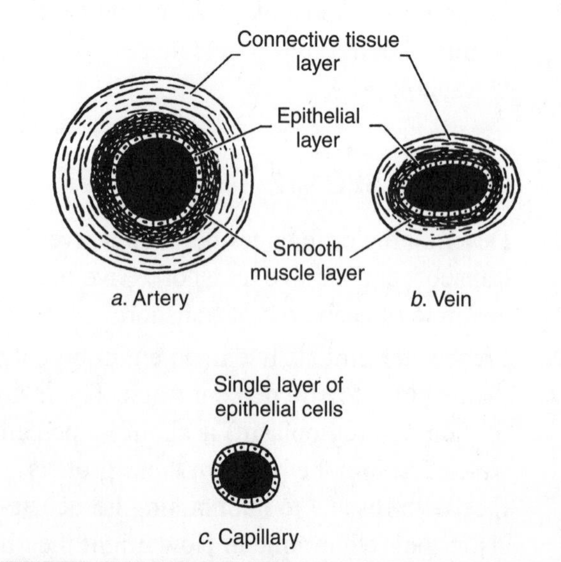

Figure 2-8. The three types of blood vessels.

QUESTIONS

MULTIPLE CHOICE

51. The primary function of the root hairs in a plant is to A. prevent excessive loss of water B. provide increased surface area for absorption C. conduct water and minerals upward D. conduct organic food materials upward and downward

52. A circulatory system in which the blood remains within vessels is called A. a closed circulatory system B. an open circulatory system C. an internal circulatory system D. an external circulatory system

OPEN RESPONSE

53. The epidermal (outermost) cells of a plant's roots can continue to absorb water even when the concentration of water in the soil is very low, even lower than that in its cells. Explain the biological process that enables the root epidermal cells to do this.

54. Use your knowledge of biology to answer the following questions comparing a single-celled organism (such as an amoeba) and a multicelled organism (such as a human):

- What are two similarities in their transport systems?
- What are two differences in their transport systems?

RESPIRATION

The life processes of all organisms require energy; and there is potential energy in the chemical bonds of organic molecules such as glucose. However, this energy cannot be used directly in cell metabolism. During *cellular respiration*, these chemical bonds are broken; the energy that is released is temporarily stored in the bonds of the energy-transfer compound called *ATP* (adenosine triphosphate). This process occurs continuously in the cells of all organisms.

Cellular Respiration

Cellular respiration involves a series of enzyme-controlled reactions in which the energy released by the breakdown of the chemical bonds in glucose is transferred to ATP. In eukaryotes, cellular respiration occurs within the mitochondria. The formation of ATP is a critical component of cellular respiration since cells cannot *directly* use the energy present in glucose. Adenosine triphosphate is often called the "energy currency" of the cell because it is immediately available to provide the energy needed for life processes. When ATP is broken down by hydrolysis (the chemical addition of water), ADP (adenosine diphosphate) and phosphate (P) are produced, and energy is released for use by the cell.

The conversion of ATP to ADP is a reversible reaction catalyzed by the enzyme ATP synthase. In living organisms, ATP is constantly being converted to ADP, and the energy released is used for the reactions of cell metabolism. The ADP is then converted back to ATP by the reactions of cellular respiration.

$$H_2O + ATP \leftarrow \text{ATP synthase} \rightarrow ADP + P + \text{energy}$$

Anaerobic Respiration. In most organisms, cellular respiration requires the presence of oxygen, and the process is known as *aerobic respiration*. In a few kinds of organisms, oxygen is not used, and the process is known as *anaerobic respiration*, or *fermentation*. Some cells, such as muscle cells, which normally use aerobic respiration, can also use anaerobic respiration in the absence of oxygen. Other cells, such as yeast and some bacteria, which use anaerobic respiration, lack the enzymes necessary for aerobic respiration. In this type of respiration, there is a net gain of only two ATP molecules for each molecule of glucose consumed.

Aerobic Respiration. In aerobic respiration, glucose is broken down completely to carbon dioxide and water by a series of enzyme-controlled reactions. Under ideal circumstances, these reactions, which take place mainly in the mitochondria, produce a net gain of 36 ATP molecules. However, the actual number of ATP molecules produced can vary from as few as 20 to as many as 38 per glucose molecule broken down, depending on various factors in the cell's environment.

$$\text{glucose} + \text{oxygen} \xrightarrow{\text{enzymes}} \text{water} + \text{carbon dioxide} + \text{ATP}$$

$$C_6H_{12}O_6 + 6O_2 \xrightarrow{\text{enzymes}} 6H_2O + 6CO_2 + 36\ \text{ATP}$$

Adaptations for Respiration

The oxygen used in aerobic cellular respiration comes from the environment, and the carbon dioxide produced must be released into the environment. Although the chemical processes of respiration are similar in most organisms, living things show a variety of adaptations for the exchange of these respiratory gases.

Protists. In single-celled organisms, such as protists, all or most of the cells are in direct contact with the environment. So the exchange of respiratory gases takes place by diffusion through the thin, moist cell membranes (Figure 2-9).

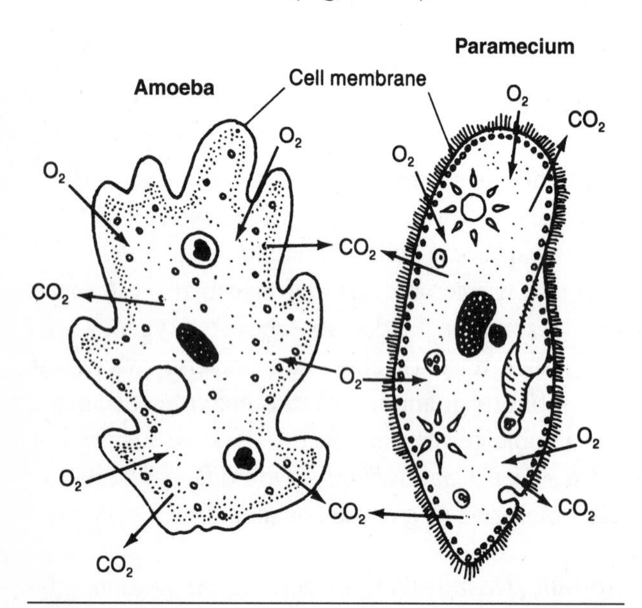

Figure 2-9. Respiration in amoeba and paramecium.

Plants. In plants, respiratory gases are exchanged through the leaves, stems, and roots. The exchange of respiratory gases occurs by diffusion through the cell membranes of internal cells, which are surrounded by intercellular spaces. The intercellular spaces open to the environment through the stomata, openings on the undersurface of the leaf.

Humans. In humans, the exchange of respiratory gases occurs by diffusion through thin, moist membranes within the lungs. Hemoglobin aids in the transport of oxygen in the blood. Carbon dioxide and oxygen are carried between the respiratory surface in the lungs and the environment by a system of air tubes.

QUESTIONS

MULTIPLE CHOICE

55. Most animals make energy available for cell activity by transferring the potential energy of glucose to ATP. This process occurs during A. aerobic respiration only B. anaerobic respiration only C. both aerobic and anaerobic respiration D. neither aerobic nor anaerobic respiration

56. In animal cells, the energy to convert ADP to ATP comes directly from A. hormones B. sunlight C. organic molecules D. inorganic molecules

57. The organelles in which most of the reactions of aerobic cellular respiration take place are the A. ribosomes B. chloroplasts C. lysosomes D. mitochondria

58. Substances that most directly control the rate of reaction during cellular respiration are known as A. enzymes B. phosphates C. monosaccharides D. disaccharides

59. Which end product is of the greatest benefit to the organism in which respiration occurs? A. glucose B. carbon dioxide C. ATP molecules D. water molecules

60. Protists obtain oxygen from their environment through A. stomata B. cell membranes C. vacuoles D. mitochondria

61. Which process requires carbon dioxide molecules? A. cellular respiration B. asexual reproduction C. active transport D. photosynthesis

62. In humans, respiratory gases are exchanged between the lungs and the environment through A. air tubes B. hemoglobin C. vacuoles D. stomata

63. Arrows *A*, *B*, and *C* in the diagram below represent the processes necessary to make the energy stored in food available for muscle activity. The correct sequence of processes represented by *A*, *B*, and *C* is

A. diffusion→synthesis→active transport

B. digestion→diffusion→cellular respiration

C. digestion→excretion→cellular respiration

D. synthesis→active transport→excretion

Food —A→ Simpler molecules —B→ Mitochondria —C→ ATP in muscle cells

Base your answers to questions 64 through 68 on the diagram below, which represents a cellular process in animals, and on your knowledge of biology.

64. The items labeled as food molecules most likely represent A. starch B. glucose C. phosphate D. chlorophyll

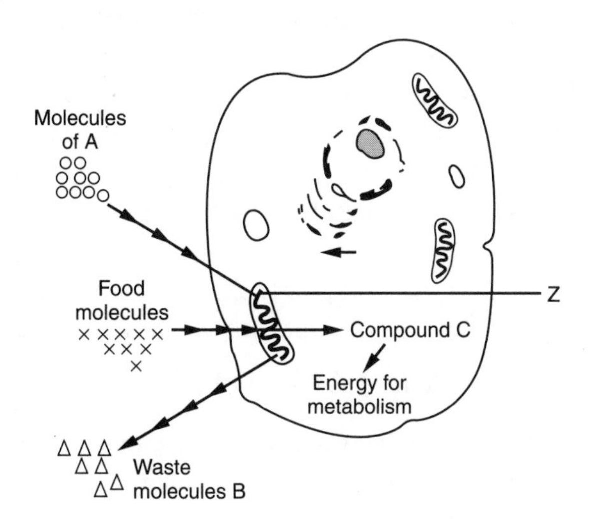

65. Compound *C* most likely represents some molecules of A. oxygen B. glucose C. ATP D. DNA

66. If this cell is carrying on aerobic respiration, *B* represents molecules of a waste product known as A. carbon dioxide B. ATP C. ethyl alcohol D. phosphate

67. If this represents a cell from the human body, the molecules of *A* are most probably A. carbon dioxide B. enzymes C. lipids D. oxygen

68. The cell organelle labeled *Z* is called a A. chloroplast B. mitochondrion C. nucleolus D. vacuole

OPEN RESPONSE

69. Why must all organisms carry out cellular respiration?

70. Briefly compare the processes of cellular respiration in aerobic and anaerobic organisms. Include the following:

- the function of oxygen;
- the net gain of ATP molecules.

Base your answers to question 71 on the following information and on your knowledge of biology.

71. A biologist was culturing some muscle cells from a mouse (an aerobic organism) in a petri dish. He was interested in measuring the amount of ATP produced by the muscle cells when the cells were supplied with glucose. At the beginning of the experiment, the cells were producing large quantities of ATP. He then added a substance called malonic acid to the cell culture, and the amount of ATP produced fell to near zero.

- Which organelles in the muscle cells were most likely affected by the malonic acid?
- Propose a testable hypothesis concerning respiration in cells treated with malonic acid.
- Predict the effect of malonic acid on an anaerobic organism and explain your prediction.

EXCRETION

Metabolic activities of living cells produce waste materials. The life process by which the wastes of metabolism are removed from the body is called *excretion*.

Wastes of Metabolism

The waste products of various metabolic processes are shown in Table 2-1. Some wastes are toxins (compounds that are poisonous to body tissues), while other wastes are nontoxic. In animals, toxic wastes are usually excreted from the body. In plants, toxic wastes are sealed off and stored, sometimes in vacuoles. Some nontoxic wastes are excreted, while others are recycled and used in metabolic activities.

***Table 2-1.* The Waste Products of Metabolism**

Metabolic Activity	*Waste Products*
Respiration	Carbon dioxide and water
Dehydration synthesis	Water
Protein metabolism	Nitrogenous wastes
Certain metabolic processes	Mineral salts

Nitrogen-containing, or *nitrogenous*, wastes are produced by the breakdown of amino acids. Different kinds of organisms produce different kinds of nitrogenous wastes, including *uric acid*, which is nontoxic; *urea*, which is moderately toxic; and *ammonia*, which is highly toxic.

Adaptations for Excretion

In single-celled or simple organisms (such as sponges), wastes pass from the cells directly into the external environment. More complex organisms have a specialized excretory system.

Protists. In general, the excretion of wastes in protists is accomplished by diffusion through the cell membrane (Figure 2-10). In freshwater protozoans, such as the paramecium, water continuously enters the cell by osmosis. In these organisms, the excess water collects in organelles called *contractile vacuoles*. The vacuoles move to the surface of the cell and expel the water back into the environment. This process involves active transport.

In freshwater protozoans, the nitrogenous waste product is ammonia. Although it is very toxic, ammonia is also very soluble in water, and thus it can be easily excreted from the cells of these organisms.

In photosynthetic protists, such as algae, some of the carbon dioxide produced by cellular respiration can be recycled and used in photosynthesis. Some of the oxygen produced by photosynthesis can be used in cellular respiration.

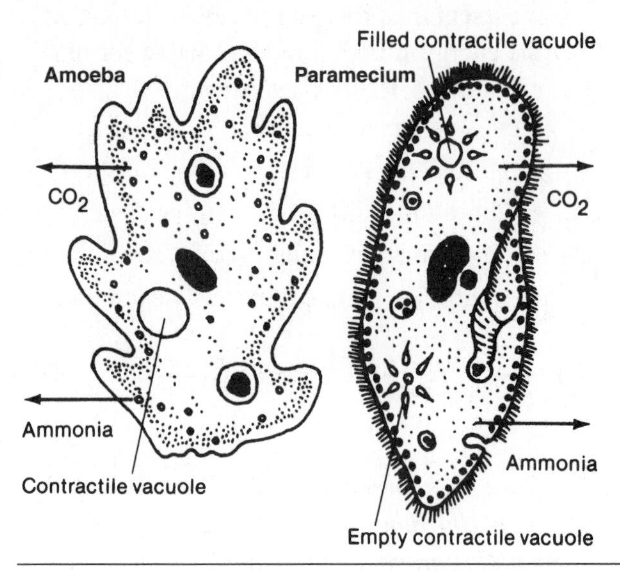

Figure 2-10. Excretion in amoeba and paramecium.

Plants. In plants, as in algae, some of the waste gases produced by photosynthesis and cellular respiration are recycled. Excess gases diffuse out of the plant through stomata on the leaves, tiny openings on the stems, and epidermal cells on the roots.

Humans. In humans, carbon dioxide is excreted by the lungs. Water, salts, and urea are excreted by the kidneys (and skin). These waste products are passed out of the body through various specialized tubes.

QUESTIONS

MULTIPLE CHOICE

72. Metabolic wastes of animals most likely include A. water, carbon dioxide, oxygen, and salts B. carbon dioxide, nitrogenous compounds, water, and salts C. hormones, water, salts, and oxygen D. glucose, carbon dioxide, nitrogenous compounds, and water

73. Which activity would most likely produce nitrogenous waste products? A. protein metabolism B. glucose metabolism C. lipid metabolism D. starch metabolism

74. The leaf structures that are closely associated with both respiration and excretion are the A. root hairs B. stomata C. waxy surfaces D. epidermal cells

75. Protists can function without an organized excretory system because their cells A. do not produce wastes B. change all wastes into useful substances C. remove only solid wastes D. are in direct contact with a water environment

76. Which statement best describes the excretion of nitrogenous wastes from paramecia? A. Urea is excreted by nephrons. B. Uric acid is excreted by nephrons. C. Urea is excreted through tiny tubules. D. Ammonia is excreted through cell membranes.

77. Most toxic products of plant metabolism are stored in the A. stomata B. vacuoles C. root cells D. chloroplasts

78. In freshwater protozoans, the organelles involved in the maintenance of water balance are A. food vacuoles B. mitochondria C. contractile vacuoles D. pseudopods

OPEN RESPONSE

79. Identify two waste products excreted by the kidneys. Why must these wastes be removed?

Base your answer to the following questions on the table below and on your knowledge of biology.

80. The table below compares three nitrogenous waste products, their toxicity levels, their solubility in water, and the habitats in which the organisms that produce them typically live.

Type of Waste	*Toxicity*	*Solubility*	*Habitat of Organism*
Ammonia	Very high	Very good	Aquatic (in water)
Urea	Moderate	Good	Land, most often
Uric acid	Low	None	Land, often desert

- What connection exists between the habitat of an organism and the toxicity of its nitrogenous waste?
- What connection exists between the solubility of each nitrogenous waste and its toxicity?
- State a possible biological benefit of the connections between waste toxicity, waste solubility, and an organism's habitat.

81. Organisms produce waste products as a result of their metabolic activities. These wastes include carbon dioxide, water, mineral salts, and nitrogenous wastes (such as ammonia, urea, and uric acid). Select any three of these four metabolic wastes and answer the following:

- How are these wastes produced in an organism?
- Why must these wastes be removed from an organism?

REGULATION

Regulation involves the control and *coordination* of life activities. In all organisms, there are chemicals that regulate life activities. In multicellular animals, there is nerve control in addition to chemical control. Both nerve control and chemical control aid organisms in their maintenance of homeostasis.

Nerve Control

Nerve control depends mainly on the functioning of **nerve cells**, or *neurons*, which are specialized for the transmission of impulses from one part of the body to another.

Structure of a Nerve Cell. The three parts of a nerve cell are the *dendrites*; the cell body, or *cyton*; and the *axon* (Figure 2-11). Dendrites are composed of many branches, but the axon (covered by the myelin sheath) has branches mainly at the end that is

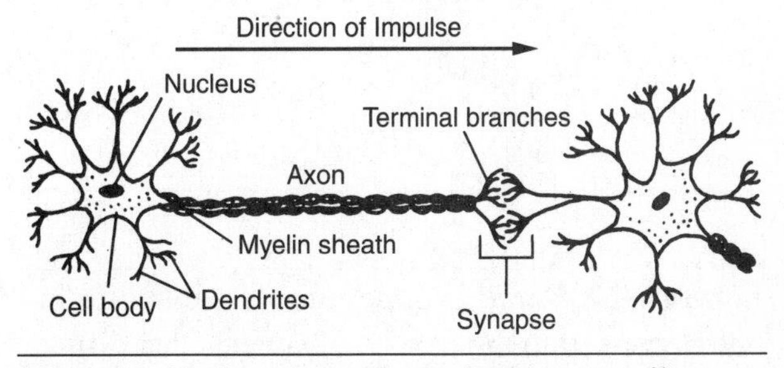

Figure 2-11. Structure of a typical nerve cell.

farthest from the cell body. Impulses are received by the dendrites and passed to the *cell body*, which contains the nucleus and other organelles. From the cell body, impulses pass along the axon to its terminal (end) branches.

Impulses. An *impulse* is a region of electrical and chemical, or *electrochemical*, change that travels over the membrane of a nerve cell. When electrochemical impulses reach the terminal branches (end brush) of an axon, they stimulate the release of chemicals called *neurotransmitters*.

Neurotransmitters and Synapses. The junction between adjacent nerve cells is called a *synapse*. At the synapse, the nerve cells do not touch; there is a small gap between them. When impulses reach the terminal branches of the axon of one nerve cell, they stimulate the release of neurotransmitters, such as acetylcholine, which diffuse across the gap of the synapse. The neurotransmitter stimulates impulses in the dendrites of the second nerve cell. In this way, impulses pass from one nerve cell to another (Figure 2-12).

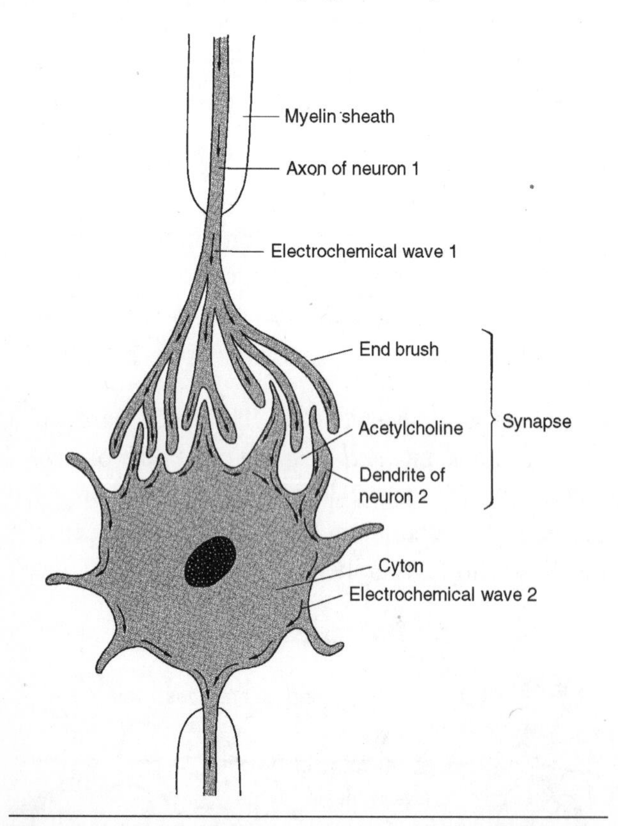

Figure 2-12. Nerve impulses are transmitted from one nerve cell to the next by chemicals that diffuse across the gap (at the synapse).

The axons of some nerve cells have junctions with a muscle or a gland. In such cases, the chemicals released by the terminal branches of the axon stimulate contraction of the muscle or secretion by the gland.

Stimulus and Receptors. Any change in the external or internal environment that initiates impulses is called a **stimulus** (plural, *stimuli*). Stimuli are detected by specialized structures called **receptors**. Each kind of receptor is sensitive to a particular kind of stimulus; for example, eyes are sensitive to light, ears to sound, and so on.

Responses and Effectors. The reaction of an organism to a stimulus is called a **response**. The response to a stimulus is carried out by *effectors*, generally the muscles or glands.

Adaptations for Nerve Control

Even the simplest animals have some type of nerve cells that transmits impulses.

Simple Animals. In some animals, there is no brain, but there is a structure called a *nerve net* that transmits messages throughout the body. In other animals, there is a primitive brain, a *nerve cord* that runs the length of the body, *peripheral nerves* that serve all parts of the body, and in some, sense receptors.

Humans. Humans have a central nervous system that consists of a highly developed brain and a dorsal (spinal) nerve cord that runs down the back. The central nervous system permits impulses to travel in one direction along definite pathways. There is also a peripheral nervous system that consists of an elaborate network of nerves. The peripheral nervous system carries signals between the central nervous system and all parts of the body. In addition, there are many highly developed sense organs, such as smell, sight, hearing, taste, and skin receptors.

Chemical Control

In both plants and animals, various aspects of their life activities are controlled by the chemicals called **hormones**.

Plant Hormones. In plants, there are no organs specialized for the production of hormones. Plant hormones are produced in greatest abundance in the cells of actively growing regions, called *meristems,* such as the tips of roots and stems and in buds and seeds. The hormones produced in these regions affect the growth and development of cells in other parts of the plant. The effects of hormones vary with their concentration and with the type of tissue they act on.

Animal Hormones. Unlike plants, many animals do have organs specialized for the synthesis and secretion of hormones. These organs, called *endocrine glands,* or ductless glands, release their secretions directly into the bloodstream. Hormones are found in a wide variety of animals, both vertebrates and invertebrates. The various hormones control the animals' metabolic activities, as well as their metamorphosis and reproduction.

QUESTIONS

MULTIPLE CHOICE

82. Animal cells that are specialized for conducting electrochemical impulses are known as A. nerve cells B. synapses C. nephrons D. neurotransmitters

83. A hawk gliding over a field suddenly dives toward a moving rabbit. The hawk's reaction to the rabbit is known as a A. stimulus B. synapse C. response D. impulse

84. Transmission of nerve impulses at synapses involves chemicals called A. hormones B. neurotransmitters C. enzymes D. nucleic acids

85. Neurotransmitters, such as acetylcholine, are initially detected by which part of a nerve cell? A. dendrites B. nucleus C. terminal branches D. mitochondrion

86. The nucleus of a nerve cell is found in the A. dendrite B. axon C. synapse D. cell body

87. Structures that detect stimuli are called A. effectors B. receptors C. synapses D. cell bodies

88. The secretions of endocrine glands are known as A. enzymes B. hormones C. pigments D. neurotransmitters

89. A chemical injected into a tadpole caused the tadpole to undergo rapid metamorphosis into a frog. This chemical was most probably a (an) A. enzyme B. neurotransmitter C. hormone D. blood protein

90. The two systems that directly control homeostasis in most animals are the A. nervous and endocrine B. endocrine and excretory C. nervous and circulatory D. excretory and circulatory

OPEN RESPONSE

91. Compare the central nervous system and the peripheral nervous system. What are the main structures and functions of each system?

92. Explain why the endocrine glands are also referred to as *ductless* glands. How is this feature related to their function?

LOCOMOTION

Locomotion is the ability to move from place to place. Among many protists and animals, locomotion improves the organism's ability to survive. It increases chances of finding food and shelter, avoiding predators and other dangers, and finding a mate.

Adaptations for Locomotion

Many protists and almost all animals are capable of some form of locomotion, or **movement**, from one place to another. Such organisms are said to be *motile*. The hydra is generally a *sessile* organism; it tends to remain in one place, fastened to another structure. However, it does have fibers that permit some limited movements.

Protists. There are three basic forms of locomotion among protists. In the amoeba, locomotion is by ameboid motion, in which the cell flows into newly formed extensions of its cytoplasm, called *pseudopods*. This causes the organism to move in the direction of the pseudopods. In the paramecium, locomotion involves *cilia*, which are short, hairlike organelles that cover the outer surface of the cell. The cilia wave back and forth in a coordinated way, moving the cell through the water. Some algae and other protozoans move by means of *flagella*, long, hairlike organelles that can pull the cell through the water (Figure 2-13).

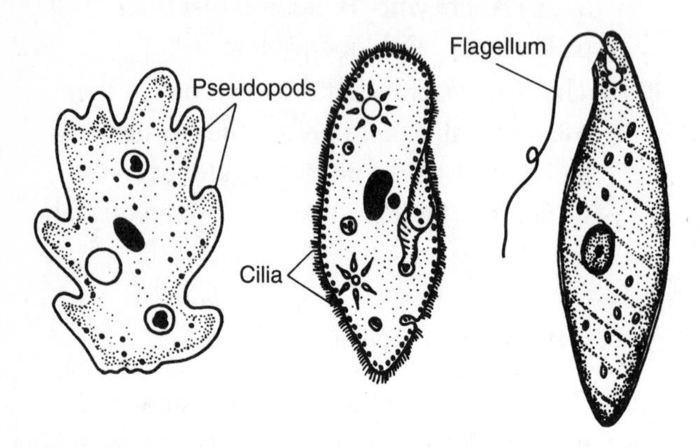

Figure 2-13. Locomotion in amoeba, paramecium, and euglena (an alga). (Not drawn to scale.)

Humans. Humans have an internal skeleton, or *endoskeleton*. Locomotion is accomplished by the interaction of muscles with jointed bones and cartilage (in the appendages), in response to nerve impulses.

QUESTIONS

MULTIPLE CHOICE

93. Locomotion increases an animal's opportunity to do all of the following *except*
A. obtain food B. find a mate and reproduce C. escape from predators D. transmit impulses

94. Which structures are *not* associated with locomotion among protists? A. flagella B. cilia C. pseudopods D. tentacles

95. Which organism is able to move due to the interaction of its muscles and skeleton?
A. amoeba B. paramecium C. human D. earthworm

OPEN RESPONSE

96. Describe four survival advantages of locomotion.

97. Some organisms are sessile; that is, they are incapable of moving from place to place. Explain how their survival might be aided in other ways.

READING COMPREHENSION

Base your answers to questions 98 through 101 on the information below and on your knowledge of biology. Source: *Science News* (April 30, 2005): Vol. 167, No. 18, p. 275.

Gene Therapy Slows Down Alzheimer's Disease

Putting extra copies of the gene for a cellular growth factor into the brains of people with Alzheimer's disease slows the degenerative condition, a new study suggests.

Alzheimer's disease kills neurons, the brain cells that orchestrate message signaling throughout the nervous system. The gene added in this study encodes nerve growth factor (NGF), a protein that keeps these cells alive and so facilitates signaling among them. The vehicle for the human gene was the patients' own skin cells. Researchers took a bit of skin tissue from each of eight people diagnosed with early Alzheimer's disease and used a non-replicating virus to transfer genes for human NGF into the skin cells. The scientists then injected these genetically modified cells into each patient's brain. However, two of the patients were excluded from the study soon after that surgery because of bleeding in their brains. Over the next two years, positron-emission tomography scans of the other patients revealed increased metabolic activity in their brains, a sign of neuron rejuvenation. An autopsy on one of the excluded patients, who died of a heart attack during the study, revealed that the implanted cells were making NGF. Nearby neurons appeared healthy.

These biological findings paralleled changes in the patients' behavior. Although standard testing indicated that the patients, on average, continued their mental declines during the 2 years after surgery, the pace of cognitive loss was only half as great as the patients had been experiencing before undergoing the gene therapy, says study coauthor Mark H. Tuszynski of the University of California, San Diego in La Jolla. The findings will appear in an upcoming *Nature Medicine*. The patients scored best on tests administered more than 6 months after the surgery, suggesting that the transplanted cells took several months to rev up their production of NGF, he says.

This marks the first time that researchers have surgically intervened in Alzheimer's disease to the benefit of patients, Tuszynski says.

Although the gene therapy slows the pace of Alzheimer's decline more than drugs currently prescribed for the disease do, he adds, the surgery is unlikely to represent a cure because it doesn't address the fundamental symptom of Alzheimer's disease: the accumulation of waxy plaques in a person's brain. "The magnitude of effect shown here is not terribly great, but any positive benefits for Alzheimer's patients would be good," say Curt R. Freed of the University of Colorado School of Medicine in Denver. "This could signal a new phase of treatment for this disorder."

However, Freed notes that skin cells used as a gene-delivery vehicle "may not be as natural to the brain as are the brain's own cells." With that in mind, Tuszynski is teaming with researchers to inject an innocuous [harmless] virus, loaded with the gene for NGF, directly into the brains of Alzheimer's patients. The scientists expect this virus to install the gene into neurons, which would then crank out NGF, says David A. Bennett of Rush University Medical Center in Chicago, who is collaborating on the project.

98. Explain the function of human NGF (nerve growth factor).

99. How was the gene for NGF introduced into the brains of Alzheimer's patients?

100. What two facts in the article show that this gene therapy does not actually cure Alzheimer's disease?

101. Describe how researchers might modify the technique to improve results in patients.

CHAPTER 3

Human Anatomy and Physiology

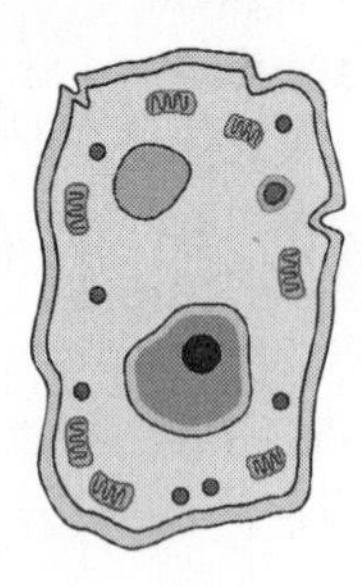

Standards 4.1, 4.2, 4.3, 4.4, 4.5, 4.7, 4.8 (Anatomy and Physiology)
BROAD CONCEPT: There is a relationship between the organization of cells into tissues, and tissues into organs. The structure and function of organs determine their relationship within body systems of an organism. Homeostasis allows the body to perform its normal functions.

LEVELS OF ORGANIZATION

In multicellular organisms, there are several levels of organization that allow living things to maintain and carry out their normal metabolic functions. The simplest level of organization is the **cell**, the basic unit of living things. Cells, in turn, are organized by function into the next higher level, the **tissues**. For example, groups of muscle cells that function together make up the muscle tissue. Tissues of the same type are arranged into **organs**, such as the stomach, lung, or heart. Organs that work together make up an **organ system**, such as the respiratory system or the digestive system. Finally, all of the organ systems work together so that the **organism** can carry out its life activities.

There is always some interaction between the organ systems of an organism. What occurs in one system has an effect on another system. Exercising, for example, affects the circulatory, respiratory, and muscular systems of a person. In order for its organ systems to continue to function effectively, a body's internal stability, or **homeostasis**, must be maintained. Changes that occur within an organism trigger adjustments in its systems, which tend to restore the original stability. For example, strenuous exercise increases the rate at which the heart delivers blood to muscle tissue (to supply the cells with nutrients and oxygen and to remove waste products). Additionally, the respiratory system speeds up its activity to bring in more oxygen and to remove carbon dioxide. After the exercise is completed, the systems soon return to their normal pre-exercise level of activity. Failure to maintain such homeostasis in an organism can cause illness or even be life threatening.

NUTRITION

The specific nutritional requirements of individual humans depend on such factors as age, gender, and activity level. However, like all other animals, humans are *heterotrophs*. This means they must take in, or *ingest*, the nutrients they need for energy and growth, which includes carbohydrates, proteins, lipids, vitamins, minerals, and water. Carbohydrates, lipids (fats), and proteins are made up of large molecules, or *macromolecules*, that must be digested before they can be absorbed and used by the cells. Recall that cells have selectively permeable membranes that limit the size and type of substances that can enter them. *Digestion* involves physical and chemical processes that break the macromolecules down to a size that is small enough to pass through the cell membranes. Vitamins, minerals, and water are made up of small molecules that can be taken in or absorbed by cells without first being digested.

Human Digestive System

The human digestive system is an organ system that consists of a one-way digestive tube called the *gastrointestinal*, or GI, tract and accessory organs (Figure 3-1). Food is moved through the GI tract by rhythmic, muscular contractions called *peristalsis*. As food moves through the tract, it is broken down physically and chemically. The accessory organs—liver, gallbladder, and pancreas—secrete enzymes and other substances that aid digestion into the digestive tract.

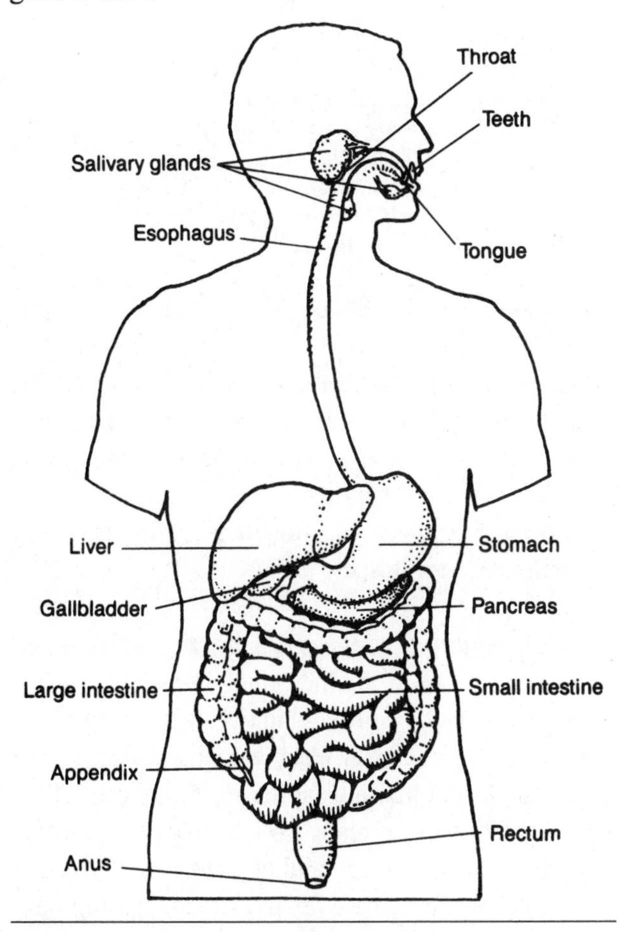

Figure 3-1. The human digestive system.

Oral Cavity. The mouth, or *oral cavity*, contains the teeth, tongue, and openings from the salivary glands. Food is ingested through the mouth, and digestion begins there. The teeth function in the physical breakdown of food into smaller pieces, which provides a larger surface area for the chemical action of digestive enzymes.

The *salivary glands* secrete saliva, a fluid that passes into the mouth through ducts. Saliva contains an enzyme, amylase, which begins the chemical digestion of starch. The tongue aids in chewing and in mixing saliva with the food by moving the food around in the mouth. The tongue also moves the food mass to the back of the mouth for swallowing. In addition, the tongue contains special cells on its surface that allow us to taste our food as sweet, bitter, salty, or sour. Along with receptors located in the nose, the tongue's receptors enable us to fully enjoy the flavor of our food. This is actually an evolutionary adaptation that stimulates the feeding response as well as helping us to avoid ingesting harmful substances, which often have an unpleasant flavor.

Esophagus. When food is swallowed, it passes into the esophagus; and peristalsis of the esophagus wall moves it downward to the stomach. Digestion of starch continues while the food is in the esophagus. The esophagus itself has no digestive function other than to connect the mouth with the stomach. They connect through a circular ring of muscle called the *sphincter*, which allows food to enter the stomach. Sometimes, this muscle weakens and allows stomach contents to flow back into the esophagus, causing a severe burning sensation. This condition is known as *gastric esophageal reflux disease*, or *GERD*. Untreated, GERD can damage the lining of the esophagus and ultimately lead to esophageal cancer.

Stomach. Food reaching the lower end of the esophagus enters the stomach, a muscular sac in which it is mixed and liquefied (physical digestion). Gastric glands in the stomach lining secrete a fluid that contains hydrochloric acid and the enzyme gastric protease (also known as *pepsin*). Hydrochloric acid provides the acidic pH required for effective functioning of gastric protease, which begins the chemical digestion of proteins. The proteins, which are large polypeptides, are broken into smaller peptides. The human stomach can hold a volume of about 1 to $1\frac{1}{2}$ liters of material. *Note:* Both physical and chemical digestion occurs in the stomach.

Small Intestine. Partially digested food moves from the stomach into the small intestine, a long, convoluted tube in which most digestion occurs. The walls of the small intestine are lined with intestinal glands that secrete several different enzymes. These enzymes digest proteins, lipids, and disaccharides. The liver, gallbladder, and pancreas secrete substances into the small intestine as well.

The *liver* produces bile, which passes into the *gallbladder*, where it is stored temporarily. From the gallbladder, *bile* passes through ducts into the small intestine, where it acts on fats, physically breaking them down into tiny droplets. This process, known as *emulsification*, increases the surface area of the fats for subsequent chemical digestion by enzymes. Bile also helps neutralize the acidic food mass from the stomach.

The **pancreas** produces and secretes a mixture of enzymes that passes through the pancreatic duct into the small intestine. Proteases, lipases, and amylase in the pancreatic secretion, together with the enzymes secreted by the intestinal glands, complete the chemical digestion of proteins, lipids, and carbohydrates in the small intestine. These macromolecules are thus broken down into their simpler building blocks: carbohydrates (such as disaccharides) → simple sugars (glucose); lipids → glycerol and fatty acids; and proteins (polypeptides) → amino acids.

The end products of digestion, which include amino acids, fatty acids, glycerol, and glucose, are absorbed through the lining of the small intestine. The intestinal lining is specially adapted for absorption. Its surface area is greatly increased by many folds and by fingerlike projections called *villi* (singular, *villus*). These features are adaptations that enable the small intestine to absorb the greatest quantity of digestive end products.

Each villus contains a lacteal and capillaries (Figure 3-2). A *lacteal* is a small vessel of the lymphatic system. Fatty acids and glycerol (the end products of fat digestion) are absorbed into the lacteals; they are transported in the lymph, which is eventually added to the blood. Glucose and amino acids are absorbed into the blood of the capillaries and transported to the liver for temporary storage. From the liver, glucose and amino acids are distributed by the blood to all the cells, as they are needed.

When excess glucose is removed from the blood in the liver, it is converted to, and stored as, *glycogen*, an insoluble polysaccharide. When the concentration of glucose in the blood drops below a certain level, the glycogen is broken down to glucose, which is then returned to the blood. The storage of excess glucose as glycogen is an adaptation for the maintenance of a constant blood glucose level. This process, which is regulated by two hormones—insulin and glucagon—illustrates how the body maintains homeostasis.

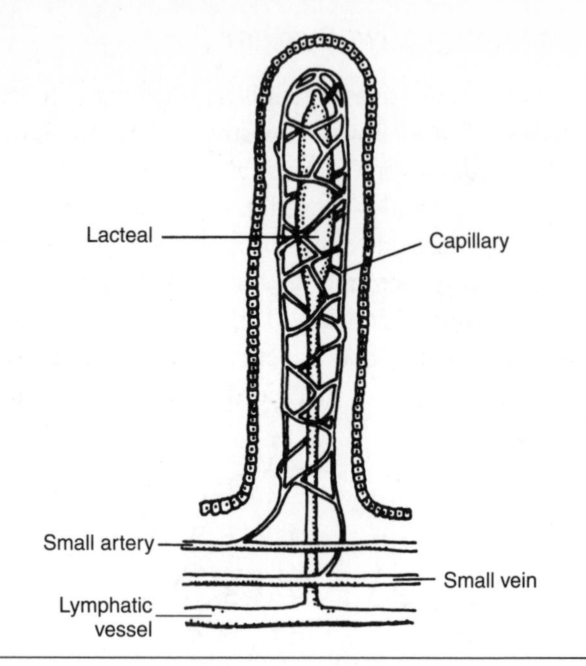

Figure 3-2. Structure of a villus.

Large Intestine. Undigested and indigestible foods and water move from the small intestine into the *large intestine*, or *colon*, which is shorter and wider than the small intestine. Water is reabsorbed from the undigested food into the capillaries in the wall of the large intestine. This reabsorption helps the body to conserve water. The remaining wastes, called *feces*, are moved through the large intestine by strong peristaltic action to the rectum, where they are stored temporarily. The feces are periodically released, or *egested*, from the body through the anus.

Mechanism of Chemical Digestion

In digestion, large, insoluble macromolecules are broken down into small, soluble molecules by the process of *hydrolysis*. Each of the many hydrolytic reactions of digestion is regulated by a specific hydrolytic enzyme. Chemically, hydrolysis is the opposite of dehydration synthesis: In hydrolysis, large molecules are split by the chemical addition of water. (Note that *hydro* means "water" and *lysis* means "breaking down.") In dehydration synthesis, small molecules are joined (through the chemical removal of water) to build, or *synthesize*, larger molecules.

In a series of chemical reactions, polysaccharides such as starch are broken down by hydrolysis into monosaccharides (simple sugars). In the presence of water and proteases (protein-digesting enzymes), proteins are broken down by hydrolysis into their amino acid subunits. Similarly, in the presence of water and lipases (lipid-digesting enzymes), lipid molecules are broken down by hydrolysis into fatty acids and glycerol.

Nutritional Requirements

A balanced diet must contain carbohydrates, proteins, and fats, as well as vitamins, minerals, and water. Ingredients labels that are printed on all packaged food items indicate how much of a particular nutrient is supplied per serving, as well as the serving's calorie content. Many U.S. cities are now requiring that restaurants and fast-food establishments post the calorie counts of items on their menus. In addition to a balanced diet, exercise is recommended to maintain proper health. (*Note:* You can see the current recommended daily allowances for all the major food groups by checking the "food pyramid" on the Web site *www.mypyramid.gov*.)

Carbohydrates. Carbohydrates serve as the major source of energy in the body. Excess carbohydrates are converted to glycogen or fat and stored in the body as an energy reserve. *Cellulose*, a complex carbohydrate found in the cell walls of fruits, vegetables, and whole grains, provides indigestible material that serves as *roughage*. Roughage, also called *fiber*, helps to move the food mass through the intestines, especially through the colon. It is thought that the quick movement of fecal (waste) matter through the colon greatly reduces one's risk of developing colorectal cancer.

Proteins. Proteins in food are broken down into their constituent amino acids, which are then used to synthesize human proteins. Twenty different amino acids are needed for the synthesis of human proteins. Twelve of these can be synthesized in the body from other amino acids; but the other eight, called the *essential amino acids,* must be obtained from the food.

All necessary amino acids must be present at the same time for protein synthesis to occur. An inadequate supply of any essential amino acid limits protein synthesis. Meat proteins generally contain all of the essential amino acids. Such foods are called *complete* protein foods. Vegetable proteins are generally *incomplete* protein foods—they lack one or more essential amino acids. However, a variety of vegetable proteins, if eaten together, can complement each other and provide all the essential amino acids. For example, a meal of rice and beans supplies complete protein.

Fats (Lipids). Fats contain relatively large amounts of potential energy and serve as an energy-storage compound in organisms. Fats (phospholipids) are also a structural component of cell membranes.

Fats are classified as saturated and unsaturated. *Saturated fats*, which are found in meats, butter, and other animal products, are solid at room temperature. Chemically, fats consist of a glycerol molecule chemically bonded to three, long carbon-chain fatty acids. Saturated fats have no double bonds between any carbon atoms and thus contain the maximum number of hydrogen atoms attached to the carbon atoms. *Unsaturated fats* contain one or more double bonds between the carbon atoms and thus could (but usually do not) hold additional hydrogen atoms. In addition, because of the double bonds, the fatty acid chain of the fat is not straight but rather bent at an angle. This prevents the fat molecules from packing tightly together; as a result, unsaturated fats (for example, oils) generally exist as liquids at room temperature. An excess of saturated fats, as well as another type of fat called a *trans fat*, in the diet is thought to contribute to cardiovascular disease. Fortunately, some forms of unsaturated fats (such as olive oil) are thought to protect against cardiovascular disease. However, it is generally considered wise to limit one's intake of all kinds of fats in order to maintain good health.

Disorders of the Digestive System

An *ulcer* is an open sore in the lining of the stomach or intestines. Ulcers may be caused by the presence of excess amounts of hydrochloric acid, which breaks down the lining of the digestive tract, or by bacterial infection. Recent medical research has determined that a species of bacteria (*Helicobacter pylori*) is one of the chief causes of ulcers. Ulcers are painful and sometimes cause bleeding.

Constipation is a condition marked by difficulty in eliminating feces from the large intestine. Constipation occurs when too much water is removed from the feces in the large intestine or when there is a reduction in peristaltic activity, slowing down the movement of waste through the large intestine. Insufficient roughage in the diet may also be a cause of constipation.

Diarrhea is a gastrointestinal disturbance characterized by frequent elimination of watery feces. This condition may result from decreased water absorption in the large intestine and increased peristaltic activity. Prolonged diarrhea may result in severe dehydration, which can even cause death.

Appendicitis is an inflammation of the appendix, a small pouch located at the beginning of the large intestine. *Gallstones* are small, hardened cholesterol deposits that sometimes form in the gallbladder. When gallstones enter the bile duct and block the flow of bile, they cause severe pain.

QUESTIONS

MULTIPLE CHOICE

1. Into which parts of the human digestive system are digestive enzymes secreted? A. mouth, esophagus, stomach B. stomach, small intestine, large intestine C. mouth, stomach, small intestine D. esophagus, stomach, large intestine
2. In humans, excess glucose is stored as the polysaccharide known as A. glycogen B. glycerol C. maltose D. cellulose
3. After a person's stomach was surgically removed, the chemical digestion of ingested protein would probably begin in the A. mouth B. small intestine C. large intestine D. liver
4. Which organ forms part of the human gastrointestinal tract? A. trachea B. esophagus C. diaphragm D. aorta
5. The intestinal folds and villi of the human small intestine function primarily to A. increase the surface area for absorption of digested nutrients B. excrete metabolic wastes C. circulate blood D. force the movement of food in one direction through the digestive tract
6. Lipase aids in the chemical digestion of A. fats B. proteins C. enzymes D. salts
7. In humans, which of the following is true of carbohydrate digestion? A. It begins in the oral cavity and ends in the esophagus. B. It begins in the oral cavity and ends in the small intestine. C. It begins in the small intestine and ends in the large intestine. D. It begins and ends in the small intestine.
8. Organisms are classified as heterotrophs if they derive their metabolic energy by A. photosynthesis B. absorbing minerals C. exercising D. ingesting nutrients
9. Glands located within the digestive tube include A. gastric glands and thyroid glands B. gastric glands and intestinal glands C. thyroid glands and intestinal glands D. adrenal glands and intestinal glands
10. The small lymphatic vessels that extend into the villi are called the A. veins B. lacteals C. glands D. capillaries
11. The principal function of physical digestion is the A. hydrolysis of food molecules for storage in the liver B. production of more surface area for enzyme action C. synthesis of enzymes necessary for food absorption D. breakdown of large molecules to smaller ones by the addition of water
12. In which organ's walls does peristalsis occur? A. liver B. pancreas C. oral cavity D. esophagus
13. A person who consumes large amounts of saturated fats may increase his or her chances of developing A. meningitis B. hemophilia C. pneumonia D. cardiovascular disease

Base your answers to questions 14 through 18 on your knowledge of biology and on the following graph, which shows how much carbohydrates, proteins, and fats are chemically digested as food passes through the human digestive tract. The letters represent body structures, in sequence, that make up the digestive tract.

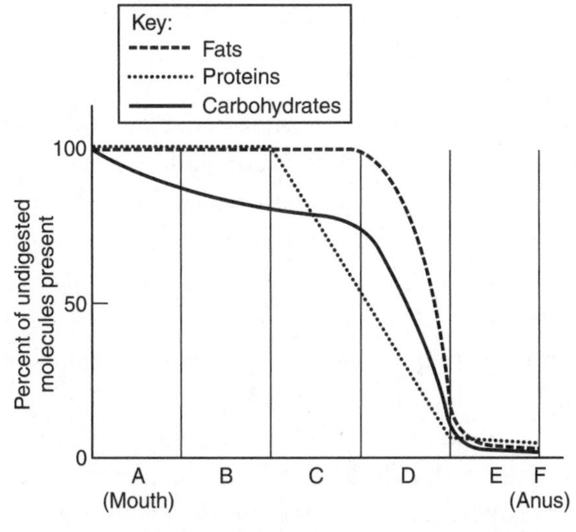

14. Proteins are digested in both A. *A* and *B* B. *B* and *C* C. *C* and *D* D. *A* and *C*

15. The organ represented by letter *C* is most probably the A. esophagus B. stomach C. small intestine D. large intestine

16. Enzymes secreted by the pancreas enter the system at A. *A* B. *B* C. *C* D. *D*

17. The final products of digestion are absorbed almost entirely in A. *F* B. *B* C. *C* D. *D*

18. Water is removed from the undigested material in A. *A* B. *B* C. *E* D. *D*

OPEN RESPONSE

19. Use your knowledge of biology to complete the following table.

Nutrient	*Digestive End Products*	*Where Chemical Digestion Begins*	*End Products Absorbed by*
Starches	Simple sugars		Villi, capillaries
Lipids		Small intestine	
Proteins	Amino acids		

Base your answers to questions 20 and 21 on the following food pyramid, which shows suggested daily servings for several types of food.

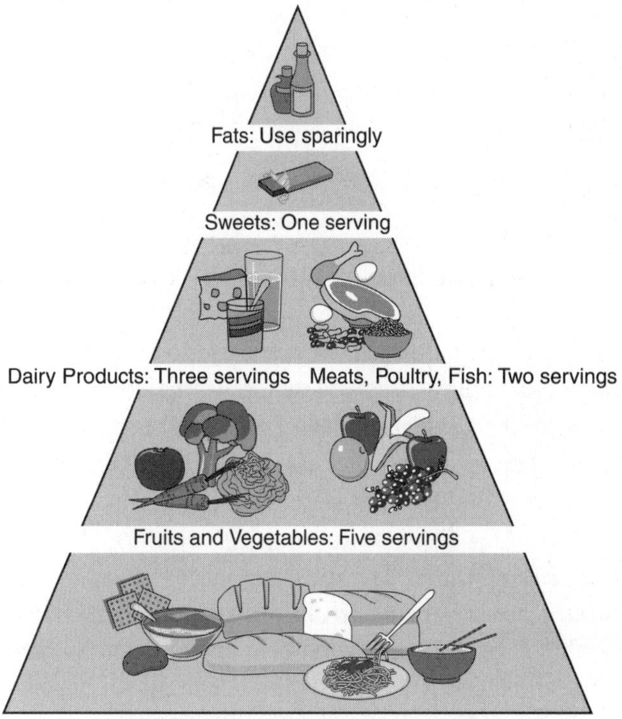

20. Make up a balanced, daily meal plan for breakfast, lunch, and dinner for one person.

21. Explain why people are advised to ingest fats "sparingly," that is, in low amounts.

22. Based on your knowledge of human biology, answer the following:

- Identify three organs of the human digestive system.
- Describe the type of digestion that occurs in each organ.

23. Explain why most foods (nutrients) eaten by humans must be digested before they can be used by the body.

TRANSPORT

Transport includes the absorption and distribution of materials throughout the body. In humans, dissolved and suspended materials are transported in the blood, which is moved throughout the body by the circulatory system.

Blood

Blood consists of a fluid called *plasma* in which red blood cells, white blood cells, and platelets are suspended (Figure 3-3).

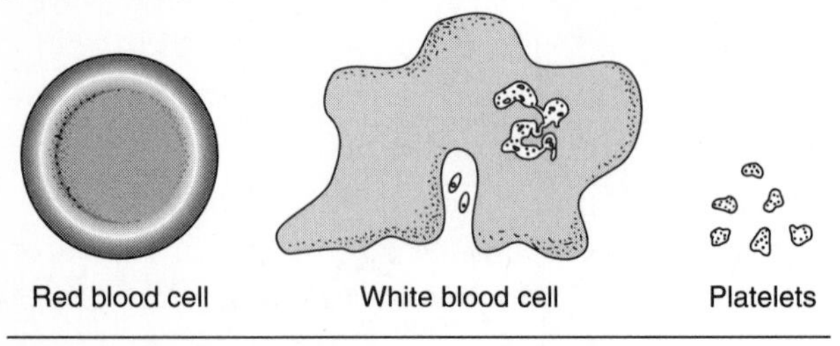

Figure 3-3. The three main types of blood cells.

Plasma. The blood *plasma* consists mostly of water. It contains many dissolved materials, including inorganic ions, wastes, nutrients, and a variety of proteins. The proteins include antibodies, enzymes, hormones, and clotting factors.

Red Blood Cells. The most numerous cells in the plasma are the *red blood cells*, which are constantly produced in the marrow of long bones (such as the arms and legs). Mature red blood cells do not have a nucleus and therefore are unable to undergo mitosis and reproduce. Within red blood cells is the iron-containing pigment hemoglobin, which chemically bonds to oxygen, thereby delivering it throughout the body.

White Blood Cells. The **white blood cells** are larger than the red blood cells and contain one or more nuclei. White blood cells are produced in the bone marrow and in lymph nodes. These special blood cells function in the immune response, defending the body against infectious agents such as bacteria and viruses. There are several types of white blood cells, including phagocytes and lymphocytes.

Phagocytes are white blood cells that engulf and destroy bacteria at the site of an infection. The phagocytes leave the capillaries, by means of ameboid motion by squeezing out through the walls of the blood vessels, and enter the body tissues. There, they engulf bacteria and other foreign matter in the same way that amoebas engulf food. After the bacteria have been engulfed, they are digested and destroyed. In some cases, the white blood cell itself dies after destroying the bacteria.

Lymphocytes are white blood cells that produce very specific protein molecules called **antibodies**. Antibodies react chemically with foreign substances or microorganisms in the blood and inactivate them. The substances that cause antibody production are called **antigens**. Most antigens are protein in nature. An antigen–antibody reaction is referred to as an *immune response*. (The immune response is discussed in greater detail later in this chapter.)

Platelets. The small cell fragments that are involved in the clotting of blood are called *platelets.* A platelet consists of cytoplasm surrounded by a cell membrane and, like red blood cells, it has no nucleus.

Blood Clotting. When an injury occurs, blood vessels break and blood is released. To stop the loss of blood, a blood clot forms, blocking the wound. Clotting involves a series of enzyme-controlled reactions. All the substances required for clotting are normally present in the blood, but clot formation does not take place unless there is a break in a blood vessel. When this occurs, blood platelets are ruptured and they release an enzyme that starts the clotting reactions. A series of complex reactions occurs in the process of blood clotting. In one of the final steps of this process, the soluble plasma protein *fibrinogen* is converted by an enzyme to the insoluble protein *fibrin*, which forms a meshwork of solid fibers across the wound. Blood cells become trapped in the fibers, forming the clot. Some people who suffer from the condition known as *hemophilia* lack one of the clotting factors (Factor VIII) in their plasma; as a result they have trouble stopping blood flow from a wound. Hemophiliacs are often transfused with Factor VIII to allow their blood to clot normally in the event of an injury that causes bleeding.

Blood-typing. Knowledge of human blood chemistry has made possible the transplanting of organs and the transfusion of blood from one person to another. In both organ transplants and blood transfusions, an immune response is stimulated if the body of the recipient recognizes foreign antigens in the tissue or blood from the donor. In organ transplants, an antigen–antibody reaction against the transplanted organ is called *rejection*. Donor tissue proteins must be carefully matched to those of the recipient to avoid rejection. There are drugs available that can suppress the immune response so that the transplanted organs are less likely to be rejected. However, suppressing the immune system comes with risk, since it can also cause a reduced response to any infection that might develop.

Blood-typing for transfusions is based on the presence or absence of antigens on the surface of red blood cells. An individual's blood type is determined by genes inherited from his or her parents. The most important blood-typing system is the ABO blood group system. In this system, the red blood cells may have on their surface a special kind of protein–carbohydrate molecule. One of these molecules is known as Antigen A and the other is called Antigen B. An individual red blood cell may have the Antigen A on its surface, in which case the individual has type A blood. If the red blood cell has Antigen B on its surface, the individual has type B blood. It is possible for a red blood cell to have both Antigens (A and B) on its surface. In this case, the individual has type AB blood. If the red blood cell has neither A nor B, he or she has type O blood (the O is actually a zero). In addition, the plasma of the blood may contain antibodies: anti-A and anti-B. Table 3-1 shows the antigens and antibodies for each type of blood.

Table 3-1. **Antigens and Antibodies of the ABO Blood Group System**

Blood Type	*Antigens on Red Cells*	*Antibodies in Plasma*
A	A	Anti-B
B	B	Anti-A
AB	A and B	Neither Anti-A nor Anti-B
O	Neither A nor B	Anti-A and Anti-B

Transport Vessels

Blood circulates throughout the human body within the blood vessels, which include *arteries*, *capillaries*, and *veins*. (Refer to Figure 2-8 on page 40.)

Arteries. Blood is carried from the heart to all parts of the body in arteries, which are thick-walled, muscular vessels that expand and contract to accommodate the forceful flow of blood from the heart. The rhythmic expansion and contraction of the arteries produced by the heartbeat aids the flow of blood to all parts of the body; it is called the *pulse*. During a medical examination, a doctor measures the force with which the blood pushes against the artery walls. This measurement is known as the *blood pressure*. Maintenance of normal blood pressure is important to proper health.

Capillaries. With increasing distance from the heart, arteries branch into smaller and smaller vessels, finally forming capillaries, tiny blood vessels with cell membranes only one cell layer thick. Capillaries are the site of exchange of materials between the blood and the body tissues. Because capillaries are so thin, materials can be exchanged through their walls by diffusion.

Veins. Blood flows from the capillaries into the veins, thin-walled vessels that carry the blood back to the heart. Veins contain flaps of tissue that act as valves (Figure 3-4). The valves allow the blood in the veins to flow in only one direction—back toward the heart.

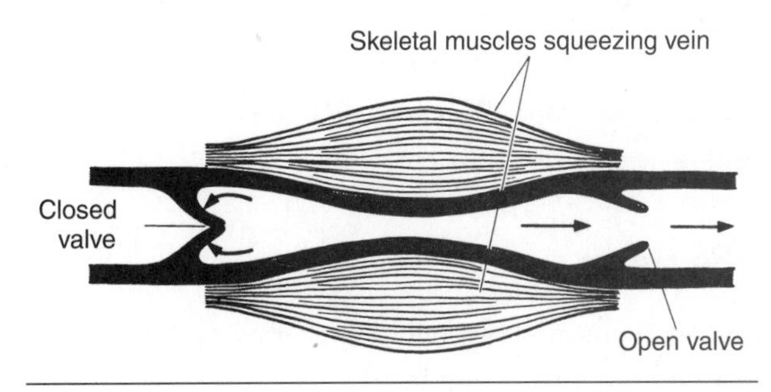

Figure 3-4. Blood flow in a vein.

Intercellular Fluid and Lymph

As blood passes through the capillaries of the body, some of the plasma is forced out of the vessels and into the surrounding tissues. This fluid, which bathes all the cells of the body, is called *intercellular fluid*, or *ICF*. Materials diffusing between the cells and the blood of the capillaries are dissolved in the ICF.

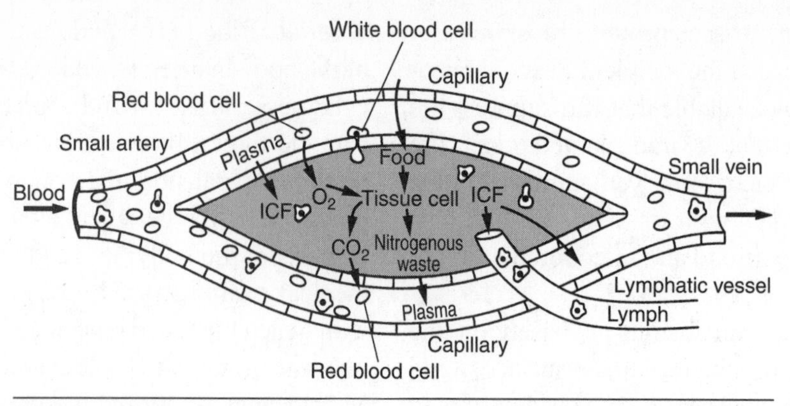

Figure 3-5. Molecules diffuse between the capillaries, intercellular fluid, and body cells.

Excess intercellular fluid is drained from the tissues by tiny *lymph vessels*, which are part of the *lymphatic system*. Once inside these vessels, the fluid is called *lymph*. The lymph vessels merge, forming larger vessels. Eventually, all lymph flows into two large lymph ducts, which empty into veins near the heart. In this way, the fluid lost from the blood is returned to the blood (Figure 3-5).

Major lymph vessels have enlarged regions called *lymph nodes* in which phagocytic cells filter bacteria and dead cells from the lymph. Some lymph vessels contain valves that, like those in the veins, keep the lymph flowing back toward the heart. The spleen, an organ located near the stomach, is also part of the lymphatic system. As lymph flows through the spleen, it is filtered of impurities.

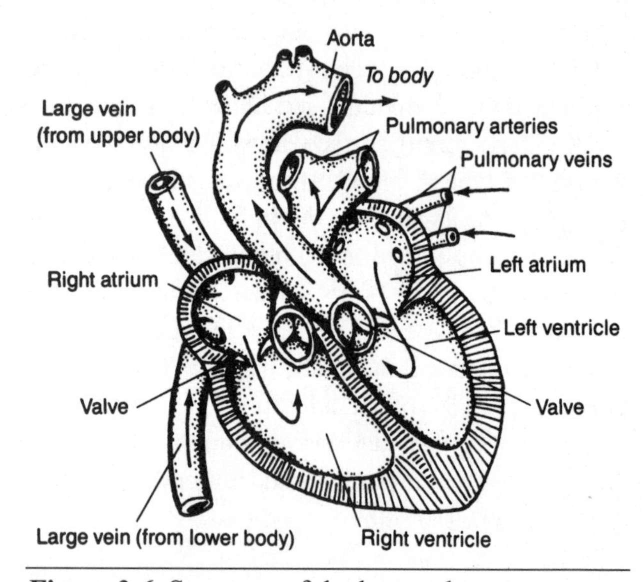

Figure 3-6. Structure of the human heart.

The Heart

Blood is pumped through the arteries of the body by the contractions of the heart.

Structure of the Heart. The heart has four chambers (Figure 3-6). The two upper chambers, the *atria* (singular, *atrium*), receive blood returning to the heart from the rest of the body. The two lower chambers, the *ventricles*, pump blood out of the heart into the arteries. The walls of the ventricles are thicker and more muscular than those of the atria—a feature related to their function of having to force blood through the thick-walled arteries.

Circulation Through the Heart. The deoxygenated (oxygen-poor) blood from the body is returned to the right atrium of the heart through two large veins—one from the upper part of the body and one from the lower part of the body. This deoxygenated blood flows down from the right atrium into the right ventricle; from there it is pumped out of the heart through the pulmonary arteries to the lungs. When in the lungs, the blood gives up carbon dioxide and picks up oxygen. The oxygenated (oxygen-rich) blood is then returned through the pulmonary veins to the left atrium of the heart. The blood then passes from the left atrium into the left ventricle, which pumps it out of the heart through the *aorta*, the largest artery in the body. (From there, the blood is transported to deliver oxygen to the body's cells.)

This one-way flow of blood through the heart is controlled by valves that prevent backflow of the blood. There are valves between the atria and the ventricles, between the right ventricle and the pulmonary artery, and between the left ventricle and the aorta.

Blood Pressure. The pressure exerted by the blood on the walls of the arteries during the pumping action of the heart is referred to as *blood pressure*. During the contraction phase of the heartbeat cycle, arterial blood pressure is highest. This is known as the *systolic pressure*. During the relaxation phase of the heartbeat cycle, blood pressure is lowest. This is called the *diastolic pressure*. Blood pressure is recorded in millimeters of mercury (mmHg) as two numbers: the systolic pressure over the diastolic pressure. For example, a reading of 110/70 indicates a healthy blood pressure, with a systolic pressure of 110 and a diastolic pressure of 70.

Pathways of Circulation

The pathway of blood between the heart and the lungs is called the *pulmonary circulation*. The circulatory pathway between the heart and all other parts of the body except the lungs is called the *systemic circulation*. The system of blood vessels that supplies the heart itself is called the *coronary circulation*.

Role of the Liver and Kidneys

Although not technically a part of the circulatory system, the liver plays an important role in removing toxins (such as alcohol) from the blood as it flows through this vital organ. Through a series of chemical reactions, the liver detoxifies harmful substances that may be present in the blood. The kidneys filter waste products out of the blood as it flows through them. Some of the wastes that are removed are salts and nitrogenous wastes resulting from cellular activities.

Disorders of the Transport System

Diseases of the heart and blood vessels are called *cardiovascular* diseases (*cardio* refers to the heart; *vascular* refers to the blood vessels). The most common form of cardiovascular disease is high blood pressure, or *hypertension*, which is characterized by elevated arterial blood pressure. This condition can be caused by a number of factors, including stress, diet, heredity, cigarette smoking, obesity, and aging. High blood pressure can damage the lining of arteries and weaken the muscle of the heart. As mentioned previously, maintenance of normal blood pressure is essential to good health. Generally, a systolic pressure of 140 or higher over an extended period of time is considered to be hypertension. High blood pressure can lead to heart disease, stroke, and kidney failure. People with chronic high blood pressure can be treated with medication, but they must also be careful about their diet (avoiding salty foods) and exercise regularly.

A blockage of the coronary artery or its branches is a *coronary thrombosis*, or heart attack. As a result of the blockage, some of the muscle tissue of the heart is deprived of oxygen and is damaged.

A narrowing of the coronary arteries may cause temporary shortages of oxygen to the heart muscle, resulting in intense pain in the chest and sometimes in the left arm and shoulder. This condition is called *angina pectoris*.

Anemia is a condition in which the blood cannot carry sufficient amounts of oxygen to the body cells. Anemia may be due to inadequate amounts of hemoglobin in the red blood cells or to too few red blood cells. One form of anemia is caused by a shortage of iron in the diet.

Leukemia is a form of cancer in which the bone marrow produces abnormally large numbers of white blood cells.

QUESTIONS

MULTIPLE CHOICE

24. Which is a characteristic of lymph nodes?
A. They carry blood under great pressure.
B. They move fluids by means of a muscular pump. C. They produce new red blood cells.
D. They contain phagocytic cells.

25. The accumulation of specific antibodies in the plasma, due to the presence of an antigen, is characteristic of A. an immune response B. angina pectoris C. a coronary thrombosis D. cerebral palsy

26. Blood is carried from the heart to all parts of the body in thick-walled vessels known as A. capillaries B. lymph vessels C. veins D. arteries

27. In the human body, which blood components engulf foreign bacteria? A. red blood cells B. white blood cells C. antibodies D. platelets

28. In humans, the exchange of materials between blood and intercellular fluid directly involves blood vessels known as A. capillaries B. arterioles C. venules D. arteries

29. An injury to a blood vessel may result in the formation of a blood clot when A. bone marrow cells decrease platelet production B. kidney tubules synthesize clotting factors C. ruptured platelets release enzyme molecules D. white blood cells release antibodies

30. Oxygen carried by the blood in capillaries normally enters the body cells by A. active transport B. osmosis C. diffusion D. pinocytosis

31. Which type of vessel normally contains valves that prevent the backward flow of blood? A. artery B. arteriole C. capillary D. vein

32. The thin-walled vessels that carry the blood back to the heart are known as A. capillaries B. lymph vessels C. veins D. arteries

33. The right ventricle is the chamber of the heart that contains A. deoxygenated blood and pumps this blood to the lungs B. deoxygenated blood and pumps this blood to the brain C. oxygenated blood and pumps this blood to the lungs D. oxygenated blood and pumps this blood to the brain

34. Which two systems are most directly involved in providing human cells with the molecules needed for the synthesis of fats? A. digestive and circulatory B. excretory and digestive C. immune and muscular D. reproductive and circulatory

Base your answers to questions 35 through 39 on your knowledge of biology and on the diagram below, which represents the exchange of materials between capillaries and cells.

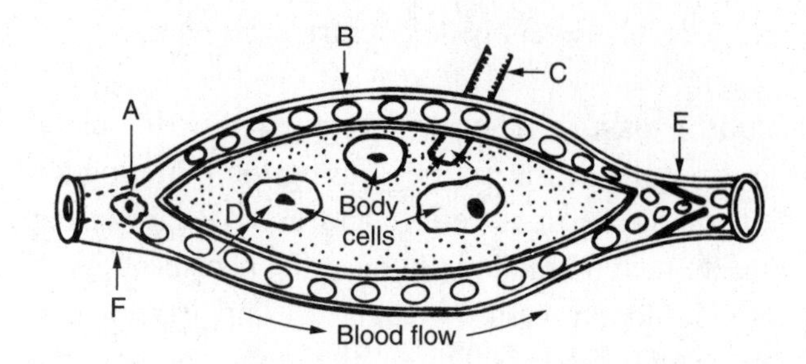

35. Blood vessel *B* has membranes that are very thin, enabling this type of vessel to A. transport hemoglobin to body cells B. transport red blood cells into the tissue spaces C. withstand the pressure of the blood coming in from veins D. easily transport substances into and out of the blood

36. A function of cell *A* is to A. carry oxygen B. engulf disease-producing bacteria C. transport digested food D. produce hemoglobin

37. A substance that diffuses in the direction indicated by *D* is most likely A. fibrin B. oxygen C. urea D. bile

38. Which vessel most likely contains the greatest amount of carbon dioxide? A. *F* B. *B* C. *C* D. *E*

39. Excess intercellular fluid (ICF) is constantly drained off by the lymphatic vessels. Which letter represents such a vessel? A. *E* B. *B* C. *C* D. *F*

Base your answers to questions 40 through 43 on the following diagram and on your knowledge of biology. The diagram represents the human heart; the arrows indicate the direction of blood flow.

40. The aorta is represented by number A. 1 B. 6 C. 8 D. 4

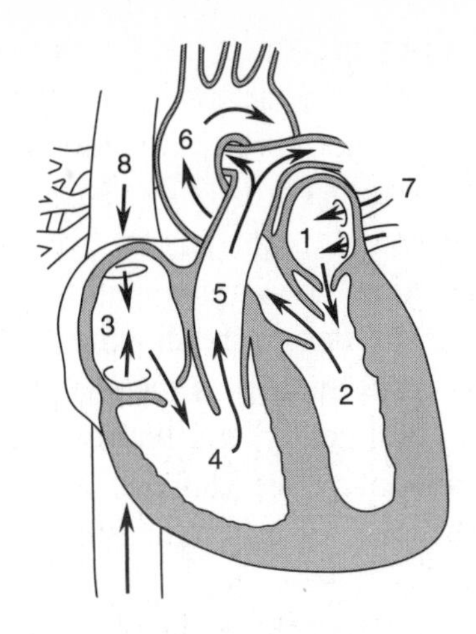

41. Deoxygenated blood returns to the heart through the structure represented by number A. 8 B. 7 C. 3 D. 5
42. The chamber that pumps blood to all parts of the body except the lungs is represented by number A. 1 B. 2 C. 3 D. 4
43. Blood passes from the heart to the lungs through the structure represented by number A. 5 B. 6 C. 7 D. 8

OPEN RESPONSE

44. Explain what effects faulty valves in the veins would have on a human's blood flow.
45. Describe the relationship that exists between the circulatory system and the lymphatic system.
46. The four major blood components are red blood cells, white blood cells, plasma, and platelets. Describe one major function for each blood component listed.
47. Generally (excluding pulmonary arteries), blood in arteries has a higher oxygen content than that of blood in veins. Give a scientifically valid explanation for this observation.
48. Why are people who are anemic (have too little hemoglobin and/or too few red blood cells) often advised to take in extra iron in their diets?
49. Using appropriate information, fill in spaces *A* and *B* in the following chart. In space *A* identify an organ in the human body where molecules diffuse into the blood. In space *B* identify a specific molecule that diffuses into the blood at this organ.

An organ in the human body where molecules diffuse into the blood	A specific molecule that diffuses into the blood at this organ
A	**B**

50. Describe the pathway of blood flow through the heart, beginning and ending with the point at which the blood returns to the heart from the body organs.
51. Compare the structure and function of the three major types of blood vessels: arteries, veins, and capillaries.

RESPIRATION

Respiration includes cellular respiration and gas exchange. The process of cellular respiration in humans is basically the same as that in other aerobic organisms. The respiratory and circulatory systems work together to deliver oxygen to the body's cells so that aerobic respiration can occur. In the cells, glucose is broken down completely to yield carbon dioxide and water, and ATP is formed from ADP and phosphate.

Anaerobic respiration occurs in human skeletal muscle during prolonged exercise when the amount of oxygen supplied by the circulatory system becomes inadequate for aerobic respiration. Under these circumstances, glucose is broken down in the muscle to lactic acid. The accumulation of lactic acid in skeletal muscle is thought to be responsible for muscle fatigue. When adequate oxygen is again available, aerobic respiration resumes and the lactic acid returns to the aerobic respiration pathway.

Human Respiratory System

The human respiratory system moves respiratory gases between the external environment and the internal surfaces for gas exchange within the lungs. The respiratory system consists of a network of passageways that permit air to flow into and out of the lungs (Figure 3-7).

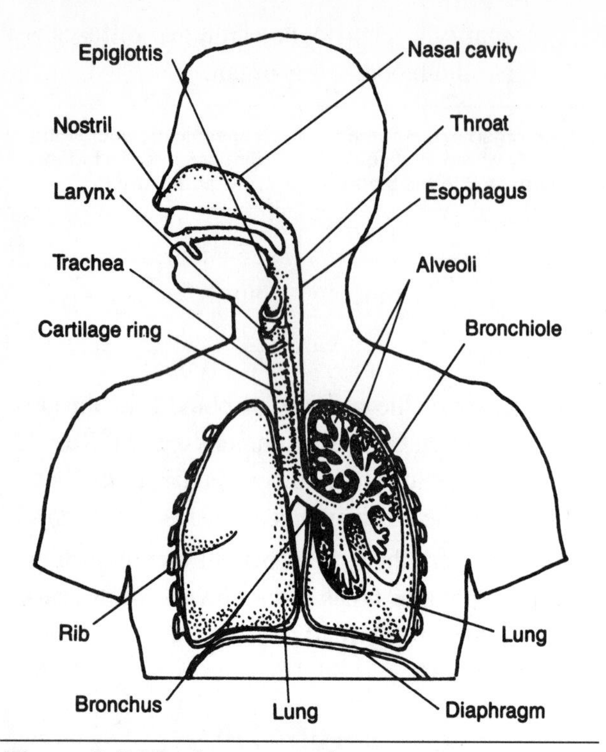

Figure 3-7. The human respiratory system.

Nasal Cavity. Air generally enters the respiratory system through the nostrils and passes into the *nasal cavity*. This cavity is lined with a ciliated mucous membrane that cleans, warms, and moistens the air. It is generally better to breathe through the nose than through the mouth because the nasal passages are more effective in filtering impurities out of the air, as well as moistening and warming it. When the nasal passages become inflamed (as during a cold), mouth breathing becomes much more important.

Pharynx. From the nasal cavity, air passes into the *pharynx,* or throat, the area where the oral cavity and nasal cavity meet. Air passes through the pharynx on its way to the trachea. It is important to note that the pharynx is a part of the body shared by two systems: the digestive and respiratory systems.

Trachea. The *trachea*, or windpipe, is a tube through which air passes from the pharynx to the lungs. The opening (from the pharynx) to the trachea is protected by a flap of tissue called the *epiglottis*. Because the pharynx is shared by both the digestive and respiratory systems, the epiglottis covers the opening of the trachea during swallowing so that food and liquids cannot enter the air passages. During breathing, the opening of the trachea is uncovered. In the top of the trachea is the *larynx*, or voice box, which functions in speech. The larynx contains cords of cartilage that vibrate and produce sound as air passes over them. This sound is the voice. The unique qualities of an individual's voice are due to the thickness and tension of the vocal cords as well as the shape and size of the pharynx, oral cavity, and nasal passages or sinuses. (A person's voice changes when he or she has a cold because of inflammation in the sinuses and pharynx.)

The walls of the trachea contain rings of cartilage that keep the trachea open so that the passage of air remains unobstructed. The trachea is lined with a ciliated mucous membrane. Microscopic particles in the inhaled air are trapped by mucus, and the beating of the cilia sweeps the mucus upward toward the pharynx (to be expelled from the body by coughing or sneezing).

Bronchi and Bronchioles. The lower end of the trachea splits, forming two tubes called the *bronchi* (singular, *bronchus*). The bronchi, like the trachea, are lined with a mucous membrane and ringed with cartilage. Each bronchus extends into a lung, where it branches into smaller and smaller tubes called *bronchioles*.

The bronchioles are lined with mucous membranes, but they lack cartilage rings. At the end of each bronchiole is a cluster of tiny, hollow air sacs called *alveoli*.

Alveoli. The lungs contain millions of alveoli (singular, *alveolus*). The alveoli membranes are thin and moist and are surrounded by capillaries. The alveoli are the functional units for gas exchange in the human respiratory system. Oxygen diffuses from the alveoli into the surrounding capillaries, while carbon dioxide and water diffuse from the capillaries into the alveoli.

Lungs. Each bronchus with its bronchioles and alveoli make up a *lung*. Humans have two lungs and each is divided into areas called *lobes*. The right lung has three lobes, while the left has only two. Each lung is situated in a membranous sac called the *pleura*.

The lungs are highly elastic and are able to expand and contract as they fill up with and empty out air.

Breathing

Air moves into and out of the lungs during *breathing*. The lungs are highly elastic, yet they contain no muscle tissue. They expand and contract in response to pressure changes in the chest cavity brought about by the actions of the rib cage and the diaphragm (a sheet of skeletal muscle). As such, the lungs themselves are actually passive during the breathing act.

During *inhalation*, the ribs push upward and outward and the diaphragm moves down, enlarging the chest cavity. The enlargement of the chest cavity reduces the pressure around the lungs, which expand, and air flows into the lungs. In *exhalation*, the ribs move inward and downward and the diaphragm moves up. The chest cavity becomes smaller, causing an increase in chest pressure, and air is forced out of the lungs (Figure 3-8).

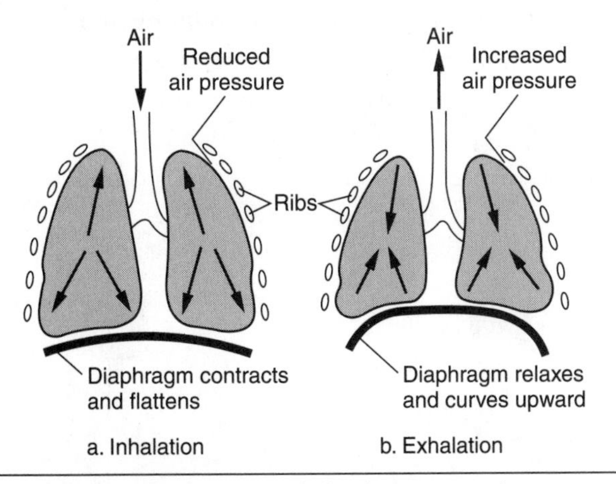

Figure 3-8. Movement of diaphragm as air goes into and out of the lungs.

Gas Exchange. The air that enters the alveoli is rich in oxygen. The blood in the capillaries surrounding the alveoli is oxygen-poor and contains the wastes of cellular respiration—carbon dioxide and water. The oxygen diffuses from the alveoli into the blood, where it enters the red blood cells and becomes loosely bound to the hemoglobin in them.

The oxygen and hemoglobin separate in the capillaries of the body tissues. The oxygen diffuses out of the capillaries, through the intercellular fluid, and into the body cells. Carbon dioxide and water diffuse out from the cells and into the blood. When the blood returns to the lungs, these wastes diffuse into the alveoli and are expelled from the body in the exhaled air.

Breathing Rate. The rate of breathing is controlled by the breathing center in the medulla of the brain. The breathing center is sensitive to the concentration of carbon dioxide in the blood. When the carbon dioxide level is high, nerve impulses from the breathing center are sent to the rib muscles and to the diaphragm to increase the breathing rate, which speeds up the rate of carbon dioxide removal from the body. As the carbon dioxide level in the blood drops, the breathing rate decreases. This regulation of carbon dioxide levels is one example of the feedback mechanisms, or *feedback loops*, by which the body maintains homeostasis. In a **feedback mechanism**, an initial change in one part of the loop (such as a high CO_2 level) stimulates a reaction in a second part of the system (an increased breathing rate). When the condition has been corrected (that is, a lower CO_2 level), the system feeds back information to the first part, shutting it off. This, in turn, shuts off the reaction in the second part.

Disorders of the Respiratory System

Bronchitis is an inflammation of the linings of the bronchial tubes. As a result of such swelling, the air passages become narrowed and filled with mucus, causing breathing difficulties and coughing. *Asthma* is an allergic reaction characterized by a narrowing of the bronchial tubes, which results in difficulty in breathing. *Emphysema* is a disease in which the membranes of the alveoli break down, decreasing the surface area for gas exchange. Emphysema is marked by shortness of breath, difficulty in breathing, and decreased lung capacity.

Chronic Obstructive Pulmonary Disease (*COPD*) is a group of diseases that are all characterized by wheezing, difficulty in breathing, and crackling sounds heard when breathing in. People with COPD suffer from the inability to get adequate amounts of oxygen to their tissues. *Pleurisy* is an inflammation of the pleural membranes surrounding the lungs, often caused by bacterial or viral infections. This disease causes great pain when a person attempts to take a deep breath. *Tuberculosis*, or *TB*, is caused by a bacterium and is contagious. TB causes nodules, or *tubercles*, to develop in the lungs and these may eventually spread to other parts of the body. These tubercles greatly interfere with the normal functioning of the lungs, and the results can be fatal.

QUESTIONS

MULTIPLE CHOICE

52. The alveoli in humans are structures most closely associated with A. gas exchange B. anaerobic respiration C. glandular secretion D. neural transmission

53. In humans, the center that detects and regulates the amount of carbon dioxide in the blood is situated in the A. cerebrum B. diaphragm C. medulla D. rib muscles

54. The exchange of air between the human body and the environment is a result of the rhythmic contractions of the rib cage muscles and the A. diaphragm B. lungs C. trachea D. heart

55. The breathing rate of humans is regulated mainly by the concentration of A. carbon dioxide in the blood B. oxygen in the blood C. platelets in the blood D. white blood cells in the blood

Base your answers to questions 56 through 60 on the diagram below, which represents part of the human respiratory system.

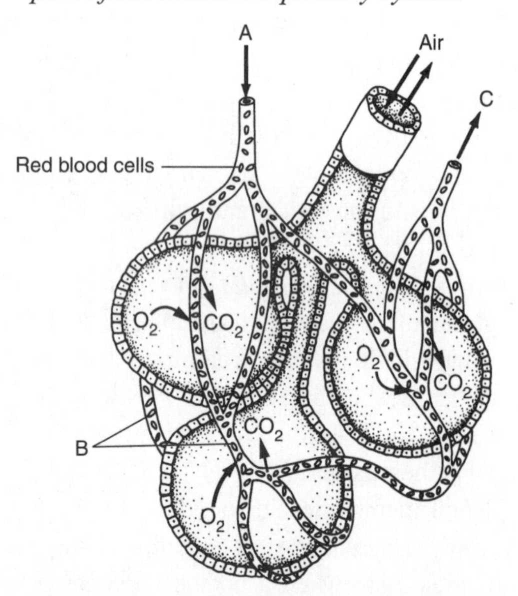

56. The blood vessels labeled *B* that are surrounding these air sacs are called A. arteries B. capillaries C. veins D. lymphatic ducts

57. These air sacs are known as A. alveoli B. bronchi C. bronchioles D. tracheae

58. The heart chamber that most directly pumps blood to the vessel network at *A* is the A. right atrium B. left atrium C. right ventricle D. left ventricle

59. The process most directly involved in the exchange of gases between these air sacs and blood vessels is called A. active transport B. pinocytosis C. hydrolysis D. diffusion

60. Compared to blood entering at *A*, blood leaving the vessel network at *C* has a lower concentration of A. oxygen B. hemoglobin and carbon dioxide C. carbon dioxide D. oxygen and hemoglobin

OPEN RESPONSE

Base your answers to questions 61 through 63 on your knowledge of biology and on the diagram below, which represents a model of the human respiratory system.

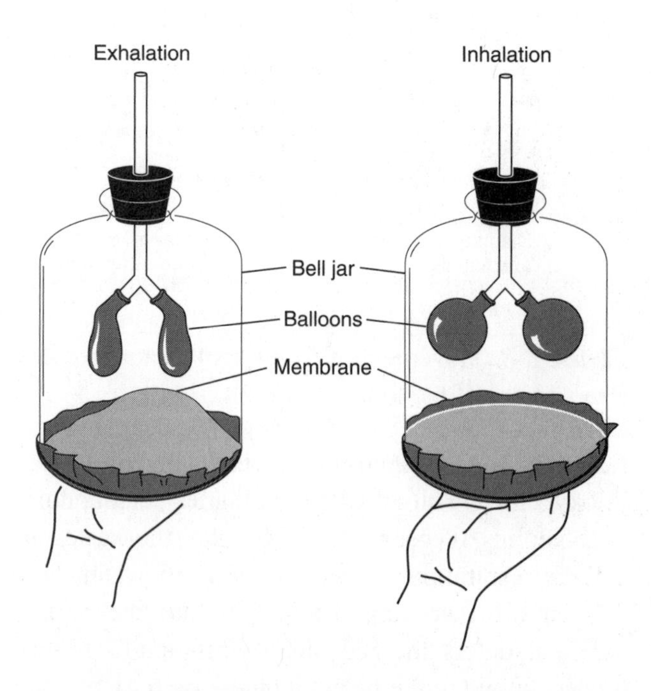

61. Explain which parts of the respiratory system the Y-tube, balloons, and rubber membrane represent.

62. Describe what happens to the balloons when the rubber membrane is pulled downward.

63. Give a scientific reason for your answer to question 62. How does this process apply to humans?
64. Breathing rate is controlled by the respiratory center in the brain, which responds to carbon dioxide levels in the blood. High levels of carbon dioxide increase the breathing rate; low levels decrease the breathing rate. Give a scientific reason for why a person's breathing rate increases during and after vigorous exercise.
65. State how the alveoli in our lungs satisfy the conditions needed to be a good respiratory surface. Include at least two conditions that allow gas exchange to take place.
66. Place the following terms in a sequence that shows the correct pathway air takes during breathing in a human (starting with taking in air from the environment): *bronchi*, *alveoli*, *nose/mouth*, *trachea*, and *bronchioles*.

EXCRETION

The metabolic wastes of humans include carbon dioxide, water, salts, and urea. Excretory wastes pass from the cells into the blood and are carried to the excretory organs that expel them from the body. The excretory organs include the lungs, liver, sweat glands, and kidneys.

Lungs. The *lungs* function in the excretion of carbon dioxide and water vapor, which are the wastes of cellular respiration.

Liver. The *liver* is a large organ that performs many functions essential to human survival. One of the excretory functions of the liver is to get rid of excess amino acids. The amino groups are removed and converted into *urea*, which is excreted (in urine) by the kidneys. The remaining amino acid molecules can be broken down by cellular respiration. The liver is also responsible (along with the spleen) for the breakdown and recycling of red blood cells.

Sweat Glands. The *sweat glands* of the skin excrete wastes, including water, salts, and a small amount of urea. These wastes pass by diffusion from capillaries into the sweat glands and then through ducts to pores on the surface of the skin (Figure 3-9). The mixture of wastes and water excreted by the sweat glands is called sweat, or *perspiration*.

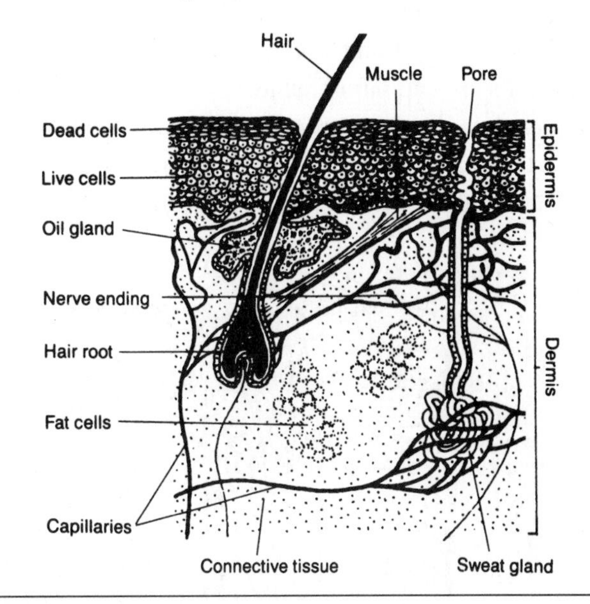

Figure 3-9. Structure of the skin.

Perspiration functions primarily in the regulation of body temperature. The evaporation of sweat from the surface of the skin occurs when heat is absorbed from skin cells, and it serves to lower the body temperature. This method of temperature regulation is another example of homeostasis.

Urinary System

The human urinary system consists of the kidneys, ureters, urinary bladder, and urethra (Figure 3-10).

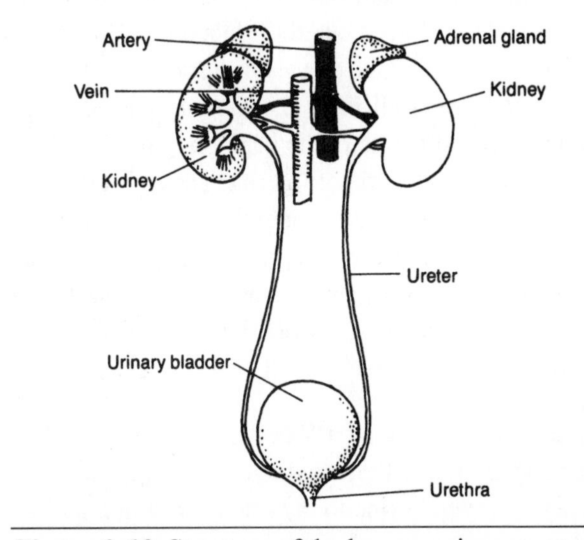

Figure 3-10. Structure of the human urinary system.

Kidneys. Human *kidneys* perform two major functions: they remove urea from the blood, and they regulate the concentrations of most of the substances in the body fluids. Blood is carried to each kidney by a large artery. Within the kidney, the artery divides and subdivides into smaller and smaller arteries and then into balls of capillaries called *glomeruli* (singular, *glomerulus*). Each glomerulus is part of a *nephron*, the functional unit of the kidney (Figure 3-11). There are about one million nephrons in each kidney.

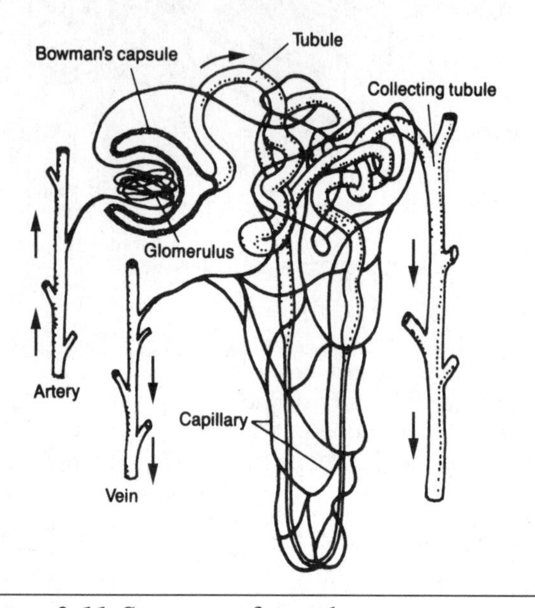

Figure 3-11. Structure of a nephron.

A nephron consists of a glomerulus surrounded by a cup-shaped structure called *Bowman's capsule.* Extending from the capsule is a long, coiled tubule that is surrounded by capillaries. As blood flows through the glomerulus, water, salts, urea, glucose, and some amino acids are forced out of the blood into Bowman's capsule. This process is called *filtration.* As these substances—referred to as the *filtrate*—pass through the long, coiled tubule of the nephron, glucose, water, amino acids, and some of the salts are reabsorbed by active and passive transport into the blood in the capillaries surrounding the tubule. The fluid that remains in the tubules consists of water, urea, and salts, and is called *urine*. Urine passes from the small tubule of the nephron into larger tubules and then to a ureter.

Ureters and Urinary Bladder. Urine flows from each kidney into a large tubule called the *ureter*. The ureters carry the urine to the urinary bladder, a muscular organ in which urine is stored temporarily.

Urethra. Urine is periodically expelled from the bladder into a tube called the *urethra*. This tube leads to the outside of the body.

Diseases of the Urinary System

Diseases of the kidneys affect the body's ability to eliminate normal amounts of metabolic wastes. *Kidney stones* form when the mineral calcium builds up in the kidney, causing a lot of pain as they pass out through the ureter. *Gout* is a condition that produces symptoms similar to arthritis and is caused by deposits of uric acid in the joints. Victims of gout suffer from severe pain and stiffness in the joints. Diets that are extremely high in protein result in the production of large amounts of urea, which the kidneys must remove from the blood. The extra strain on the kidneys in eliminating these wastes may result in a kidney disorder, or *malfunction*. Renal failure occurs when the kidneys fail to function.

QUESTIONS

MULTIPLE CHOICE

67. Which human body system includes the lungs, liver, skin, and kidneys? A. respiratory B. digestive C. transport D. excretory

68. In humans, the filtrate produced by the nephrons is temporarily stored in the A. glomerulus B. alveolus C. gallbladder D. urinary bladder

69. What is the principal waste from excess amino acids in humans? A. salt B. urea C. uric acid D. carbon dioxide

70. In humans, the organ that breaks down red blood cells and amino acids is the A. kidney B. liver C. gallbladder D. small intestine

71. The main components of urine, besides water, are A. amino acids and fatty acids B. urea and salts C. ammonia and bile D. hydrochloric acid and bases

72. In humans, urine is eliminated from the bladder through the A. urethra B. ureter C. nephron D. collecting tubule

73. The basic structural and functional excretory units of the human kidney are known as A. neurons B. nephrons C. alveoli D. ureters

74. The excretory organ that is also associated with the storage of glycogen is the A. stomach B. lung C. kidney D. liver

Base your answers to questions 75 through 77 on your knowledge of biology and on the diagram below, which illustrates a nephron and its capillaries.

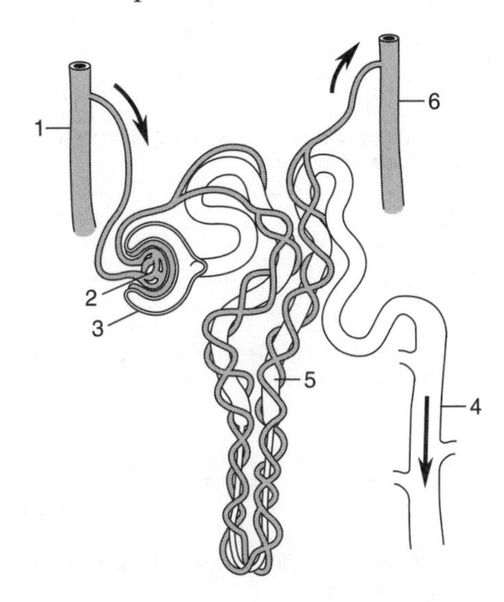

75. Into which structure does the filtrate first pass? A. 5 B. 6 C. 3 D. 4

76. In which area is water being reabsorbed? A. 5 B. 2 C. 3 D. 4

77. In which area does urine collect? A. 1 B. 2 C. 6 D. 4

OPEN RESPONSE

78. Briefly describe how each of the following functions as an excretory organ: the liver; the skin; and the lungs.

Base your answers to questions 79 and 80 on the following diagram and information. The diagram represents a nephron from which samples of fluid were extracted. The samples were recovered from the areas labeled A and B in the diagram. The concentrations of five substances in the fluid extracted from both sites were compared and the results are listed in the table.

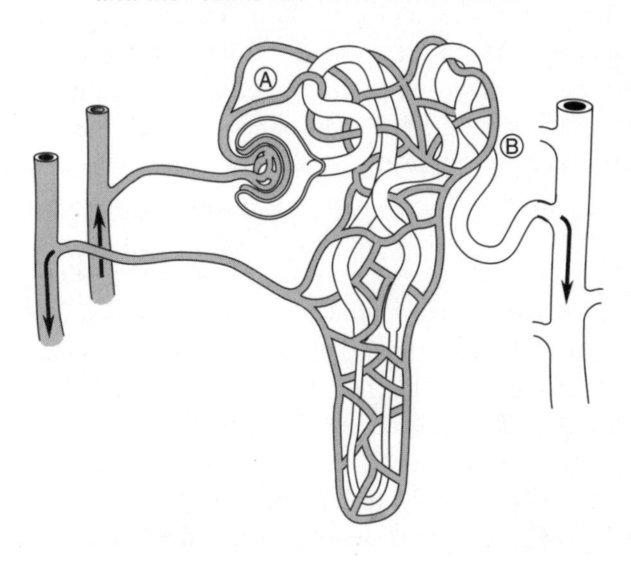

Substance	*Concentration at A*	*Concentration at B*
Water	High	Low
Urea	Moderate	High
Glucose	Moderate	Zero
Amino acids	High	Zero
Salts	Low	High

79. Explain the change in the concentrations of water and urea from area *A* to area *B*.

80. Why are there no amino acids present in the fluid extracted from area *B*, yet the concentration at area *A* was high?

NERVOUS SYSTEM

Regulation in humans involves the interaction of the nervous and endocrine systems with each other and with other systems in the body. The endocrine and nervous systems are similar in that they both secrete chemicals and both play a major role in the maintenance of homeostasis. In general, they differ in that the responses of the nervous system are much more rapid and of shorter duration than those of the endocrine system, whose actions take much longer to develop but may last for many years or for the lifetime of the individual, once begun.

Nerve Cells

The nervous system is made up of **nerve cells**, or *neurons*, which are adapted for the transmission of impulses. A typical nerve cell consists of several important parts: the cell body, or *cyton*, which includes the nucleus; numerous *dendrites* that branch out from the cell body to receive stimuli from their environment; the *axon*, which is a long, thin extension covered by the *myelin sheath* of specialized *Schwann cells*; and the axon terminals, or *end branches* (end brush). Between the end branches of one neuron and the dendrites of the next is a tiny space called the *synapse*. Thus, although nerve cells transmit and receive impulses, they do not physically touch one another. (Refer to Figure 2-12 on page 46.)

The nervous system contains three different types of nerve cells, which differ both in structure and function; these are the sensory neurons, motor neurons, and interneurons. *Sensory neurons* transmit impulses from the sense organs, or receptors, to the brain and the spinal cord. Sense organs include the eyes, ears, tongue, nose, and skin. *Motor neurons* transmit impulses from the brain and spinal cord to the *effectors*, that is, to the muscles and the glands. *Interneurons* are found in the spinal cord and brain; they transmit nerve impulses from sensory neurons to motor neurons.

Transmission of Nerve Impulses

Neurons possess the property of *irritability*; that is, they are capable of being stimulated and then they are capable of undergoing changes. When a neuron is not conducting an impulse, its membrane is said to be at *resting potential*. Experiments have shown that a difference in electrical potential exists across the membrane of an axon. The area outside the membrane (extracellular fluid) is positively charged while the area inside (cytoplasm) is negatively charged. These charges across the membrane arise from ions (charged atoms) that are present on either side. On the outside of the cell, there is a large concentration of positively charged sodium ions (Na^+), while inside the neuron there are positively charged potassium ions (K^+) and negatively charged chloride ions (Cl^-). The membrane itself is normally impermeable to sodium ions; so even though there is a high concentration of Na^+ outside, these ions remain mostly outside and the membrane potential is maintained. When a nerve cell is stimulated by a strong disturbance in its environment, it undergoes a sudden change in permeability. The sodium ions flood into the cell, reversing the membrane's electrical charge. The membrane is then said to be *depolarized* (Figure 3-12).

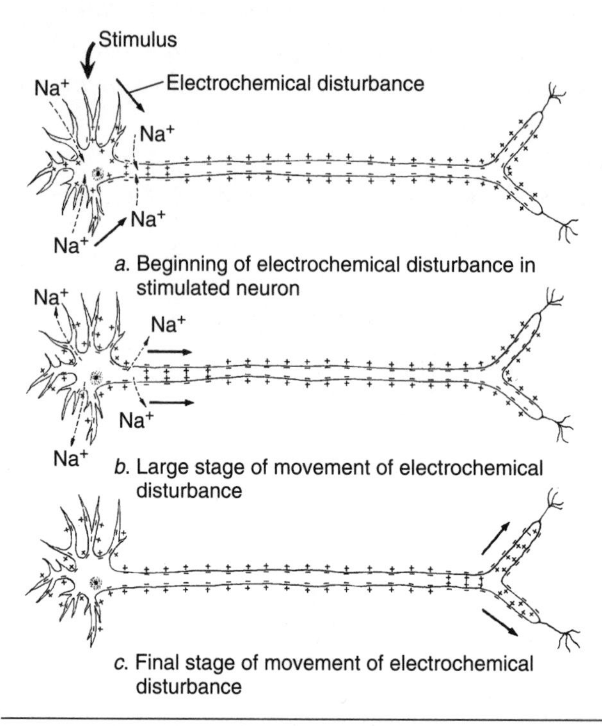

Figure 3-12. A nerve impulse is the movement of electrical changes (depolarization of the cell membrane) along the length of the neuron.

This depolarization generates an *action potential* in the membrane that causes the impulse to travel along the entire nerve cell, from dendrites to cell body to axon terminals. The depolarization that occurs in one portion of the axon membrane moves along the entire length of the axon. This wave of depolarization is known as the *nerve impulse*. When the impulse reaches the axon terminals, special vesicles in the terminal branches secrete a neurotransmitter (such as acetycholine) into the synapse. This chemical stimulates the next neuron in line so that the impulse can continue in that cell. The receiving neuron has special receptor molecules on its dendrites to which the neurotransmitter can attach and generate a new impulse. Soon after the neurotransmitter is secreted, it is destroyed by an enzyme or reabsorbed by the nerve cell so that it does not continue to function in the synapse.

Once the impulse has passed a particular section of the axon membrane, the axon again undergoes changes. The Na^+ ions are actively pumped out of the neuron and the K^+ ions re-enter it. This process is carried out by special transport proteins in the neuron's membrane and it is known as the *sodium–potassium pump*. It uses the energy of ATP to pump against the concentration gradient and is, therefore, a form of active transport. The pump mechanism restores the resting potential of the membrane to what it was before the stimulus. Then, after the resting potential has been restored, the neuron can again be stimulated to transmit a new impulse.

Nerves

The nerve cells, or parts of nerve cells, are bound together in bundles called *nerves*. There are three kinds of nerves: *sensory nerves*, which contain only sensory neurons; *motor nerves*, which contain only motor neurons; and *mixed nerves*, which contain both sensory and motor neurons.

Central Nervous System

The two main divisions of the human nervous system are the *central nervous system* (CNS), which includes the brain and spinal cord, and the *peripheral nervous system* (PNS), which includes all the nerves outside the central nervous system.

The Brain. The *brain* is a large mass of nerve cells located in the cranial cavity. It is surrounded and protected by the bones of the skull and by several membranes known as the *meninges*. The three major parts of the brain are the cerebrum, the cerebellum, and the medulla (Figure 3-13). Each controls different functions of the body.

In humans, the *cerebrum* is the largest part of the brain. It is the center for thought, memory, judgment, reasoning, the senses, and learning. The cerebrum receives and interprets messages from the sense organs; and it initiates all voluntary, or conscious, movements.

The *cerebellum* is located below and behind the cerebrum. It coordinates all motor activities and is involved in maintaining the body's balance.

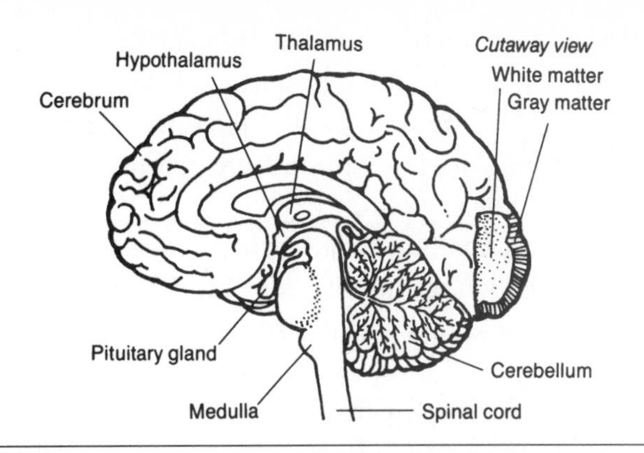

Figure 3-13. Structure of the human brain.

The *medulla* is located at the base of the brain and connects the brain and the spinal cord. The medulla controls many important involuntary activities in the body, including breathing, heartbeat, blood pressure, and peristalsis.

The Spinal Cord. The medulla of the brain is continuous with the *spinal cord*, which is surrounded and protected by the vertebrae of the backbone, or *spinal column*, and the spinal meninges. The spinal cord coordinates activities between the brain and other body structures. Impulses from sense receptors throughout the body are transmitted by sensory neurons to the spinal cord. In the spinal cord, impulses are transmitted by interneurons to the brain. Impulses from the brain are carried by motor neurons through the spinal cord and then to the appropriate effectors, which are muscles or glands that respond to the stimulus. The spinal cord is also the center for reflex activity. For most reflexes, the brain is not directly involved in their activity. In fact, the reflex action usually occurs before the brain is consciously aware that it has happened.

Peripheral Nervous System

The peripheral nervous system includes all neurons, both sensory and motor, outside the central nervous system. These nerve cells carry impulses between the central nervous system and the rest of the body. The two main divisions of the peripheral nervous system are the somatic nervous system and the autonomic nervous system.

The *somatic nervous system* includes all the nerves that control the movements of the voluntary muscles of the body, as well as the sensory neurons that transmit impulses from sense receptors to the central nervous system.

The *autonomic nervous system* consists of the nerves that control the activities of smooth muscle, cardiac muscle, and glands. The activities of this system, which are involuntary (that is, they happen without conscious control), include regulation of the heartbeat and circulation, respiration, and peristalsis.

Behavior and the Nervous System

All animals, including humans, have behaviors that help them maintain homeostasis and aid their survival. These behaviors are controlled by the nervous system. Some behaviors are inborn, or *innate*, while others are learned.

Habits. A *habit* is a kind of learned behavior that becomes automatic through repetition. The repetition establishes pathways for nerve impulse transmission that permit a rapid, automatic response to a particular stimulus. For example, when we learn to write, we are trained to dot the letter *i* and cross the letter *t*. After this behavior is learned, it is done as a habit without conscious thought.

Reflexes. An automatic, inborn response to a particular stimulus is called a *reflex*. In a reflex response, impulses follow a set pathway called a *reflex arc* (Figure 3-14). In this pathway, impulses pass from a receptor to a sensory neuron to an interneuron (in the spinal cord) to a motor neuron to an effector. Although impulses may also pass from an interneuron to the brain, the reflex response is controlled by the spinal cord and, as mentioned above, occurs without the involvement of the brain. Reflexes are generally protective in nature, allowing a rapid response to a potentially dangerous stimulus. For example, if you accidentally touch a hot surface, such as a stovetop, you pull your hand away from the source of heat without any conscious thought. Sensory neurons in the receptor (your hand) receive the painful stimulus. It is rapidly transmitted to the spinal cord, to the interneurons, which in turn transmit the impulse to motor neurons. The motor neurons carry an impulse to a muscle, which quickly contracts and pulls your hand away from the hot surface. You become consciously aware of what has happened (including the pain!) when the brain receives impulses from interneurons in the spinal cord.

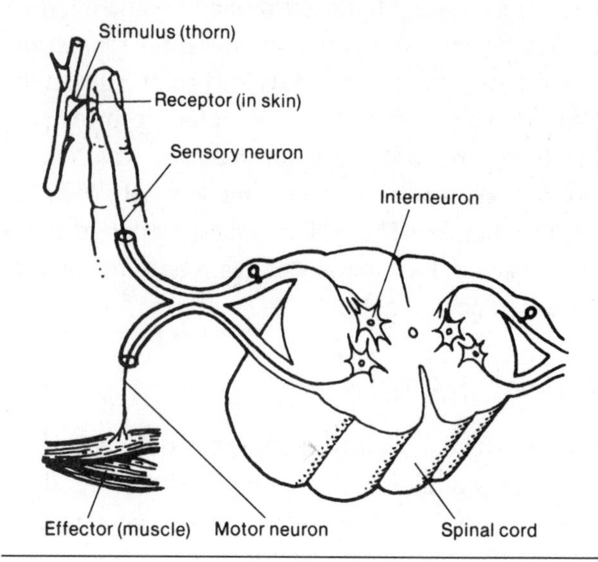

Figure 3-14. A typical reflex arc.

Disorders of the Nervous System

Cerebral palsy is a group of diseases caused by damage to the parts of the brain that control voluntary movement. This damage occurs during embryonic development. *Meningitis* is an inflammation of the membranes (meninges) that surround the brain and spinal cord. Meningitis may be caused by viral or bacterial infections, and symptoms include headache, muscle stiffness, fever, and chills. A *stroke* is a disorder in which the brain is damaged as a result of a *cerebral hemorrhage* (a broken blood vessel) or a blood clot (in a blood vessel) in the brain. *Polio* is a disease that affects the central nervous system; it may result in paralysis. Polio is caused by a virus and can be prevented by immunization.

Alzheimer's is a degenerative disease in which neurons in the brain are gradually destroyed. It is thought that certain types of plaques develop in the neurons of the brain as the disease progresses. This fatal illness generally strikes older people and begins with such symptoms as forgetfulness, mood swings, and unusual behavior. As the disease progresses, the person becomes less and less capable of handling simple daily tasks such as dressing, bathing, and eating on his or her own. Because a large segment of the American population is aging, this disease is becoming more and more commonplace.

QUESTIONS

MULTIPLE CHOICE

81. The major function of a motor neuron is to A. transmit impulses from the spinal cord to the brain B. act as a receptor for environment stimuli C. transmit impulses from sense organs to the central nervous system D. transmit impulses from the central nervous system to muscles or glands

82. Nerves are composed of bundles of A. muscle cells B. neurons C. phagocytes D. bone cells

83. Which part of the human central nervous system is involved primarily with sensory interpretation and thinking? A. spinal cord B. medulla C. cerebrum D. cerebellum

84. The somatic nervous system contains nerves that run from the central nervous system to the A. muscles of the skeleton B. heart C. smooth muscles of the gastrointestinal tract D. endocrine glands

85. If the cerebellum of a human were damaged, which of the following would probably result? A. inability to reason B. difficulty in breathing C. loss of sight D. loss of balance

86. Which is the correct route of an impulse in a reflex arc?

A. receptor → sensory neuron → interneuron → motor neuron → effector

B. effector → receptor → motor neuron → sensory neuron → interneuron

C. sensory neuron → effector → motor neuron → receptor → interneuron

D. motor neuron → sensory neuron → interneuron → effector

87. The brain and the spinal cord make up the A. autonomic nervous system B. peripheral nervous system C. central nervous system D. somatic nervous system

88. Impulses are transmitted from receptors to the central nervous system by A. receptor neurons B. sensory neurons C. interneurons D. motor neurons

Base your answers to questions 89 through 92 on the following diagram of the human brain.

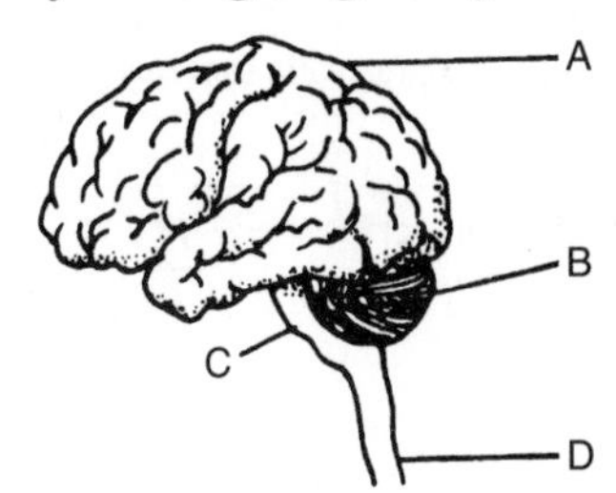

89. Injury to which part would most likely result in loss of memory? A. *A* B. *B* C. *C* D. *D*

90. Which part of the brain controls the involuntary movements of the digestive system? A. *A* B. *B* C. *C* D. *D*

91. Which part of the brain is involved with balance and the coordination of body movements? A. *A* B. *B* C. *C* D. *D*

92. Sight and hearing are functions of the structure labeled A. *A* B. *B* C. *C* D. *D*

Base your answers to questions 93 and 94 on the diagram below and on your knowledge of biology.

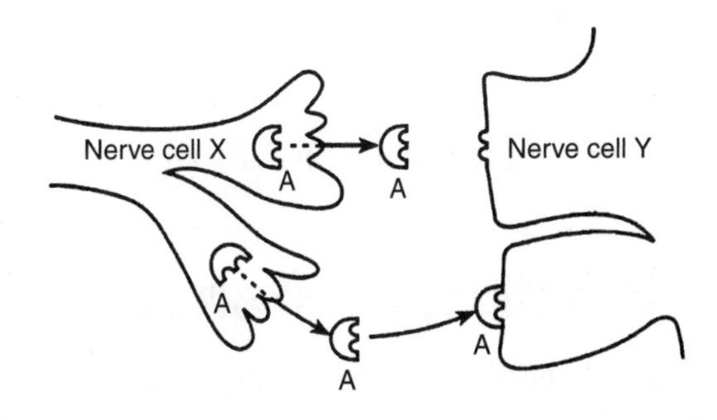

93. The process represented in the diagram best illustrates A. cellular communication B. muscle contraction C. extraction of energy from nutrients D. waste disposal

94. Which statement best describes the diagram? A. Nerve cell *X* is releasing receptor molecules. B. Nerve cell *Y* is signaling nerve cell *X*. C. Nerve cell *X* is attaching to nerve cell *Y*. D. Nerve cell *Y* contains receptor molecules for substance *A*.

OPEN RESPONSE

95. Use the following terms to complete the chart below, which outlines the human nervous system: *somatic nervous system, brain, cerebrum, cerebellum, autonomic nervous system, medulla,* and *spinal cord.*

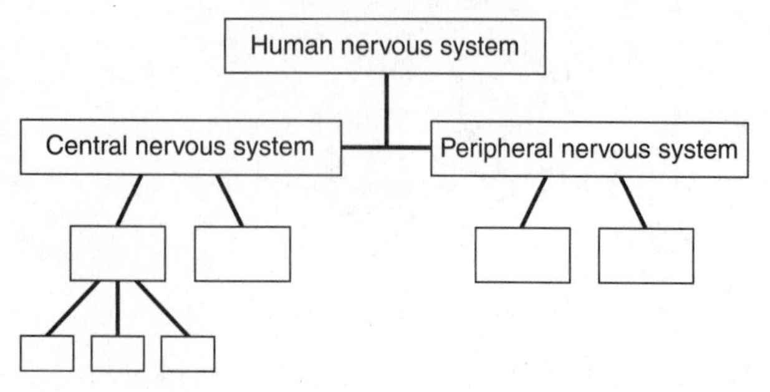

96. Briefly describe the main functions of the three major regions of the human brain: the cerebrum; the cerebellum; and the medulla.

97. Suppose that the enzyme that normally destroys a neurotransmitter does not function properly. How would this affect the nervous system?

98. Briefly explain how a reflex arc works to protect the human body from a potentially dangerous stimulus. Provide an example.

99. The human nervous system is one of the most complex in the world. Give one example of how our brains have led to the progress that we consider to be "advanced" compared to that of other animals.

ENDOCRINE SYSTEM

The human endocrine system is made up of the endocrine glands, which secrete **hormones** directly into the blood. The hormones are transported by the circulatory system to the organs and tissues on which they act.

Endocrine Glands

The glands of the human endocrine system are the hypothalamus, pituitary, thyroid, parathyroids, adrenals, islets of Langerhans, and gonads (ovaries and testes) (Figure 3-15).

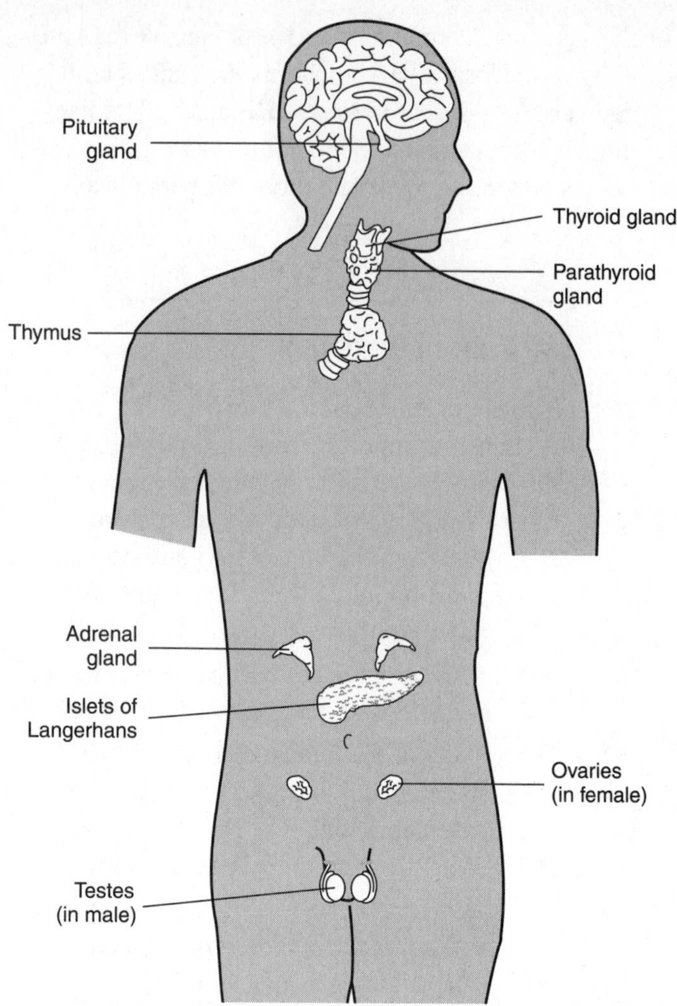

Figure 3-15. Structure of the human endocrine system.

Hypothalamus. Hormone-secreting cells are located in a small part of the brain called the *hypothalamus*. These secretions are referred to as *releasing factors* because they cause to the pituitary gland to release its hormones.

Pituitary Gland. Many hormones are secreted by the *pituitary gland*, which is located at the base of the brain. Some pituitary hormones regulate the activities of other endocrine glands as well.

Growth-stimulating hormone is a pituitary hormone that has widespread effects in the body, in addition to stimulating the growth of long bones. *Thyroid-stimulating hormone* (*TSH*) is a pituitary hormone that stimulates the thyroid gland to begin its secretion of the hormone thyroxin. *Follicle-stimulating hormone* (*FSH*) is a pituitary hormone that stimulates the development of follicles and ova (eggs) within the ovaries of females. In males, it influences sperm production.

Thyroid Gland. The iodine-containing hormone *thyroxin* is produced by the *thyroid gland*, which is located in the neck. Thyroxin regulates the rate of metabolism in the body cells and is essential for normal mental and physical development. In order to produce sufficient amounts of thyroxin, it is important to have an adequate supply of iodine in the diet. Fish, other seafoods, and iodized table salt are good sources of this essential mineral.

Parathyroid Glands. Embedded in the back of the thyroid gland are the *parathyroid glands*, which secrete the hormone *parathormone*. Parathormone controls calcium metabolism. Calcium is required for normal nerve function, blood clotting, and the growth of teeth and bones, as well as the proper functioning of muscle tissue.

Adrenal Glands. One *adrenal gland* is located on the top of each kidney. The outer layer of the adrenal glands is the adrenal cortex; the inner layer is the adrenal medulla.

The *adrenal cortex* secretes two types of steroid hormones. One type stimulates the conversion of fats and proteins to glucose, thereby increasing the level of glucose in the blood. The other type stimulates the reabsorption of sodium from the kidney tubules into the bloodstream. The concentration of sodium in the blood affects blood pressure and water balance.

The *adrenal medulla* secretes the hormone *adrenaline*, which increases the blood glucose level and accelerates the heartbeat and breathing rates. Adrenaline is released in times of stress and heavy exercise.

Islets of Langerhans. The small groups of endocrine cells that are found throughout the pancreas are called the *islets of Langerhans*. These endocrine cells secrete the hormones insulin and glucagon. The islets are composed of two different types of cells: alpha cells and beta cells. The *alpha cells* secrete the hormone glucagon, while the *beta cells* secrete the hormone insulin.

The hormone **insulin** causes the absorption of glucose from the blood into the body cells, thereby lowering the blood glucose level. It also stimulates the conversion of glucose to glycogen in the liver and in skeletal muscle.

The hormone *glucagon* increases the blood glucose level by stimulating the conversion of glycogen to glucose in the liver and skeletal muscle. The glucose then passes from these organs and tissues back into the blood. As you can see, these two hormones have precisely opposite effects. In this way, insulin and glucagon help the body maintain homeostasis by keeping the blood glucose level within healthy limits.

The Gonads. The male and female *gonads*—the testes and ovaries—both function as endocrine glands. The **testes** (singular, *testis*) secrete the male sex hormone **testosterone**, which stimulates the development of the male reproductive organs and secondary sex characteristics, such as body hair, a deeper voice, and muscular tissues. Testosterone also stimulates the production of **sperm**. The **ovaries** secrete the female sex hormones **estrogen** and **progesterone**. Estrogen influences the development of the female reproductive organs and secondary sex characteristics, such as breasts and wider hips. Estrogen also stimulates the production of **egg** cells. Progesterone stimulates the thickening of the uterine lining in preparation for the implantation of an **embryo** (the fertilized egg cell). These two hormones, along with secretions of the pituitary gland, are also responsible for the changes that occur in a woman's body during the menstrual cycle.

Hormones and Negative Feedback

The secretion of hormones by the endocrine glands is regulated by *feedback mechanisms*. In many cases, the level of one hormone in the blood can either stimulate (positive feedback) or inhibit (negative feedback) the production of a second hormone. The blood level of the second hormone in turn stimulates or inhibits the production of the first hormone. For example, the relationship between the pituitary gland's secretion of thyroid-stimulating hormone (TSH), which stimulates the thyroid's secretion of the hormone thyroxin (to regulate metabolism) is a classic type of a negative feedback loop, because thyroxin inhibits further production of TSH.

When the concentration of thyroxin in the blood drops below a certain level, the pituitary is stimulated to secrete TSH. This hormone, in turn, then stimulates the secretion of thyroxin by the thyroid. When the blood thyroxin concentration reaches a sufficient level, the further secretion of TSH by the pituitary is inhibited. In this way, the body can regulate thyroxin levels—just as it regulates carbon dioxide levels—and maintain homeostasis, or **stability**.

Disorders of the Endocrine System

A *goiter* is an enlargement of the thyroid gland that is most commonly caused by a lack of iodine in the diet. This condition can be treated by surgery and/or medication. *Diabetes* is a disorder that affects blood glucose levels. There are actually two types of diabetes: Type I (insulin dependent) and Type II (insulin independent). In Type I diabetes, the islets of Langerhans do not secrete adequate amounts of insulin into the bloodstream and, as a result, the blood glucose level is elevated. People with this condition need to take insulin by injection or orally. Type II diabetes develops later in life. In this case, a person's cells lose the ability to respond to insulin even though there may be an adequate supply in the blood. The condition is probably caused by the malfunction of an insulin receptor on the cell membrane, which prevents the insulin from binding to the cell. Individuals can successfully be treated with diet and medication that lowers blood sugar levels. In either case, untreated diabetes can cause serious effects such as blindness, kidney failure, and loss of limbs (from poor circulation).

Disorders in the pituitary gland may affect the release of *human growth hormone*, resulting in a negative effect on a person's growth. Recent advances in recombinant DNA technology have allowed the synthesis of human growth hormone, as well as insulin. These hormones (made by genetically engineered bacteria) can be used to replace or supplement the insufficient amount of hormone being produced by the person with the disorder. *Hormone replacement therapy* refers to the administration of hormones (usually female hormones to post-menopausal women) under the supervision of a physician. While there are some health benefits to this therapy for women whose natural production of estrogen and progesterone has diminished, there are risks as well.

It is normal for male gonads to secrete small quantities of female hormones and for female gonads to secrete small amounts of male hormones. In addition, the adrenal glands release both male and female hormones. However, malfunctions of the gonads may cause oversecretion of hormones of the opposite gender. For example, a woman may become masculinzed if her ovaries or adrenal glands oversecrete male hormones. Oversecretion of female hormones by the testes or adrenal glands can have a feminizing effect on men. In recent years, there has been much discussion about professional athletes who use male steroids (hormones) to artificially enhance their performance or build muscles. This is a very dangerous practice; the misuse of these steroids can have serious consequences, such as sterility and cancer.

QUESTIONS

MULTIPLE CHOICE

100. The part of the brain that is most directly related to the endocrine system is the
A. cerebrum B. medulla C. hypothalamus
D. cerebellum

101. Which structure secretes the substance it produces directly into the bloodstream?
A. gallbladder B. salivary gland C. adrenal gland D. skin

102. The hormones insulin and glucagon are produced by the A. thyroid B. pituitary
C. pancreas D. liver

103. Which hormone lowers blood sugar levels by increasing the rate of absorption of glucose by the body cells? A. follicle-stimulating hormone B. insulin C. parathormone D. adrenalin

104. A person was admitted to the hospital with abnormally high blood sugar level and a very high sugar content in his urine.

Which gland most likely caused this condition by secreting lower than normal amounts of its hormone? A. pancreas B. parathyroid C. salivary D. thyroid

105. Which hormone stimulates activity in the ovaries? A. testosterone B. thyroid stimulating hormone C. insulin D. follicle stimulating hormone

106. A person's rate of metabolism is regulated by a hormone secreted by the A. parathyroids B. thyroid C. pancreas D. adrenals

107. Estrogen, which influences the development of secondary sex characteristics, is secreted by the A. pituitary B. adrenals C. parathyroids D. ovaries

108. In humans, the level of calcium in the blood is regulated by the A. pancreas B. thyroid C. adrenals D. liver

109. The mechanism that regulates the secretion of hormones by endocrine glands is called A. peristalsis B. active transport C. feedback D. filtration

110. Insufficient iodine in the diet may cause goiter, a disorder of the A. adrenal glands B. pancreas C. pituitary gland D. thyroid gland

OPEN RESPONSE

Base your answers to questions 111 through 114 on the graph below, which shows the levels of glucose and insulin present in a person's blood after eating a meal.

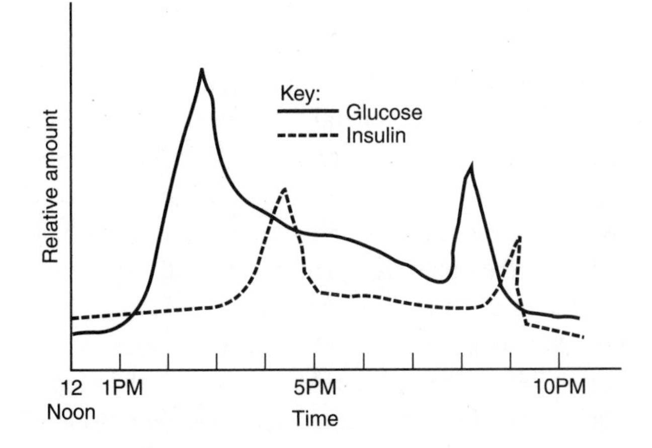

111. At approximately what times did the glucose level spike in this person?

112. What effect does insulin seem to have on the blood glucose level?

113. Describe the relationship between the levels of glucose and the levels of insulin.

114. What is the most probable reason for the time lag between the spikes in glucose level and the spikes in insulin level?

115. Use your knowledge of biology to answer the following questions about how a negative feedback mechanism works:

- How do the pituitary gland and thyroid gland affect each other?
- How does this feedback mechanism help in the maintenance of homeostasis?
- What hormones produced by each gland are part of this feedback mechanism?

116. The diagrams below represent some of the systems that make up the human body. Select one of the pairs of systems and:

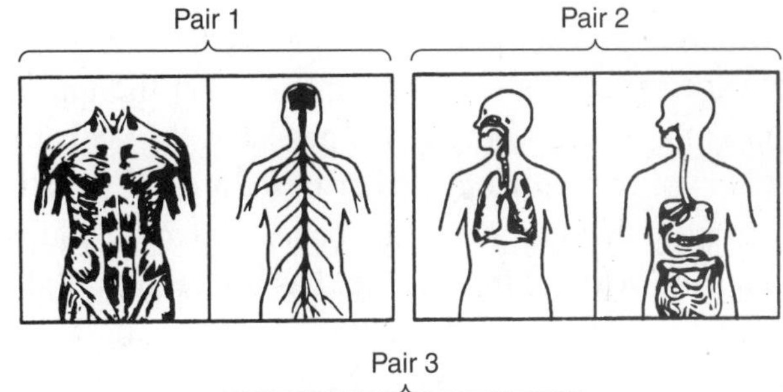

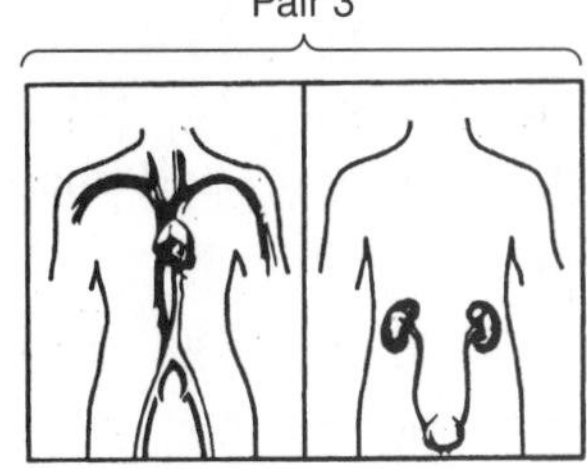

- identify each system in the pair you selected;
- state one function of each system in the pair;
- explain how the two systems work together to help maintain homeostasis.

LOCOMOTION

Locomotion, or *movement*, in humans involves the interaction of bones, cartilage, muscles, tendons, and ligaments.

Bones

The human skeleton is made up mainly of bones of various shapes and sizes. All bones are made of *bone tissue*, which is quite hard. Bone cells secrete the mineral calcium phosphate, which strengthens the bones and makes them rigid. Bones provide support and protection for the soft parts of the body (for example, the lungs and heart are protected by the rib cage and sternum); they are the sites of attachment for muscles; and they act as levers at the joints, enabling the body to move when the attached muscles contract. The production of new red blood cells and white blood cells occurs in the marrow of various bones.

Cartilage

In addition to bone, the human skeleton contains *cartilage*, a type of flexible, fibrous, elastic connective tissue. In embryos, most of the skeleton is made of cartilage. (This is most pronounced at the top of the skull, where soft cartilage persists until the baby is almost one year old.) After birth, a child's cartilage is gradually replaced by bone, so that by adulthood almost all of the cartilage has been replaced. In adults, cartilage is found at the ends of ribs, between vertebrae, at the ends of bones, and in the nose, ears, and trachea. Cartilage provides cushioning and flexibility at joints, and support and pliability in structures such as the nose and ears.

Joints

The places in the skeleton where the bones are connected to each other are called *joints*. Joints make movement of the skeleton possible. There are several kinds of movable joints in the human body. *Hinge joints*, which can move back and forth, are in the elbow and knee. *Ball-and-socket joints*, which are capable of circular movements, are found in the shoulder and hip. The neck has a *pivot joint*, which can move in a half circle. The bones of the skull are joined in *immovable joints*. These bones are fused together and provide good protection for the brain.

Muscles

Unlike other body tissues, muscle tissue has the capacity to *contract*, or shorten. All movement in the body involves muscle tissue. There are three types of muscle tissue in the human body: skeletal muscle, smooth muscle, and cardiac muscle (Figure 3-16).

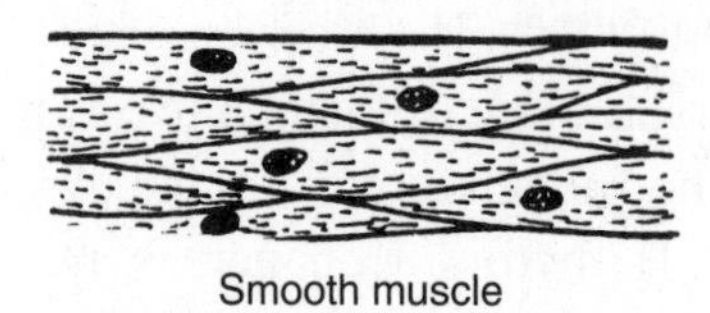

Smooth muscle

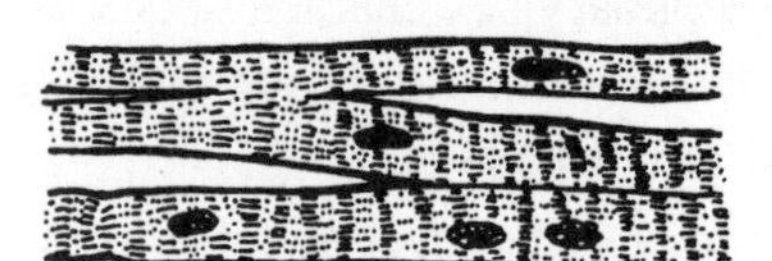

Cardiac muscle

Skeletal muscle

Figure 3-16. Smooth, cardiac, and skeletal muscle.

Skeletal Muscle. The voluntary muscles attached to the bones of the skeleton are made of *skeletal muscle* tissue. Muscle tissue of this type appears *striated*, or striped, when viewed with a microscope, and is also known as *striated muscle*. The contraction of skeletal muscle is controlled by the nervous system, which makes coordinated movements possible.

Skeletal muscles generally operate in antagonistic pairs: the contraction of one muscle of the pair extends the limb, while contraction of the other muscle flexes the limb. Figure 3-17 shows the muscles (and bones) of the upper arm. The triceps is the *extensor*, while the biceps is the *flexor*. When the biceps contracts, the triceps relaxes, and the arm flexes, bending at the elbow. When the triceps contracts, the biceps relaxes, and the arm is extended. Skeletal muscle is unusual in that there do not appear to be distinct cells. Rather, a continuous *multinucleated* (meaning "many nuclei") cytoplasm persists through most muscle tissue, or muscle *fibers*.

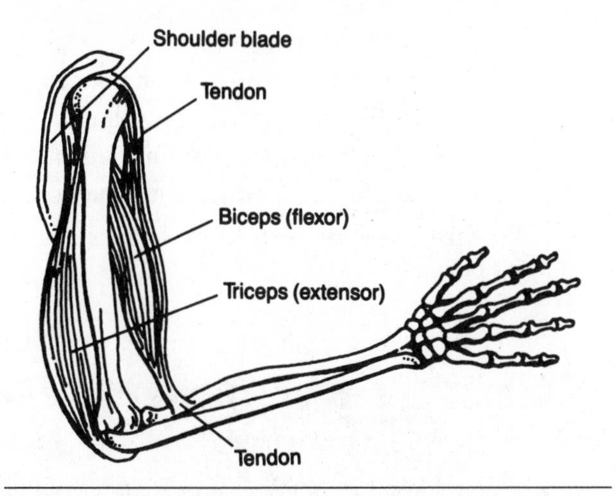

Figure 3-17. Muscles and bones of the upper arm.

Smooth Muscle. When viewed with a microscope, *smooth muscle* tissue does not appear striated. This type of muscle, which is also called *visceral muscle*, is found in the walls of the digestive organs and arteries, as well as in other internal organs. Smooth muscles are not under voluntary control; these muscles are controlled by the medulla. Peristalsis in the small and large intestine is an example of this type of muscle action.

Cardiac Muscle. The *cardiac muscle* tissue is found only in the heart. Although it appears striated when viewed with a microscope, cardiac muscle tissue is not under voluntary control, and its structure is different from that of skeletal muscle tissue. Like smooth muscle, the cardiac muscle is controlled by the medulla.

Tendons and Ligaments

Muscles are attached to bones by tough, inelastic, fibrous cords of connective tissue called *tendons*. Bones are connected together at movable joints by *ligaments,* which are composed of tough, elastic connective tissue.

Disorders of Locomotion

Arthritis is an inflammation of the joints, which can be very painful and make movement difficult. *Tendonitis* is an inflammation of a tendon, usually where it is attached to a bone. A very common site for this condition is in the elbow. Tendonitis occurs most commonly in athletes. *Osteoporosis* affects many elderly people, especially post-menopausal women. This is a condition in which there is a gradual loss of bone mass and density, leading to weakened, fragile bones that can break easily. A diet rich in calcium and vitamin D throughout one's lifetime can help prevent this disease. Although it can occur in both genders, osteoporosis is more common in women. Diagnostic medical exams such as bone density scans can detect the early stages of osteoporosis so that preventive measures can be taken (such as calcium supplements and bone-strengthening exercises).

QUESTIONS

MULTIPLE CHOICE

117. Which type of muscle tissue found in the walls of the human stomach is most closely associated with the process of peristalsis? A. striated B. cardiac C. voluntary D. smooth

118. Bones are attached to each other at movable joints by A. elastic ligaments B. cartilaginous tissues C. smooth muscles D. skeletal muscles

119. Which is *not* a major function of cartilage tissues in a human adult? A. giving pliable support to body structures B. cushioning joint areas C. adding flexibility to joints D. providing skeletal levers

120. Which type of connective tissue makes up the greatest proportion of the skeleton of a human embryo? A. ligaments B. cartilage C. tendons D. bone

121. Which structure contains pairs of opposing skeletal muscles? A. stomach B. small intestine C. heart D. hand

122. Which statement most accurately describes human skeletal muscle tissue? A. It is involuntary and striated. B. It is involuntary and lacks striations. C. It is voluntary and striated. D. It is voluntary and lacks striations.

123. In the human elbow joint, the bone of the upper arm is connected to the bones of the lower arm by flexible connective tissue called A. tendons B. ligaments C. muscles D. neurons

For each phrase in questions 124 through 128, select the human body structure in the list below that is best described by that phrase.

Human Body Structure
A. Bones
B. Cartilage tissues
C. Ligaments
D. Smooth muscles
E. Tendons
F. Voluntary muscles

124. Cause peristalsis in the digestive tract A. *B* B. *C* C. *D* D. *F*

125. Serve as extensors and flexors A. *A* B. *D* C. *E* D. *F*

126. Serve as levers for body movements A. *A* B. *B* C. *C* D. *E*

127. Bind the ends of bones together A. *B* B. *C* C. *D* D. *E*

128. Attach the muscles to bones A. *B* B. *C* C. *D* D. *E*

OPEN RESPONSE

Refer to Figure 3-17, which shows the bones and (upper arm) muscles of the human arm to answer questions 129 through 131.

129. What happens to the arm when the biceps contracts?

130. What happens to the arm when the triceps contracts?

131. Why are the biceps and triceps considered an opposing pair of muscles?

132. Briefly compare the functions of the following paired structures of the muscular and skeletal systems: smooth muscle to skeletal muscle; tendon to ligament; and bone to cartilage.

133. How do the skeletal and muscular systems work together to produce locomotion?

134. State two advantages that locomotion gives to an organism. Explain how they aid survival.

135. Suppose you were a physician and an elderly patient came into your office. You notice a bent-over posture and difficulty walking. What might your diagnosis of this patient be? How would you confirm your diagnosis? What medical advice might you give to this patient?

HOMEOSTASIS: A DYNAMIC EQUILIBRIUM

Under normal circumstances, an organism is able to maintain homeostasis in relation to both its internal and external environments. This maintaining of a **dynamic equilibrium** means that, despite the fact that external environmental conditions may change, an organism responds by taking corrective actions that maintain healthy conditions within its body.

For example, to maintain a normal temperature of about 37°C, the human body can make simple adjustments to keep the temperature within a safe range. If the body is too cold, small blood vessels in the skin may constrict in order to direct blood flow to the vital, internal organs. In addition, the body may shiver to generate more heat. If the body is overheated, blood vessels near the skin surface can dilate (open wider) to promote blood flow to the skin in order to lose heat to the surrounding air. The skin may also produce perspiration (sweat) as a means of lowering the body temperature.

Homeostasis and Feedback Mechanisms

As discussed in the sections on breathing rate and on hormones, many homeostatic adjustments in humans involve interactions called *negative feedback mechanisms*. An initial change in one part of the loop stimulates a reaction in another part of the system, which responds until stability has been reached. In this way, homeostasis is maintained. If conditions in the body change, the feedback system is triggered into action again.

A very common negative feedback loop involves the pituitary gland, the thyroid gland, and their hormones. As already described, if the level of thyroxin is too low, the pituitary gland secretes thyroid-stimulating hormone (TSH), causing the thyroid to increase its production of thyroxin. As the levels of thyroxin increase, the pituitary gland stops secreting TSH and the thyroid slows down its secretion of thyroxin. In this way, the levels of thyroxin are maintained within normal limits (Figure 3-18).

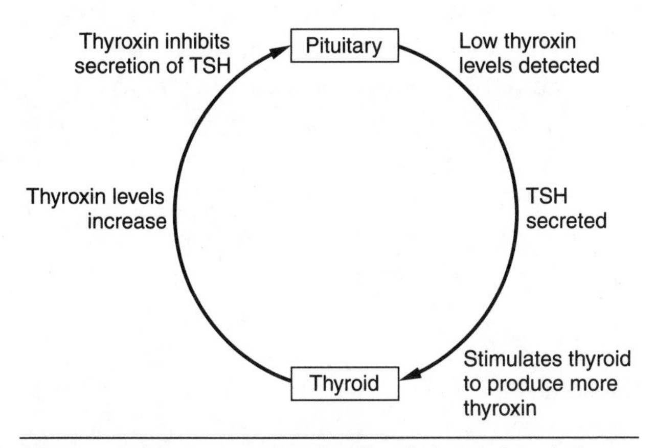

Figure 3-18. A negative feedback loop, involving the pituitary and thyroid glands.

WHEN HOMEOSTASIS FAILS: ILLNESS AND DISEASE

Causes of Disease

Any breakdown in an organism's ability to maintain or restore equilibrium can result in illness, disease, or even death. The causes of disease are many and varied. Diseases that are caused by factors inside the body are usually inherited and due to defective genetic traits. Diseases that are caused by factors outside the body, and that can be passed from one organism to another, are called *infectious* diseases. Factors that cause such diseases include microorganisms, or **microbes**, that are harmful. These disease-causing microbes, called **pathogens**, may include bacteria, fungi, protozoa, and worms, as well as nonliving particles of protein and nucleic acid, called *viruses*.

Another type of infectious particle is called a *prion*. Like viruses, these misfolded forms of proteins have the unusual ability to increase in number when inside a living organism. It is thought that prions, which form clumps, or *plaques*, that kill brain cells, cause normal proteins to misfold and become abnormal. A well-known disease caused by prions is bovine spongiform encephalopathy, or *mad cow disease*, a fatal illness that destroys the nervous system in cattle. Sheep also may suffer from a nervous-system disease caused by prions, called *scrapie*. A degenerative nervous disorder known as Creutzfeldt-Jakob disease, or *kuru*, which has infected some people, is also caused by prions.

Sometimes, unhealthy habits and/or risky behaviors can jeopardize health and lead to illness. Poor nutrition, cigarette smoking, and abuse of alcohol and drugs can all result in serious illness and a breakdown of homeostasis. For example, excessive consumption of alcoholic beverages can cause cirrhosis of the liver, a fatal disease.

Cancer. Disease may also occur when certain cells in the body behave abnormally due to a genetic mutation. Such cells can divide uncontrollably and result in the growth of *tumors*. Tumors are either benign (not spreading) or malignant (spreading). Uncontrolled growth, or *metastasis*, of malignant cells is known as **cancer**. When cancer cells spread throughout the body, they interfere with the functioning of normal cells. In such cases, the cancer can become life threatening. Although cancer may occur spontaneously, certain factors are known to increase the risk of developing it. Tobacco smoking, unhealthful diet, genetic factors, and exposure to **radiation** and certain chemicals called *carcinogens* are all thought to play a part in causing cancer.

Symptoms of Disease. Some diseases show their symptoms as soon as they begin to develop or soon after they are triggered by a pathogen. An example is influenza (the "flu"), which is caused by viruses. Other diseases may take several days, weeks, or even years before their symptoms appear. **AIDS** (***a****cquired* ***i****mmuno****d****eficiency* ***s****yndrome*) and cancer are examples of diseases that may develop in the body for years before their symptoms appear.

QUESTIONS

MULTIPLE CHOICE

136. The term that describes a body's overall ability to maintain homeostasis is A. negative loop system B. low maintenance C. dynamic equilibrium D. infectious

137. Pathogens may include all of the following *except* A. fungi B. protozoa C. bacteria D. plants

138. Viruses differ from other pathogens in that *only* viruses A. contain a true nucleus B. can reproduce on their own every 20 minutes C. consist only of protein and nucleic acid D. are able to infect healthy cells

139. Which of the following represents a correct cause-and-effect sequence? A. cirrhosis of the liver → excessive alcohol consumption B. low thyroxin levels → increase in TSH secretion C. symptoms of disease → exposure to pathogen D. dilation of blood vessels in skin → overheating of the body

140. Infectious particles known as prions, which cause nervous system diseases such as scrapie and *kuru*, consist of misshaped pieces of A. bacteria B. fungi C. proteins D. viruses

141. Tobacco smoking, unhealthful diet, genetic factors, and exposure to radiation and carcinogens are all thought to play a part in causing A. gout B. diabetes C. cancer D. tuberculosis.

OPEN RESPONSE

Base your answer to the following question on the diagram below and on your knowledge of biology.

142. What term or phrase does letter *X* most likely represent?

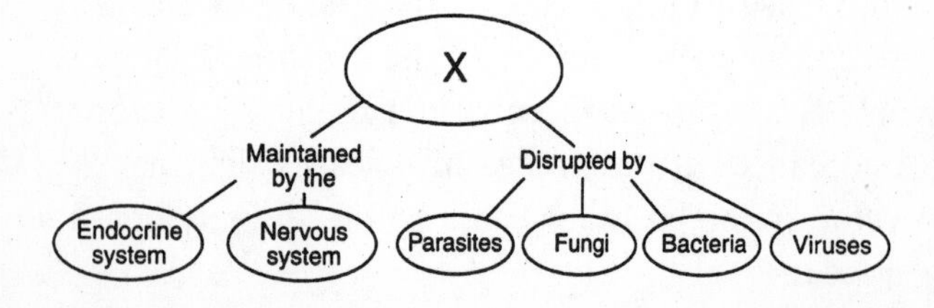

143. Select two of the following risk factors—drug abuse; poor nutrition; genetic factors; radiation; tobacco smoking—then:

- define each of the two risk factors you have selected;
- explain how each risk factor can interfere with proper functioning of the immune system.

Base your answers to questions 144 through 146 on the diagram of a negative feedback loop, shown below, and on the following data: Structure A releases a substance, X, that can stimulate structure B to release its substance, Y.

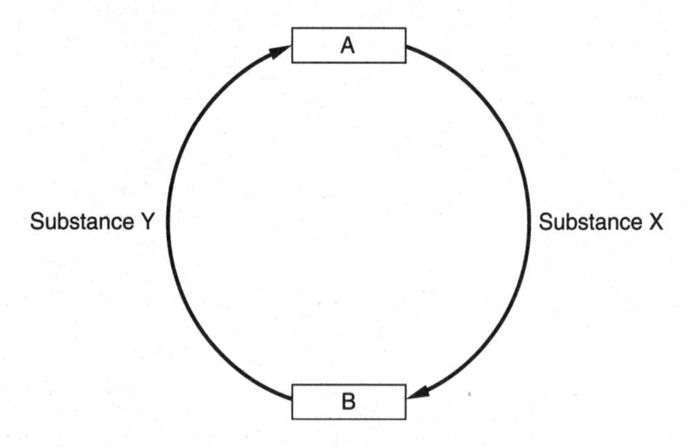

144. Under what conditions would structure *A* probably release substance *X*?

145. Explain what happens when the levels of substance *Y* get too high.

146. What type of substances are *X* and *Y* most likely to be?

THE IMMUNE SYSTEM: PROTECTION AGAINST DISEASE

The human body is well protected against invading pathogens. The first line of defense prevents harmful microorganisms from getting into the body by blocking their entry. The skin, when unbroken, provides an effective physical barrier to nearly all pathogenic organisms. Secretions such as tears, saliva, and mucus provide an effective physical and chemical barrier; they contain enzymes that destroy pathogens or help trap and flush them out of the body.

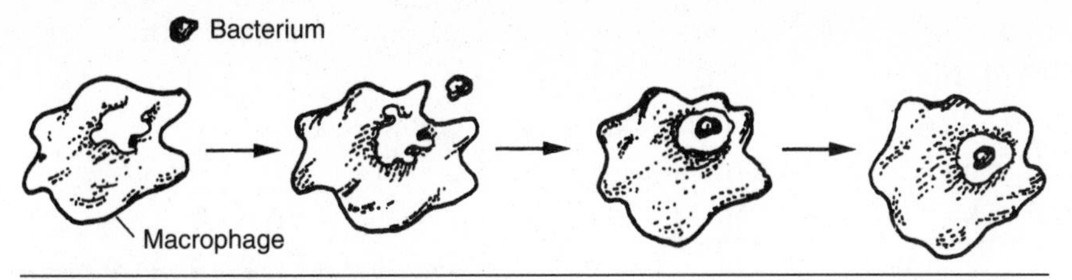

Figure 3-19. The macrophage is a type of white blood cell that engulfs and destroys invading pathogens.

Nevertheless, some pathogens manage to elude the first line of defense and gain entry. They may do so through breaks in the skin (cuts and scrapes) or through the eyes and natural openings in the body, such as the mouth and nostrils. Once inside, these invaders are confronted by the **immune system**, the body's primary defense mechanism. Invaders may be destroyed by being engulfed by special cells or by being chemically marked for destruction and elimination.

Functions of the Immune System

How does the immune system function? All cells have very specific proteins on their plasma membrane surfaces. The immune system is able to recognize proteins on cells that are foreign and to distinguish them from its own body's proteins. These invading foreign proteins are referred to as *antigens*.

Specialized Blood Cells. The human immune system consists of specialized white blood cells (leukocytes) and lymphatic organs such as the spleen, thymus, and tonsils. The system also has a number of *lymph nodes* that participate in defense mechanisms. Some white blood cells, called *macrophages*, engulf and digest pathogens (Figure 3-19). After destroying the pathogens, these white cells often die, too.

White blood cells called *T cells* are specialized to kill pathogens or mark them for destruction. Other white blood cells, called *B cells*, produce very specific *antibodies* against the pathogens. Antibodies have a chemical structure that precisely matches the shape of the antigen with which they react. Once the match has been made, the pathogen is destroyed by the antibodies. Some of the antibody-producing blood cells remain in the body's immune system as "memory cells." These specialized cells can quickly mount an attack if the body is invaded again by the same pathogen (Figure 3-20).

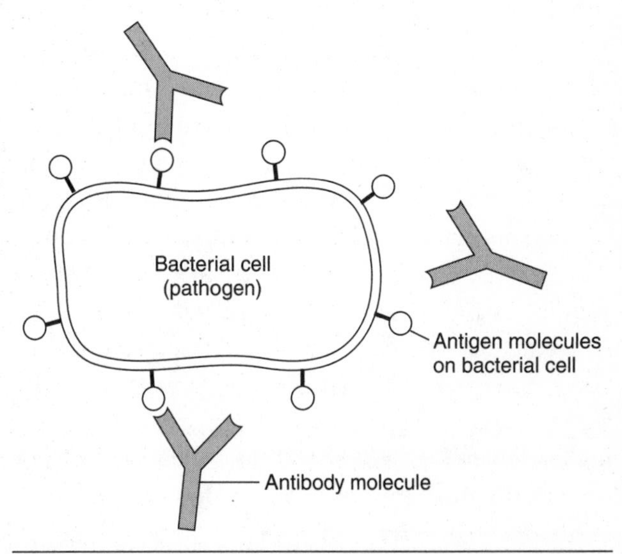

Figure 3-20. The immune system produces antibodies that are specific to the invading foreign antigens.

Vaccinations

A *vaccine* is a weakened pathogen or its antigen that is injected into an organism. **Vaccinations** are given to people and animals to provide *immunity* against particular pathogens. Once recognized by the immune system, the invading antigen causes antibodies that are specific to it to be made. The cells that produce the antibodies remain as memory cells in the person. Thus, vaccinations provide *active acquired immunity* (see following section). If the actual pathogen invades the body at a later time, these memory cells can launch an immediate response and attack the invaders, often before they have a chance to cause any disease symptoms. Some examples of vaccines that are given to people include those for the flu (influenza), MMR (measles, mumps, rubella), and hepatitis. In the United States, all children must be vaccinated for several diseases before they are permitted to attend school. Some examples of vaccines that are given to animals include distemper shots and rabies shots for pet dogs and cats.

Types of Immunity

There are several types of immunity to disease that a person may have or acquire. When a person contracts a disease (or receives a vaccination) that he or she later recovers from, his or her immune system creates antibodies that are specific to the pathogens or antigens that caused the disease. These antibodies remain in the bloodstream even after the disease symptoms have ceased. If the person becomes infected again with the same type of pathogen or antigen, the antibodies already present attack the invaders so that he or she will either recover faster or not even become ill a second time. This type of immunity is called **acquired immunity**, and it can be either *active* or *passive*.

In **active immunity**, the person is exposed to the pathogen or antigen directly. Active immunity may result from direct exposure to a pathogen or antigen from the environment or in the form of a vaccine. In either case, the person produces his or her own antibodies against the disease.

In **passive immunity**, antibodies are transferred from another source to the person. For example, during fetal development, antibodies cross the placenta from the mother to her fetus, thus providing immunity. Additionally, if a mother breast-feeds her infant, antibodies also enter the baby through the mother's milk. Providing antibodies through an injection may also induce passive immunity. For example, if a person is bitten by a rabid animal, an injection containing antibodies to the rabies virus is given to the person to prevent the development of rabies, a fatal disease.

Innate immunity in a person is determined genetically. This type of immunity is present at birth and has no relationship to exposure to pathogenic organisms or antigens. There are certain illnesses that people are naturally immune to but that may infect other animals. For example, people do not get all of the same diseases that dogs or cats can develop.

PROBLEMS IN THE IMMUNE SYSTEM

Overreactions of the Immune System

In some individuals, the immune system overreacts to certain stimuli or antigens that are harmless to most other people. Unfortunately, these severe reactions cause, rather than prevent, suffering and illness for the person.

Allergic Reactions. An **allergic reaction** is a strong response to *allergens* in pollen, animal fur, mold, insect stings, foods, and so on. The sufferer may experience sneezing, watery and itchy eyes, a runny nose, hives, coughing, and/or swelling. These uncomfortable symptoms are triggered by the immune system's release of substances called *histamines*. Although these allergy symptoms are inconvenient, they are responses made by the body in an attempt to expel the invading antigen. In some cases, the swelling may be so severe in the sinuses or throat that it interferes with breathing. An extreme type of allergic reaction, known as *anaphylactic shock*, occurs in some people in response to bee or wasp stings (and even in response to certain medications). This condition causes severe swelling and can be truly life threatening.

Autoimmune Diseases. In very rare cases, the immune system accidentally targets some of the body's own cell proteins as antigens. Once the immune system has identified an antigen as foreign, the cells bearing that protein are attacked as if they were invading foreign pathogens. This reaction produces a condition known as an **autoimmune disease** (*auto* meaning "self"). Examples of such serious diseases include rheumatoid arthritis (which causes inflammation and pain in the joint membranes) and lupus erythematosus (which causes painful swelling of the skin and joints, fever, rash, hair loss, fatigue, and sensitivity to light).

Immune Response to Transplants. People who receive transplanted organs, such as a heart, liver, or kidney (due to a *malfunction* of their own organ), may also experience problems with their immune response. Because the organ is recognized as foreign, the immune system may launch an attack against it, causing the body to reject the new organ. Physicians attempt to match the protein chemistry of the organ donor as closely as possible with that of the recipient, in order to minimize the risk of organ rejection. In addition, *immunosuppressant* drugs may be used to lessen the immune response. However, use of these medications can leave the transplant recipient quite vulnerable to infection by various microbes.

A Damaged or Weakened Immune System

HIV (***h**uman **i**mmunodeficiency **v**irus*), the agent that causes AIDS, damages the immune system by destroying specific T cells known as *helper T cells*. This leaves the affected person with a severely limited immune response. For that reason, AIDS is called an *immunodeficiency disease*. In fact, AIDS sufferers are prone to and frequently die from a variety of diseases that a healthy person's immune system could probably conquer (especially with the use of medicine), rather than from the virus itself.

Finally, as a person gets older, his or her immune system gradually weakens in its ability to respond to pathogens or cancerous cells. Consequently, older adults may be more prone than younger individuals to becoming ill or developing (malignant) tumors. Fatigue, stress, substance abuse, and poor nutrition can also contribute to a weakened immune response.

QUESTIONS

MULTIPLE CHOICE

147. Which cells are important components of the human immune system? A. red blood cells B. liver cells C. white blood cells D. nerve cells

148. A blood test showed that a person had increased levels of antibodies. This may indicate that the person has A. an infection B. diabetes C. low blood pressure D. an enlarged thyroid

149. Antibodies are produced by the body's A. T cells B. lymph node cells C. B cells D. liver cells

150. Substances that trigger a defensive response by the immune system are called A. antibodies B. antigens C. lymph nodes D. macrophages

151. A similarity between antibodies and enzymes is that both A. are lipids B. are produced by liver cells C. can make blood vessels dilate D. have very specific shapes and functions

152. Which statement does *not* describe an example of a feedback mechanism that maintains homeostasis? A. The guard cells close the openings in leaves, preventing excess water loss from a plant. B. White blood cells increase the production of antigens during an allergic reaction. C. Increased physical activity increases heart rate in humans. D. The pancreas releases insulin, helping humans to keep blood sugar levels stable.

153. Vaccines are given to people in order to A. disrupt their homeostasis B. immunize them against certain diseases C. inject T cells and B cells into them D. test if they can destroy the pathogen

154. A person's sneezing, coughing, and watery eyes right after exposure to cat hair are all indications of A. an autoimmune disease B. an infection caused by the cat C. an allergic reaction D. early warning signs of cancer

155. The use of a vaccine to stimulate the immune system to act against a specific pathogen is valuable in maintaining homeostasis because A. once the body produces chemicals to combat one type of virus, it can more easily make antibiotics B. the body can digest the weakened microbes and use them as food C. the body will be able to fight invasions by the same type of microbe in the future D. the more the immune system is challenged, the better it performs

Answer the following question based on your knowledge of biology and on the diagram below, which represents what can happen when homeostasis in an organism is threatened.

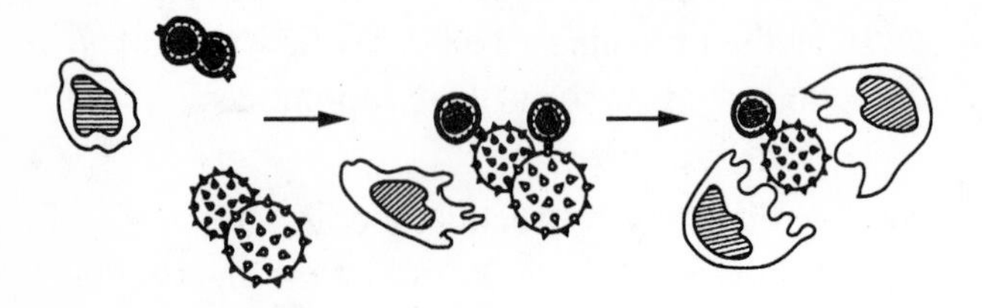

156. Which statement provides a possible explanation for the event shown? A. Ruptured blood platelets release an enzyme that starts the clotting reactions. B. Specialized cells tag and/or engulf invading microbes during an immune response. C. Embryonic development of specialized cells occurs during pregnancy. D. Cloning removes abnormal cells produced during differentiation.

157. Which statement describes an example of active acquired immunity? A. Humans generally do not get equine encephalitis, a disease of horses. B. After having mumps as a child, an adult does not generally have a recurrence of the disease. C. A patient receives an antibiotic to fight off a respiratory infection. D. Breast milk provides many antibodies to a nursing infant.

158. Antibodies that cross the placenta from mother to baby during fetal development provide the child with A. innate immunity B. active acquired immunity C. passive acquired immunity D. induced passive immunity

OPEN RESPONSE

Base your answers to questions 24 through 26 on the following graph, which shows the relationship between exposure to an antigen and the antibody response that followed.

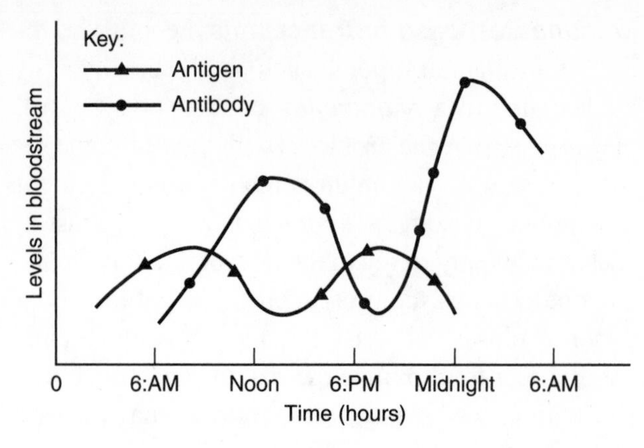

159. At what times did the antigen reach maximum levels in the bloodstream?

160. What relationship exists between the antigen levels and antibody levels in this graph?

161. The second peak of the antibody level is much greater than the first peak. Explain why.

Base your answers to question 27 through 29 on the paragraph below and on your knowledge of biology.

A boy contracted the viral disease chicken pox when he was a first grader. His doctor kept him out of school for two weeks until he recovered from the illness. Two years later, when his younger brother came down with chicken pox, the older boy did not catch it again, even though they shared a bedroom and were in close contact with one another.

162. What type of immunity to the chicken pox virus did the older boy develop? State one reason to support your answer.

163. How did the older boy's immune system protect him against chicken pox when he contracted the virus in the first grade?

164. Why didn't the older boy catch chicken pox again when his younger brother had it two years later?

165. A pharmaceutical company is proposing that its new product, *Immunoblast*, can help strengthen a person's immune system. Design an experiment in which you could test the effectiveness of this new product. Include the following steps:

- state the problem you are investigating;
- propose a suitable hypothesis;
- write the experimental procedure you would follow;
- list the data you would collect to test your hypothesis.

166. Each year, before the start of the flu season, older adults are advised to get a flu shot, or vaccination, to protect them. Answer the following questions:

- Why is the flu shot recommended more often for seniors than for younger adults?
- How does the vaccine protect people from the flu?
- Why is a new flu vaccination needed every year?

167. Describe two ways that the risk of organ rejection can be minimized in a transplant patient. Discuss one problem that is associated with one of the methods you have described.

168. AIDS is an infectious disease that has reached epidemic proportions. Briefly describe the nature of this disease and be sure to include:

- the type of pathogen that causes AIDS;
- the specific body system that is attacked by that pathogen;
- the effect on the body when this system is weakened by AIDS;
- *two* ways to prevent or control the spread of infectious diseases such as AIDS.

READING COMPREHENSION

Base your answers to questions 169 through 172 on the information below and on your knowledge of biology. Source: *Science News* (April 30, 2005): Vol. 167, No. 18, p. 285.

When the Stomach Gets Low on Acid

A shortage of stomach acid can lead to cancer, possibly as a result of bacterial overgrowth and chronic inflammation, a study in mice indicates.

Too much stomach acid is a well-studied problem that can cause more than simple gastritis, an inflammation of the stomach lining. Excess acid can lead to heartburn and cause chronic inflammation of the esophagus, esophageal scarring, and even cancer.

Turning the tables, scientists recently found that too little stomach acid might cause its own problems, including pneumonia.

In the new study of low stomach acid, Juanita L. Merchant, a gastroenterologist at the University of Michigan in Ann Arbor, and her colleagues studied 20 mice, half of which were genetically engineered to lack gastrin, the hormone that orchestrates stomach-acid secretion. Six of the mice lacking gastrin developed stomach tumors at 12 months of age, but none of the normal mice did, the researchers report in the March 31 *Oncogene*.

The mice lacking gastrin also had fewer stomach-lining cells die off, which is a normal, tumor-suppressing action. In this process, the body detects runaway cell growth and sends the aberrant cells into suicide mode. The gastrin-deficient mice lacked RUNX3, a protein that in normal mice can activate such programmed cell death. Merchant hypothesizes that inflammation brought on by excess bacterial growth might suppress RUNX3 production.

It's too early to draw a parallel between acid-deficient mice lacking all gastrin from birth and people who regularly take acid-blocking drugs for acid-reflux disease, Merchant says.

169. How might a lack of stomach acid lead to stomach cancer?

170. State three health problems caused by excess stomach acid.

171. Explain the connection between gastrin and stomach acid.

172. What role does RUNX3 play in preventing tumors in normal mice?

CHAPTER 4 Reproduction and Development

Standards 2.6, 2.7 (Cell Biology)

BROAD CONCEPT: Cells have specific structures and functions that make them distinctive. Processes in a cell can be classified broadly as growth, maintenance, and reproduction.

Standard 4.6 (Anatomy and Physiology)

BROAD CONCEPT: There is a relationship between the organization of cells into tissues, and tissues into organs. The structure and function of organs determine their relationship within body systems of an organism. Homeostasis allows the body to perform its normal functions.

The survival of a species depends on reproduction, that is, the production of new individuals. There are two ways that organisms can reproduce: *asexually* and *sexually*. In **asexual reproduction**, only one parent is involved, and the new organism develops from a cell or cells of the parent organism. In **sexual reproduction**, there are two parents, and each one contributes DNA in a specialized **sex cell** to the new organism. The two sex cells, one from each parent, fuse to form the first cell of the new generation.

THE CELL CYCLE

Cells, as units of life, have their own cycle of formation, growth, function, division, and/or death. This sequence of events is known as the **cell cycle**. While the cycle is a continuous process, biologists recognize several observable events that occur: interphase, mitosis, and cytokinesis. The first of these events, *interphase*, has three separate phases. The first phase is known as G1 (Gap 1 or Growth 1). During this growth phase, the cell carries out its normal metabolic activities, such as digestion, synthesis of proteins, and removal of wastes. The second part of interphase is known as S (Synthesis). It is during this phase that the cell prepares for its upcoming division. During S, the cell *replicates*, or makes a copy of, all of its genetic material (DNA). The next phase is called G2 (Gap 2 or Growth 2). During G2, the cell continues to grow and carry out its metabolic functions, as well as organize the extra structures (such as microtubules) that will play a role in its division. The cell's actual division occurs during the M (Mitosis) phase, which refers to the equal division of its nucleus, followed in most cases by the C (Cytokinesis) phase, which is the division of its cytoplasm into two new cells.

The cell cycle occurs in most cells that are capable of undergoing division. However, there are some types of cells in which the cell cycle is stopped in one of its phases; such cells do not undergo division. Nerve cells are an example of this type of cell. Some cells do not undergo the C phase of the cell cycle, which results in a cell that has many nuclei within the same cytoplasm. Skeletal muscle tissue in animals and some plant tissues are composed of this type of cell. In addition, the different phases of the cell cycle may vary greatly in the length of time needed to complete each phase. Normally, in most cells, G1 is the longest lasting phase. The cell cycle can be abbreviated as follows: G1 → S → G2 → M → C → G1, and so on.

MITOSIS

All cells arise from other cells by cell division, during which DNA in the nucleus duplicates, or **replicates**, and the cytoplasm divides in two, forming two cells. The process of **mitosis** (a nuclear process) is the orderly series of changes that results in the division of the duplicated sets of chromosomes and the formation of two new nuclei that are identical to each other and to the nucleus of the original parent cell. The division of the cytoplasm, called **cytokinesis**, occurs during and after mitosis, and it results in the formation of two new, identical daughter cells. The effect of this is that all the cells that come from a single cell are genetically identical to it and to each other; they are all *clones*. The process of mitosis ensures that all cells that result will have the same number of chromosomes as the parent cell had, and that is characteristic of the species. For example, in humans, all cells that arise from mitosis will have 46 chromosomes, the characteristic number for humans. Thus, mitosis maintains the correct chromosome number from generation to generation.

Events of Mitosis

Mitosis consists of four distinct stages: prophase, metaphase, anaphase, and telophase. During the period between cell divisions, the chromosome material is dispersed in the nucleus in the form of *chromatin*, long strands of DNA. While still in the S phase (of interphase), before the chromosomes become visible as distinct units, the chromatin replicates. Then, during *prophase*, the nuclear membrane disintegrates and disappears and the chromatin contracts, forming a visible set of double-stranded chromosomes. Each replicated chromosome consists of two identical strands, or *sister chromatids*, joined by a *centromere* (Figure 4-1).

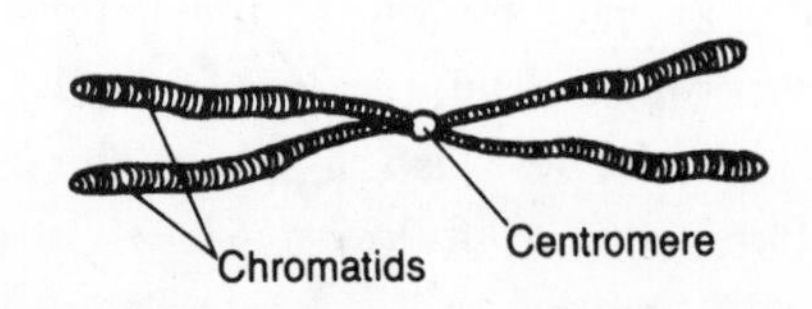

Figure 4-1. A double-stranded chromosome.

Next, during *metaphase*, a network of fibers called the *spindle apparatus* forms. In animal cells, two small organelles called *centrioles* move to the opposite ends, or *poles*, of the cell, where they appear to be involved in the formation of the spindle apparatus. Plant cells generally lack centrioles, but the spindle apparatus forms without them, and the movement of chromosomes is similar to that in animal cells.

During *anaphase*, the double-stranded chromosomes become attached to the spindle apparatus and line up along the cell's center, or equator. The two chromatids of each double-stranded chromosome separate and are pulled apart (due to shortening of the spindle fibers). Finally, during *telophase*, the chromatids move to opposite poles of the cell. A nuclear membrane forms around each of the two sets of single-stranded chromosomes, thus forming two daughter nuclei—identical to each other and to the original nucleus (Figure 4-2).

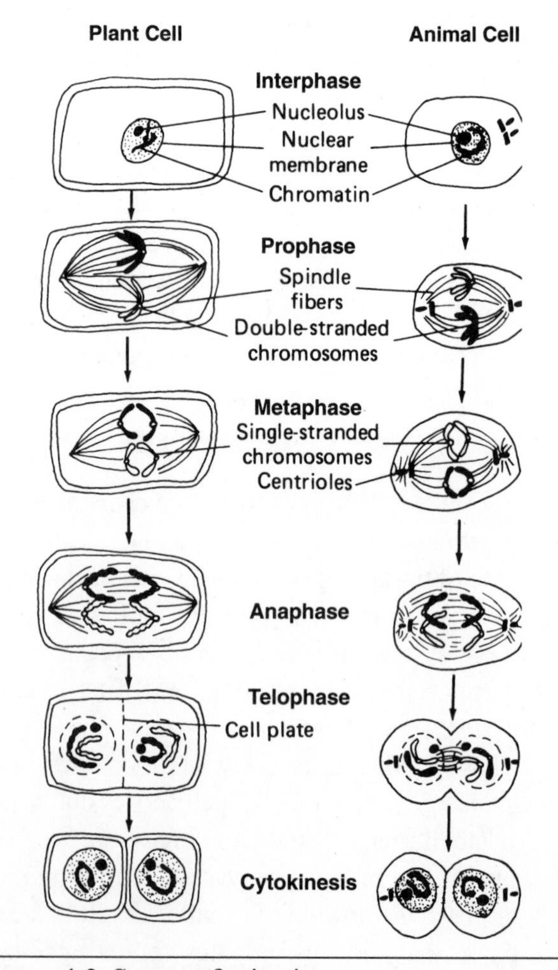

Figure 4-2. Stages of mitosis.

Division of the Cytoplasm: Cytokinesis

In animal cells, the cytoplasm is divided when the cell membrane "pinches in" at the cell's center, separating the two nuclei and dividing the cytoplasm into approximately equal halves.

In plant cells, the cytoplasm is divided when a *cell plate* forms across the center of the cell. The cell plate then forms the new cell walls.

Uncontrolled Cell Division

In multicellular organisms, cells sometimes undergo abnormal and rapid divisions, resulting in growths called *tumors*, which invade surrounding tissues and organs and interfere with their normal activities. Such migrating tumor cells are linked to a group of diseases known collectively as *cancer*.

QUESTIONS

MULTIPLE CHOICE

1. Each of the two daughter cells that results from the normal mitotic division of the original parent cell contains A. the same number of chromosomes but has genes different from those of the parent cell B. the same number of chromosomes and has genes identical to those of the parent cell C. one-half the number of chromosomes but has genes different from those of the parent cell D. one-half the number of chromosomes and has genes identical to those of the parent cell

2. The following list describes some of the events associated with normal cell division.

a. nuclear membrane formation around each set of newly formed chromosomes

b. pinching in of cell membrane to separate daughter nuclei and divide cytoplasm

c. replication of each chromosome to form sets of double-stranded chromosomes

d. movement of single-stranded chromosomes to opposite ends of the spindle fibers

What is the normal sequence in which these events occur?

A. $a \rightarrow b \rightarrow c \rightarrow d$ B. $c \rightarrow b \rightarrow d \rightarrow a$
C. $c \rightarrow d \rightarrow a \rightarrow b$ D. $d \rightarrow c \rightarrow b \rightarrow a$

3. What is the result of normal chromosome replication? A. Lost or worn-out chromosomes are replaced. B. Each daughter cell is provided with twice as many chromosomes as the parent cell. C. The exact number of centrioles is produced for spindle fiber attachment. D. Two identical sets of chromosomes are produced.

4. Normally, a complete set of chromosomes is passed on to each daughter cell as a result of A. reduction division B. mitotic cell division C. meiotic cell division D. nondisjunction

5. In nondividing cells, the chromosome material is in the form of A. chromatids B. centrioles C. spindle fibers D. chromatin

6. Organelles that play a role in mitotic division in animal cells but not in plant cells are A. centrioles B. chromatids C. cell plates D. chromosomes

7. In plant cells, after the cytoplasm divides, a cell plate forms across the center of the cell and forms the new A. cell membranes B. chromosomes C. cell walls D. centrioles

OPEN RESPONSE

8. Colchicine is a drug that prevents chromosomes from separating during cell division. Describe how colchicine might affect daughter cells produced by a cell during mitosis.

9. Red blood cells lose their nuclei when they become fully mature. How does this explain the fact that red blood cells cannot undergo mitosis?

10. Compare the process of mitosis in a plant cell and in an animal cell.

11. Using the Internet, research how a tumor forms. How is mitosis related to the formation of a tumor? How is a benign tumor different from a malignant tumor?

TYPES OF ASEXUAL REPRODUCTION

Asexual reproduction is the production of new organisms without the joining of nuclei from two specialized sex cells. In asexual reproduction, the new organism develops by mitotic cell divisions, and the offspring are genetically identical to the parent. There is no genetic variation among the offspring produced asexually.

Binary Fission

The form of asexual reproduction that occurs most commonly in single-celled organisms, such as the amoeba and paramecium, is *binary fission* (Figure 4-3). In this type of reproduction, the nucleus divides by mitosis, and the cytoplasm divides, forming two daughter cells of equal size. These newly formed cells are smaller than the parent cell, but they contain the same number of chromosomes.

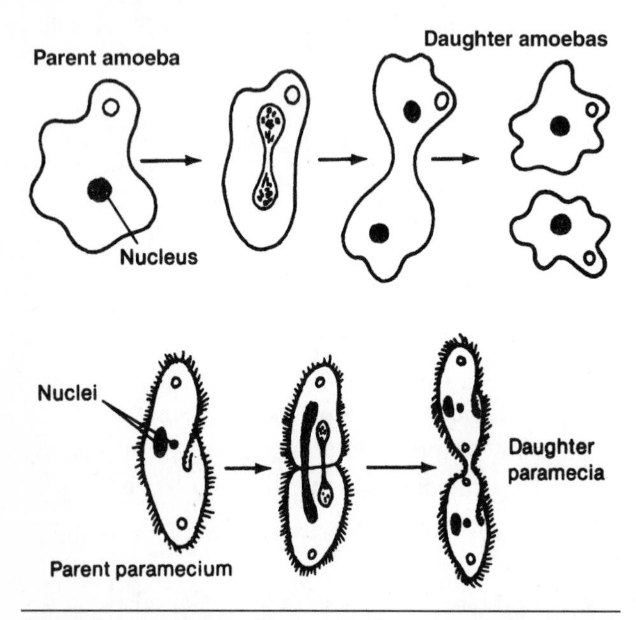

Figure 4-3. Binary fission in the amoeba (top) and the paramecium (bottom).

Budding

Yeasts and some other simple organisms carry on a form of asexual reproduction called *budding*, which is basically similar to binary fission. However, in budding, the division of the cytoplasm is unequal, so that one of the daughter cells is larger than the other. The daughter cells may separate, or they may remain attached, forming a colony (Figure 4-4).

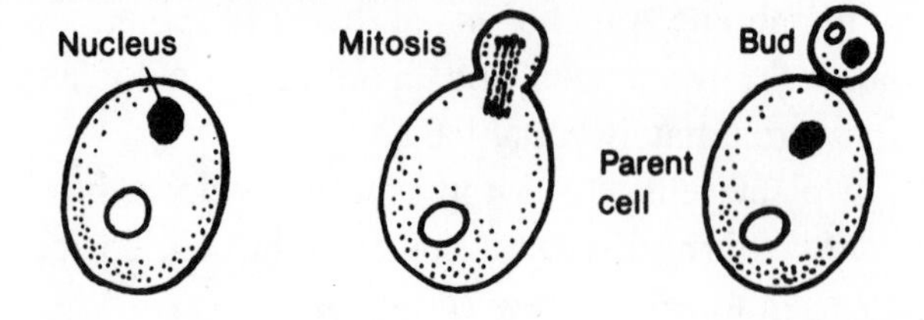

Figure 4-4. Budding in yeast.

In multicellular organisms such as the hydra, budding refers to the production of a multicellular growth, or *bud*, from the body of the parent (Figure 4-5). The bud is produced by mitotic cell division, and it develops into a new organism. The new organism may detach from the parent, or it may remain attached, forming a colony.

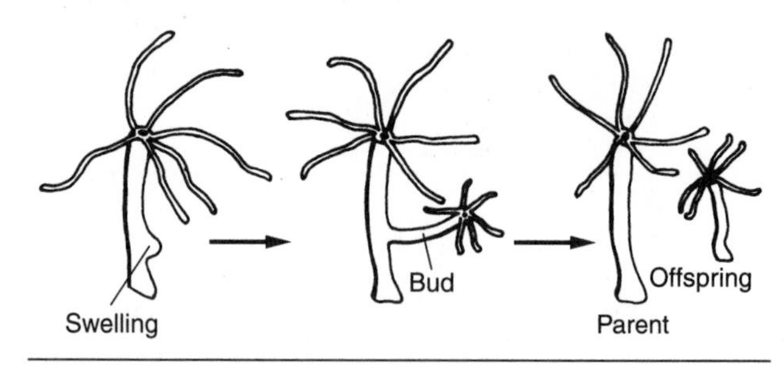

Figure 4-5. Budding in hydra.

Sporulation

In some multicellular organisms, such as bread mold, specialized cells called *spores* are produced in large numbers by mitosis. This process is called *sporulation*. Spores are generally surrounded by a tough coat, which enables them to survive harsh environmental conditions. Each spore may then develop into a new organism when environmental conditions become favorable.

Regeneration

The process of *regeneration* refers to the replacement, or regrowth, of lost or damaged body parts. For example, a lobster may regenerate a lost claw. In some cases, an entire new animal can develop from a part of the parent organism. A new sea star can develop from one arm and part of the central disk of an existing sea star (which then regenerates the missing arm). In this case, regeneration is a type of asexual reproduction.

Invertebrates generally show a greater capacity for regeneration than vertebrates do, probably because they have many more unspecialized cells and parts than vertebrates do.

Vegetative Propagation

In plants, *vegetative propagation* involves various forms of asexual reproduction in which new plants develop from the roots, stems, or leaves of the parent plant. Examples include new plant growth from bulbs, tubers, cuttings, and runners (Figure 4-6).

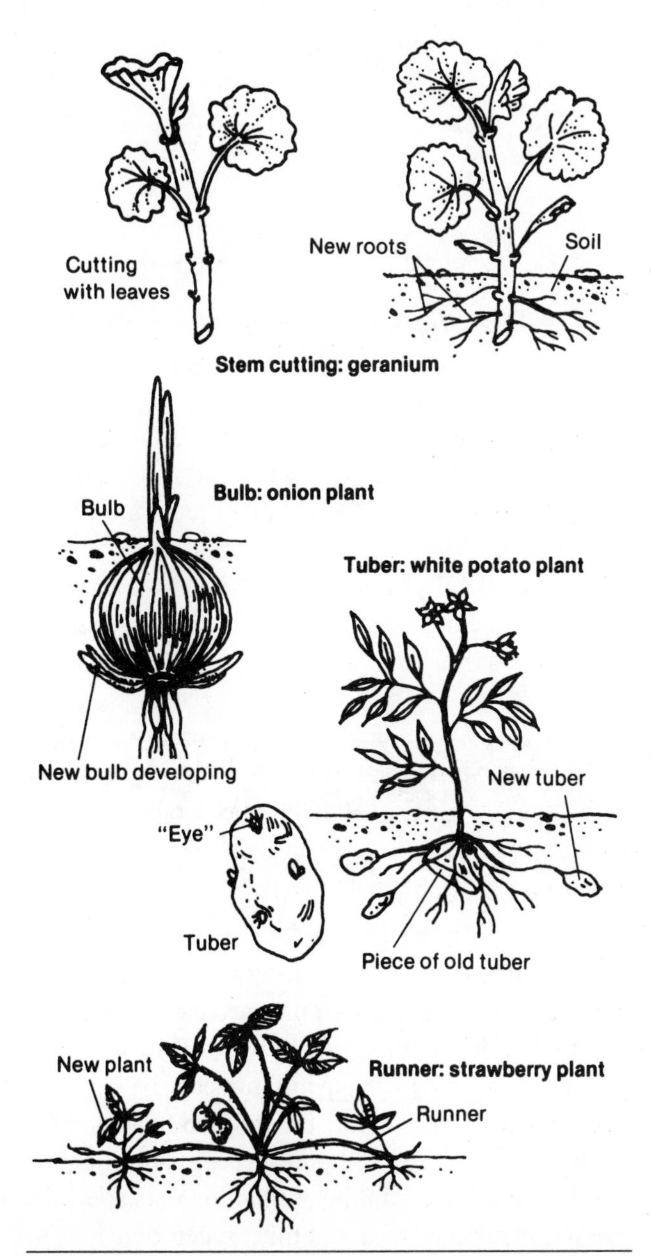

Figure 4-6. Forms of vegetative propagation.

QUESTIONS

MULTIPLE CHOICE

12. Compared to the parent cell, a daughter cell produced as a result of binary fission
A. has one-half as many chromosomes
B. has twice as many chromosomes
C. is the same size, but has fewer chromosomes D. is smaller, but has the same number of chromosomes

13. A form of asexual reproduction that occurs in yeast is A. binary fission B. budding C. vegetative propagation D. spore formation

14. What is a type of asexual reproduction that commonly occurs in many species of unicellular protists? A. external fertilization B. tissue regeneration C. binary fission D. vegetative propagation

15. A type of asexual reproduction in which new plants develop from the roots, stems, or leaves of an existing plant is called A. binary fission B. sporulation C. regeneration D. vegetative propagation

16. A form of asexual reproduction found in bread mold involves the production of large numbers of specialized cells, each surrounded by a tough coat. This process is called A. binary fission B. budding C. sporulation D. regeneration

17. Compared to vertebrates, invertebrate animals exhibit a higher degree of regenerative ability because they A. produce larger numbers of sex cells B. produce larger numbers of spindle fibers C. possess more chromosomes in their nuclei D. possess more undifferentiated cells

18. What specific type of reproduction is shown below in the diagrams of an amoeba?
A. vegetative propagation B. binary fission C. budding D. meiosis

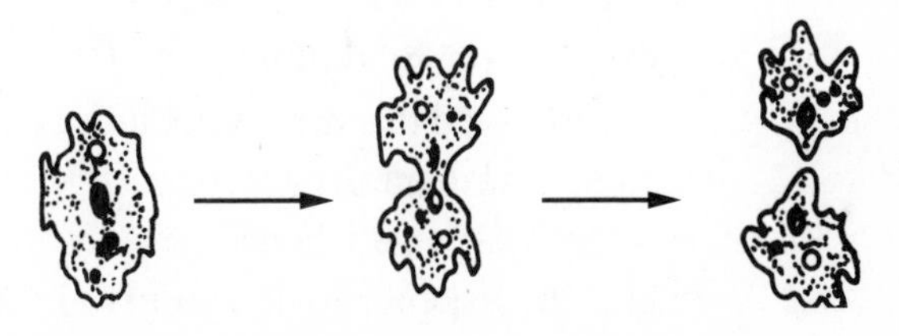

19. The chromosome content of a skin cell that is about to form two new skin cells is represented in the diagram below.

Which diagram best represents the chromosomes that would be found in the two new skin cells produced as a result of this process? A. 1 B. 2 C. 3 D. 4

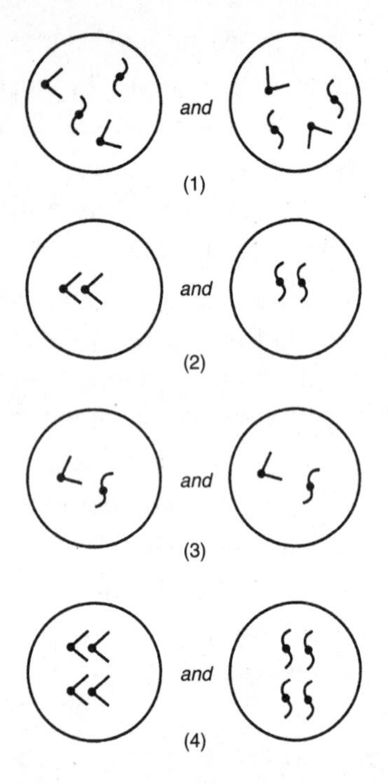

OPEN RESPONSE

20. What role does mitosis play in asexual reproduction?

21. The amoeba is a single-celled organism that reproduces asexually by mitosis. Explain why all the offspring of a single amoeba can be considered clones.

22. In what ways are regeneration and vegetative propagation similar? Why are the offspring identical to the parent in both processes?

23. A scientist noted that a paramecium culture he had in his laboratory reproduced more rapidly than average when kept in a sunny corner of the room. He also observed that other paramecium cultures kept in darker parts of the room reproduced more slowly. Use your knowledge of biology to answer the following:

- What testable question might the scientist ask based on his observations?
- State one possible hypothesis to explain the scientist's observations.
- State a procedure that the scientist could use to test the hypothesis.

24. The diagram below illustrates asexual reproduction in bread mold.

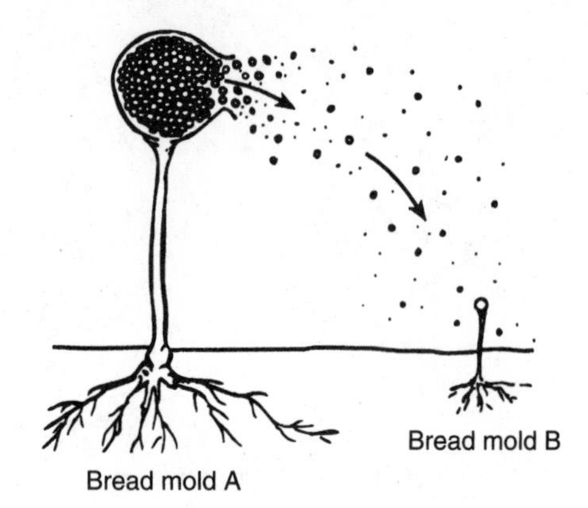

Reproductive structures known as spores were released from bread mold *A*. One of these spores developed into bread mold *B*. State how the genetic information in the nuclei of cells in bread mold *B* compares to the genetic information in the nuclei of cells in bread mold *A*.

MEIOSIS AND SEXUAL REPRODUCTION

In organisms that reproduce sexually, specialized sex cells, or **gametes**, are produced by *meiosis*, a special kind of cell division that reduces the chromosome number in the resulting cells to half of the number that was present in the parent cell. One type of gamete, the *sperm cell*, is produced by the male parent, while the other type of gamete, the *egg cell*, is produced by the female parent. The fusion of the nuclei of the sperm cell and the egg cell is called **fertilization**. The resulting cell, which is called the **zygote**, undergoes repeated mitotic cell divisions to form the *embryo*.

Chromosome Number

All members of a given species have a characteristic number of chromosomes in each of their body cells. This *diploid*, or $2n$, *chromosome number* normally remains constant from generation to generation. For example, all human body cells have 46 chromosomes, fruit flies have 8, and garden peas have 14.

The chromosomes of a body cell are actually in the form of *homologous pairs*. The two chromosomes of each homologous pair are similar in size and shape, and control the same traits. Thus, in human body cells there are 23 pairs of homologous chromosomes (23 from the mother and 23 from the father); in fruit flies there are 4 pairs; and in garden peas there are 7 pairs.

Mature sperm and egg cells contain half the diploid number of chromosomes—they contain one member of each homologous pair. Half the diploid chromosome number is called the *haploid*, or $1n$, *chromosome number*. Mature sex cells (gametes) contain the haploid (also called *monoploid*) number of chromosomes; every other cell in the body contains the diploid number.

In sexually mature individuals, haploid egg cells and sperm cells are formed in the gonads (ovaries and testes) by **meiosis**, the process of *reduction division.*

Meiosis

Meiosis occurs only in maturing sex cells and consists of two consecutive nuclear and cytoplasmic divisions but only one chromosome replication. The first meiotic division produces two cells, each containing the haploid number of double-stranded chromosomes. The second meiotic division results in the formation of four cells, each containing the haploid number of single-stranded chromosomes.

As a result of meiosis, a single primary sex cell with the diploid chromosome number gives rise to four cells, each with the haploid (n) chromosome number. These cells mature into gametes—either sperm cells or egg cells.

Meiosis is a source of genetic variations because it provides new combinations of chromosomes for the resulting gametes. A gamete receives only one member of each pair of homologous chromosomes from the $2n$ primary sex cells. The sorting and distribution of these chromosomes during formation of the gametes is random.

Gametogenesis

The process by which sperm and eggs are produced is called *gametogenesis*. It involves meiotic cell division and cell maturation. Gametogenesis occurs in specialized paired sex organs, or *gonads*. The male gonads are the testes; the female gonads are the ovaries. In most animals, the sexes are separate; that is, each individual has either testes or ovaries. However, some animals, such as the hydra and the earthworm, have both male and female gonads; such animals are called *hermaphrodites*. It is important to note that in most hermaphrodites, self-fertilization is not possible. The animal must mate with another (hermaphroditic) individual. This is yet another source for genetic variation within the species.

Spermatogenesis. The production of sperm is called *spermatogenesis* (Figure 4-7). The process begins with meiosis in primary sperm cells, which are diploid. As a result of meiosis, each primary sperm cell develops into four haploid cells of equal size. As they mature, these cells lose most of their cytoplasm and develop a long, whiplike flagellum that is used for locomotion.

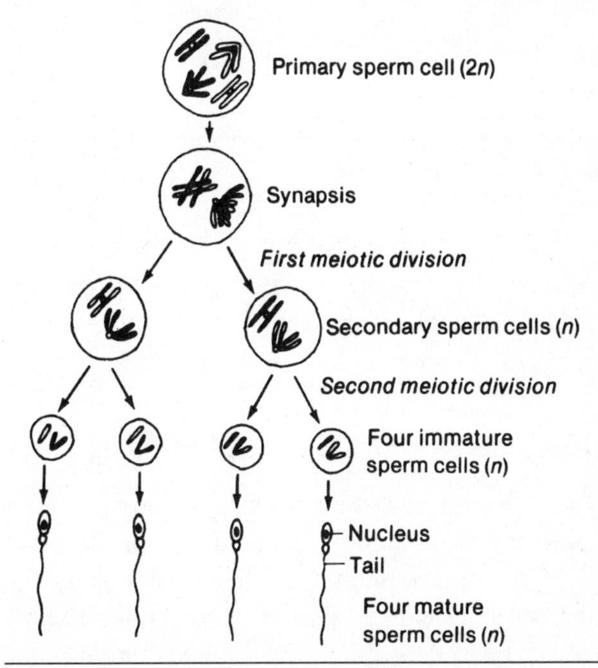

Figure 4-7. Spermatogenesis: the production of mature sperm cells.

Oogenesis. Egg cells are produced by *oogenesis* (Figure 4-8). In oogenesis, a primary egg cell undergoes meiosis. The chromosomal changes are the same as those that occur in spermatogenesis (from $2n$ to n). However, in oogenesis, division of the cytoplasm is unequal. The first meiotic division produces one large cell and one small one called a *polar body*. The larger cell then undergoes the second meiotic division, forming an egg cell and another polar body. The first polar body may also undergo a second meiotic division, forming two polar bodies. Oogenesis results in the production of one large, haploid egg cell and three small polar bodies.

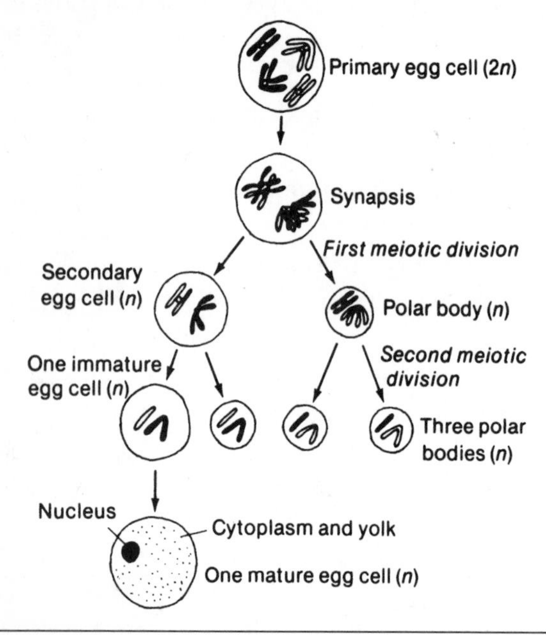

Figure 4-8. Oogenesis: the production of one mature egg cell.

The polar bodies disintegrate. The advantage of the unequal cytoplasmic division is that the egg cell is provided with a large supply of stored nutrients in the form of yolk.

Comparison of Mitosis and Meiosis

Recall that mitosis maintains the chromosome number from generation to generation. The daughter cells produced by mitotic cell division have the same number and kinds of chromosomes as the original parent cell. A cell with the $2n$ (diploid) chromosome number produces daughter cells with the $2n$ (diploid) chromosome number. Mitosis produces the body cells for growth and repair of tissues. It is also associated with asexual reproduction.

In contrast, as a result of meiotic cell division, the daughter cells have one-half the number of chromosomes of the original cell (see Table 4-1). A cell with the $2n$ (diploid) chromosome number produces daughter cells with the n (haploid) chromosome number. Meiosis occurs only in the gonads during the production of gametes. *Note:* The $2n$ chromosome number is restored during fertilization.

***Table 4-1.* A Comparison of Mitosis and Meiosis**

Mitosis	*Meiosis*
Double-stranded chromosomes line up in middle of cell in single file.	Double-stranded chromosomes line up in middle of cell in double file.
Results in diploid ($2n$) number of chromosomes in daughter cells.	Results in haploid (n) number of chromosomes in daughter cells.
Occurs in most cells of the body.	Occurs only in maturing sex cells.
Results in very few genetic variations because chromosomes remain the same.	Results in many genetic variations because of random sorting and crossing-over of chromosomes.

QUESTIONS

MULTIPLE CHOICE

25. Haploid gametes are produced in animals as a result of A. meiosis B. mitosis C. fertilization D. fission

26. In human males, the maximum number of functional sperm cells that is normally produced from each primary sex cell is A. one B. two C. three D. four

27. Sexually reproducing species show greater variation than asexually reproducing species due to A. lower rates of mutation B. higher rates of reproduction C. environmental changes D. sorting of chromosomes during gametogenesis

28. In animals, polar bodies are formed as a result of A. meiotic cell division in females

B. meiotic cell division in males C. mitotic cell division in females D. mitotic cell division in males

29. During the normal meiotic division of a diploid cell, the change in chromosome number that occurs is represented as A. $4n \rightarrow n$ B. $2n \rightarrow 4n$ C. $2n \rightarrow 1n$ D. $1n \rightarrow \frac{1}{2}n$

30. In a species of corn, the diploid number of chromosomes is 20. What would be the number of chromosomes found in each of the normal egg cells produced by this species? A. 5 B. 10 C. 20 D. 40

31. A human zygote is normally produced from two gametes that are identical in A. size B. method of locomotion C. all of their traits D. chromosome number

32. Organisms that contain both functional male and female gonads are known as A. hybrids B. hermaphrodites C. protists D. parasites

33. In sexually reproducing species, the number of chromosomes in each body cell remains the same from one generation to the next as a direct result of A. meiosis and fertilization B. mitosis and mutation C. differentiation and aging D. homeostasis and dynamic equilibrium

34. The diagrams below represent the sequence of events in a cell undergoing normal meiotic cell division.

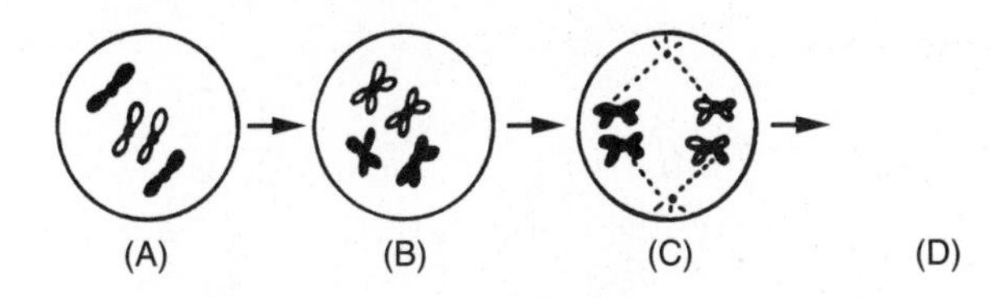

Which diagram most likely represents stage *D* of this sequence? A. diagram 1 B. diagram 2 C. diagram 3 D. diagram 4

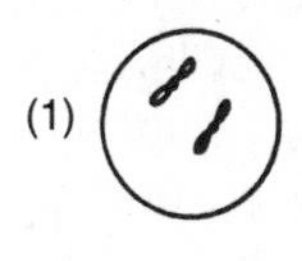

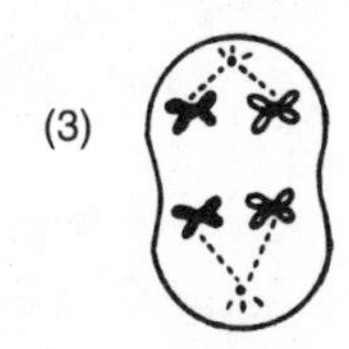

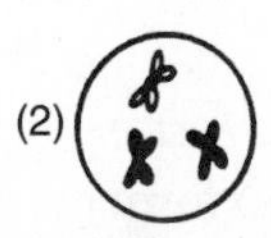

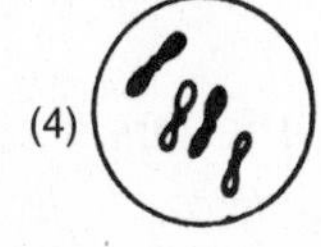

OPEN RESPONSE

Base your answers to questions 35 through 38 on your knowledge of biology and on the diagram below, which represents a diploid cell about to undergo meiosis. The shapes inside the cell represent homologous chromosomes.

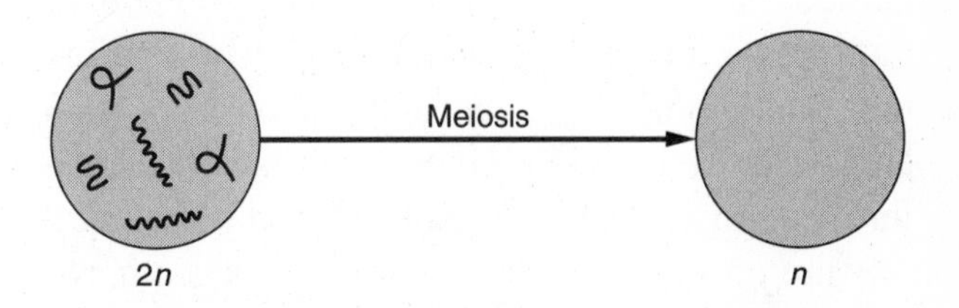

35. Copy the diagram into your notebook. In the empty circle, draw one of the resulting daughter cells produced by the diploid cell at the end of meiosis.

36. What is the diploid chromosome number of the original cell?

37. What will be the chromosome number of a daughter cell produced by this cell?

38. Name an organ in which this cell might be found. Explain your answer.

39. Briefly compare the processes of mitosis and meiosis. What is the function of each process?

40. Compare the processes of sperm and egg production in terms of the following:
- where each process occurs;
- the relative numbers of gametes produced by each process.

41. Discuss why, in sexual reproduction, it is necessary for the gametes to have the haploid number of chromosomes rather than the diploid number. How does the process of meiosis ensure that the gametes will be haploid?

42. Explain how the daughter cells produced during meiosis may be genetically different from one another even though they result from the same original diploid cell. Why is this variation important? Why are cells produced by mitosis *not* genetically different from one another?

FERTILIZATION AND DEVELOPMENT

Fertilization is the union of a haploid (n) sperm nucleus with a haploid (n) egg nucleus to form a diploid ($2n$) cell, the *zygote*, which is the first cell of a new organism. It is important to understand that normal gametes can contain only the haploid number of chromosomes (one member of each homologous pair). When gametes unite in fertilization, the homologous chromosomes from each gamete are brought together in the zygote. Thus, fertilization restores the diploid species number of chromosomes. Since chromosomes from two *different* individuals are joining together, the offspring will be genetically different from either parent, although it may share several common characteristics with both parents. Meiosis and fertilization both help to bring about genetic variation among offspring.

External Fertilization

The union of a sperm and an egg outside the body of a female animal is called *external fertilization.* External fertilization generally occurs in a watery environment and is characteristic of reproduction in frogs, most fish, and many other aquatic vertebrates.

In external fertilization, large numbers of eggs and sperm are released into the water at the same time to increase the chances that fertilization will take place and to help ensure that at least some of the fertilized eggs will develop, avoid being eaten, and survive to adulthood.

Internal Fertilization

The union of a sperm and an egg inside the moist reproductive tract of the female is called **internal fertilization**. Reproduction in most terrestrial, or land-dwelling, vertebrates, including birds and mammals, is characterized by internal fertilization.

In internal fertilization, relatively few eggs are produced at one time, since the chances that fertilization will occur are much greater with internal fertilization than with external fertilization.

Stages of Development

The early stages of embryonic development are similar in all animals. The process known as **development** begins when the zygote undergoes a rapid series of mitotic cell divisions called *cleavage.*

Cleavage. During cleavage, there is no increase in the size of the embryo—just an increase in the number of cells it contains (see Figure 4-9). Cell growth and specialization begin after cleavage.

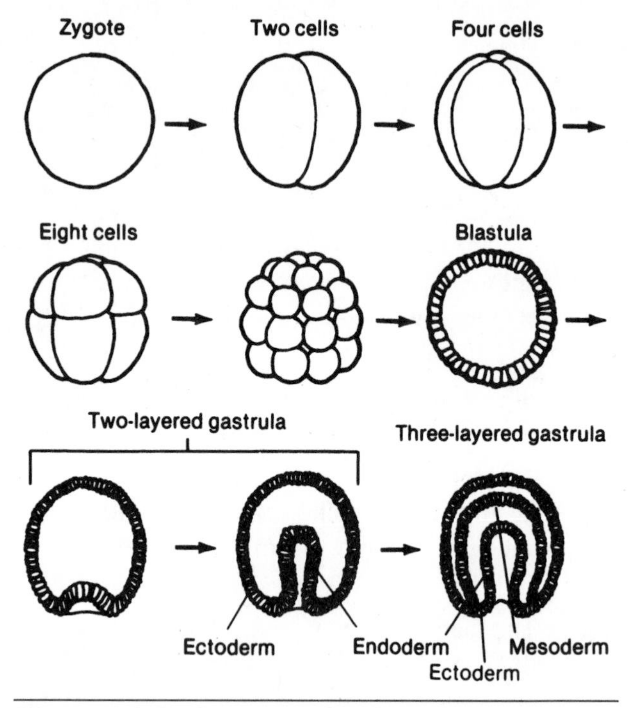

Figure 4-9. Early stages of embryonic development.

Blastula Formation. The mitotic divisions of cleavage result in the formation of the *blastula*, a hollow ball made up of a single layer of cells.

Gastrulation. As mitotic divisions continue, one side of the blastula pushes inward, or indents, in a process called *gastrulation.* The resulting embryonic stage, called a *gastrula*, consists of an inner layer, or *endoderm*, and an outer layer, or *ectoderm.* A third layer, called the *mesoderm*, forms between the endoderm and ectoderm. The endoderm, mesoderm, and ectoderm are called the *germ layers.*

Differentiation and Growth. The germ layers undergo changes, or **differentiation**, to form the various tissues, organs, and organ systems of the developing animal (see Table 4-2). During differentiation, some portions of the DNA within individual cells are active, while other portions are inactive. There are numerous methods by which sections of DNA are "switched" on or off. As a result, each cell develops into a specific type of cell, based upon which portion of its DNA is active and coding within its nucleus. For example, a stomach cell differentiates in response to signals from the active DNA in that cell, which deals with the functions of a stomach cell. Similarly, within a liver cell, different parts of the DNA—those that code for the functions of a liver cell—are active. Although each and every body cell in an animal contains the entire set of DNA for that animal, only certain parts of the DNA are active in any particular cell, depending on the type of cell it is.

Embryonic development involves growth as well as differentiation. *Growth* includes both an increase in the size of the embryonic cells and an increase in the number of cells.

***Table 4-2.* Tissues and Organs Formed from the Embryonic Germ Layers**

Embryonic Layer	*Organs and Organ Systems*
Ectoderm	Nervous system; skin
Mesoderm	Muscles; circulatory, skeletal, excretory, and reproductive systems
Endoderm	Lining of digestive and respiratory tracts; liver; pancreas

External Development

Embryonic development may occur outside or inside the body of the female. Growth of the embryo outside the female's body is called *external development.*

The eggs of many fish and amphibians are fertilized externally and develop externally in an aquatic environment. The eggs of birds and many reptiles (and even a few mammals) are fertilized internally but develop externally, encased in tough, protective shells to prevent their drying out.

Internal Development

Growth of the embryo inside the female's body is called **internal development**. In most mammals, both fertilization and development are internal. The eggs of mammals have little yolk and are very small compared with the eggs of reptiles and birds. In all mammals, the young are nourished after birth by milk from the mother's mammary glands.

Placental Mammals. Most mammals are placental mammals in which the embryo develops in the **uterus**, or womb, of the female and receives food and oxygen and gets rid of wastes through the placenta. The **placenta** is a temporary organ that forms within the uterus from embryonic and maternal tissues; it is rich in both embryonic and maternal blood vessels. Dissolved materials pass between the mother and the embryo through the blood vessels in the placenta—food and oxygen pass from the mother to the embryo, while wastes pass from the embryo to the mother. The blood of the mother and the embryo never mix. Mammals that are *marsupials*, such as kangaroos, do not have a placenta. Instead, they give birth to incompletely developed embryos, which continue to develop and receive nutrition externally from mammary glands within the mother's pouch.

QUESTIONS

MULTIPLE CHOICE

43. In the early development of a zygote, the number of cells increases, without leading to an increase in size, in the process of A. ovulation B. cleavage C. germination D. metamorphosis

44. An embryo's three germ layers are formed during A. gastrulation B. fertilization C. blastula formation D. growth

45. In most species of fish, the female produces large numbers of eggs during the reproductive cycle. This would indicate that reproduction in fish is most probably characterized by A. internal fertilization and internal development B. internal fertilization and external development C. external fertilization and internal development D. external fertilization and external development.

46. Which type of fertilization and development do birds and most reptiles have? A. internal fertilization and internal development B. internal fertilization and external development C. external fertilization and internal development D. external fertilization and external development

47. The embryos of some mammals, such as the kangaroo and the opossum, complete their development externally. What is the source of nutrition for their last stage of development? A. milk from maternal mammary glands B. diffusion of nutrients through the uterine wall C. food stored in the egg yolk D. solid foods gathered and fed to them by the mother

48. In mammals, the placenta is essential to the embryo for A. nutrition, reproduction, growth B. nutrition, respiration, excretion C. locomotion, respiration, excretion D. nutrition, excretion, reproduction

49. Which characteristic of sexual reproduction specifically favors the survival of terrestrial animals? A. fertilization within the body of the female B. male gametes that may be carried by the wind C. fusion of gametes in the outside environment D. fertilization of eggs in the water

Base your answers to questions 50 through 53 on your knowledge of biology and on the diagram below, which represents the early stages of embryonic development.

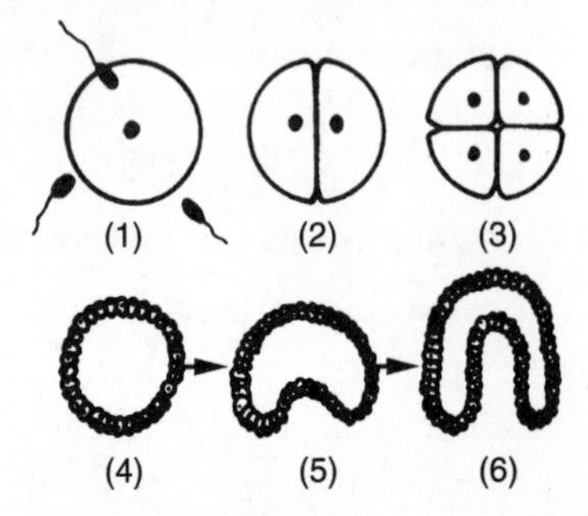

50. The structures labeled *2* and *3* are formed as a direct result of A. meiosis B. gastrulation C. cleavage D. differentiation

51. The structure in stage *4* represents a A. zygote B. blastula C. gastrula D. follicle

52. The cells of the outer embyonic layer (ectoderm) give rise to the A. digestive system and liver B. excretory system and muscles C. circulatory system and gonads D. nervous system and skin

53. Which cells are *not* represented in any of the diagrams? A. endoderm B. mesoderm C. ectoderm D. gastrula

54. The arrows in the diagram below illustrate processes in the life of a species that reproduces sexually. Which processes result directly in the formation of cells with half the amount of genetic material that is characteristic of the species? A. 1 and 2 B. 2 and 3 C. 3 and 4 D. 4 and 5

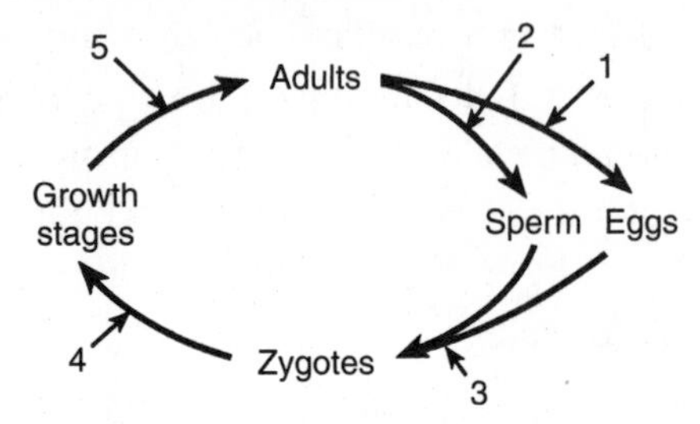

55. The development of specialized tissues and organs in a multicellular organism's embryo occurs as a result of A. cloning B. differentiation C. meiosis D. cleavage

OPEN RESPONSE

Refer to the following four terms, which describe the early stages of embryonic development, to answer questions 56 through 58: gastrula, cleavage, zygote, blastula.

56. List these terms in the correct order, from earliest to latest stage of embryonic development.

57. For each term listed, draw and label a simple sketch to illustrate that stage of embryonic development.

58. Briefly describe what occurs during each of these embryonic stages.

59. Animals that are characterized by external fertilization produce many times more gametes (sperm and eggs) than do animals that have internal fertilization. Give two reasons for this observation.

60. Stem cells are cells in which the *entire* DNA is active. Explain how a stem cell is able to differentiate into any type of cell needed by the body.

HUMAN REPRODUCTION AND DEVELOPMENT

Male Reproductive System

The male reproductive system functions in the production of sperm cells, male sex hormones, and in the placement of sperm into the female reproductive system.

Sperm Production. The sperm-producing organs, the *testes*, are located in an outpocketing of the body wall called the *scrotum* (Figure 4-10). The temperature in the scrotum, which is 1°C to 2°C cooler than normal body temperature, is best suited for the production and storage of sperm.

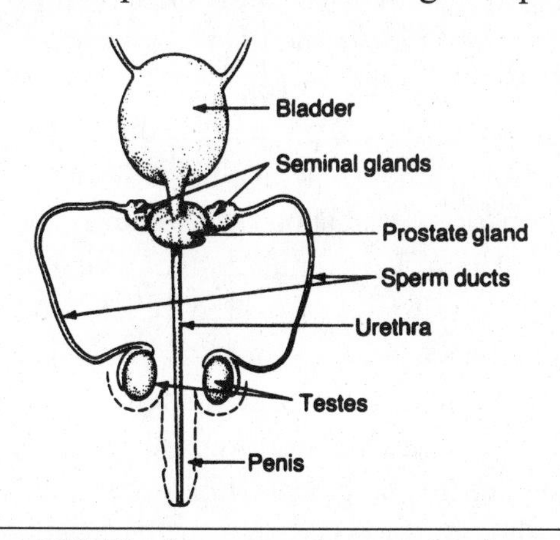

Figure 4-10. The human male reproductive system.

From the testes, the sperm pass through a series of ducts into which liquid is secreted by various glands. The liquid serves as a transport medium for the sperm cells and is an adaptation for life on land. The liquid and sperm together are called *semen*. Semen passes to the outside of the body through the urethra, a tube through the penis. The *penis* is used to deposit the semen in the female reproductive tract.

Hormone Production. The testes produce the male sex hormone testosterone, which regulates the maturation of sperm cells. Testosterone also regulates the development of male secondary sex characteristics, including body form, beard development, and deepening of the voice.

Female Reproductive System

The female reproductive system functions in the production of egg cells and female sex hormones.

Egg Production. The female reproductive organs, the *ovaries*, are located within the lower portion of the body cavity (Figure 4-11). In the ovaries, each egg cell is present in a tiny sac called a *follicle*. About once a month, a follicle matures and bursts, and the egg within it is released from the surface of the ovary, a process called *ovulation*. The egg cell then passes into the oviduct, or *fallopian tube*, which leads to the *uterus*. If sperm are present, fertilization may occur. If the egg is fertilized, it passes into the uterus, where embryonic development may occur. If the egg is not fertilized, it degenerates.

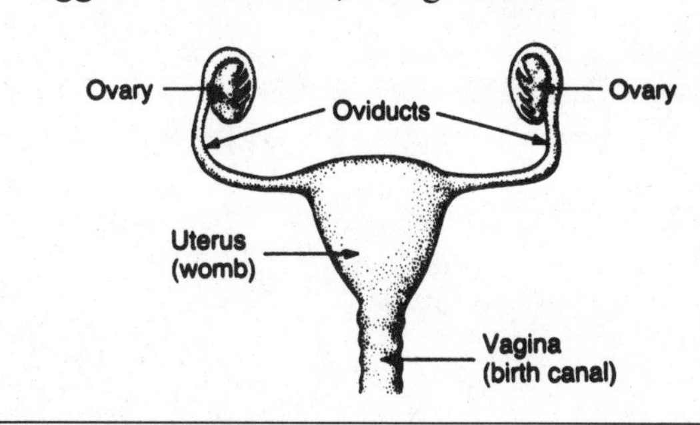

Figure 4-11. The human female reproductive system.

The lower end of the uterus, the *cervix*, opens to a muscular tube called the *vagina*, or *birth canal*. When embryonic development is complete, the baby leaves the body of the mother through the vagina.

Hormone Production. The ovaries produce the female sex hormones estrogen and progesterone. These hormones regulate the maturation of egg cells, as well as the development of secondary sex characteristics, including the development of the mammary glands and the broadening of the pelvis. Estrogen and progesterone are also involved in the menstrual cycle and **pregnancy**.

The Menstrual Cycle

The series of events that prepares the uterus for pregnancy is called the *menstrual cycle*. The cycle begins with the thickening of the lining of the uterine wall. The lining also becomes vascularized (filled with blood vessels). If fertilization does not occur, the thickened uterine lining breaks down and the material is expelled from the body during menstruation. The cycle then begins again.

The menstrual cycle begins at *puberty*, the stage at which the individual becomes capable of reproducing. It is temporarily interrupted by pregnancy and sometimes by illness, and ceases permanently (after approximately 40 years) at *menopause*. The cycle is regulated by the interaction of hormones, and lasts approximately 28 days.

The menstrual cycle consists of four stages (Figure 4-12):

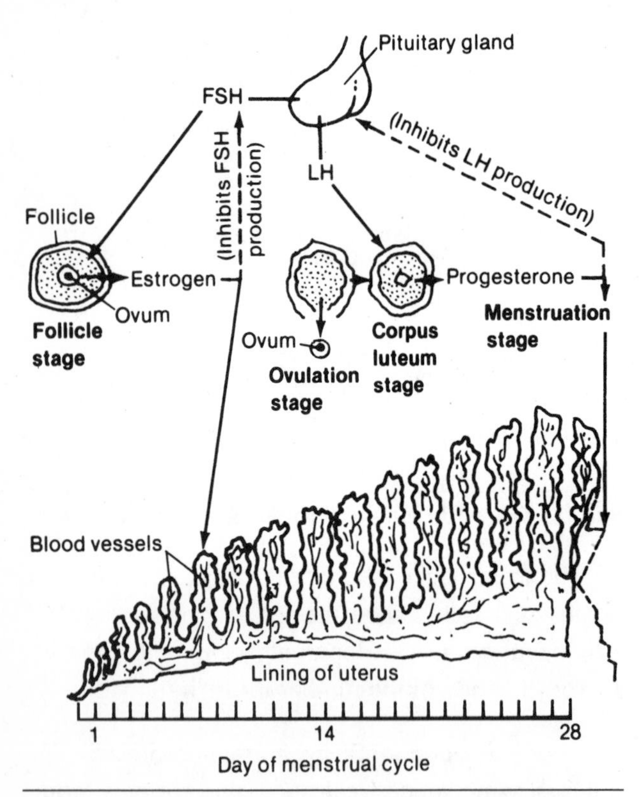

Figure 4-12. Stages of the human menstrual cycle.

(a) During the follicle stage, an egg matures and the follicle secretes estrogen, which stimulates the thickening of the uterine lining. This stage lasts about 14 days.

(b) About midway in the cycle, ovulation occurs. The egg is released from the ovary and enters the oviduct.

(c) Following ovulation, the *corpus luteum* forms from the ruptured follicle. The corpus luteum secretes progesterone, which continues the vascularization of the uterine lining started by estrogen. This stage lasts about 12 days.

(d) If fertilization does not occur, the egg cell and the thickened uterine lining break down, and the extra tissue, together with some blood and mucus, pass out of the body through the vagina. The shedding of the uterine lining is called menstruation. This stage lasts about four to five days.

Hormones of the Menstrual Cycle

The menstrual cycle is controlled by hormones that are released by the hypothalamus, pituitary gland, and ovaries.

During the follicle stage, the pituitary gland, under the influence of hormones from the hypothalamus, secretes FSH (follicle-stimulating hormone), which in turn stimulates the follicle to secrete estrogen. Estrogen stimulates ovulation and initiates vascularization of the uterine lining.

Increased blood estrogen levels inhibit the production of FSH by the pituitary, and the secretion of LH (luteinizing hormone) by the pituitary increases. Ovulation occurs at about this time in the cycle. After ovulation, LH stimulates the formation of the corpus luteum from the ruptured follicle. The corpus luteum secretes progesterone, which enhances the vascularization of the uterine lining.

If fertilization does not occur, the high levels of progesterone in the blood inhibit the production of LH by the pituitary. The drop in LH level causes a drop in the progesterone level. The lining of the uterus thins out, and at about the twenty-eighth day of the cycle, the shedding of the uterine lining, or menstruation, begins. The blood flow of menstruation is caused by the breakage of many small blood vessels.

The relationship between the ovarian hormones estrogen and progesterone and the pituitary hormones FSH and LH is an example of a *negative feedback mechanism*.

Fertilization and Development

If fertilization does occur in the oviduct, the zygote undergoes cleavage to form a blastula. Six to ten days later, the blastula becomes implanted in the uterine lining. Gastrulation usually occurs after implantation. The germ layers of the gastrula begin to differentiate and grow, resulting in the formation of specialized tissues and organs. The placenta and umbilical cord form, enabling the embryo to obtain nutrients and oxygen and to dispose of metabolic wastes. An amnion (membrane-enclosed sac) filled with fluid provides a watery environment that cushions the embryo, which helps to protect it from injury.

In Vitro Fertilization

Fertilization that occurs outside the body of the female (that is, by means of laboratory techniques) is known as *in vitro* (meaning "in glass") fertilization. After fertilization, the early embryo is implanted into the uterus, where development is completed.

Multiple Births

Sometimes two or more embryos develop in the uterus simultaneously. Fraternal twins develop when two eggs are released from the ovary at the same time and both are fertilized. The two eggs are fertilized by two different sperm cells. Fraternal twins may be of the same sex or opposite sexes. Identical twins develop when a zygote separates into two equal halves early in cleavage. Each half develops into an offspring. Since identical twins develop from the same zygote, they have identical genetic make-ups and are always of the same sex.

Birth and Development

The time between fertilization and birth is referred to as the *gestation period.* In humans, the gestation period is about nine months. After the first three months of gestation, the embryo is referred to as a **fetus**. At the end of the gestation period, the secretion of progesterone decreases and another hormone from the pituitary causes strong muscular contractions of the uterus. The amnion bursts, and the baby is pushed out of the mother's body through the vagina.

During *postnatal development* (development after birth), humans pass through different stages, including infancy, childhood, puberty, adulthood, and old age. Puberty begins at early adolescence. In males, puberty usually occurs between the ages of 12 and 18. In females, it usually occurs between the ages of 10 and 14.

Aging is a series of complex structural and functional changes in the body that occur naturally with the passage of time. The causes of aging are not fully understood. However, it now appears that aging may result from an interaction of both genetic and environmental factors. The aging process ends in death, which may be described as an irreversible cessation of brain function.

QUESTIONS

MULTIPLE CHOICE

61. Which of the following structures *least* affects the human female menstrual cycle?
A. pituitary B. ovary C. pancreas
D. corpus luteum

62. A woman gave birth to twins, one girl and one boy. The number of egg cells involved was A. 1 B. 2 C. 3 D. 4

63. A diagram of the human female reproductive structures is shown below.

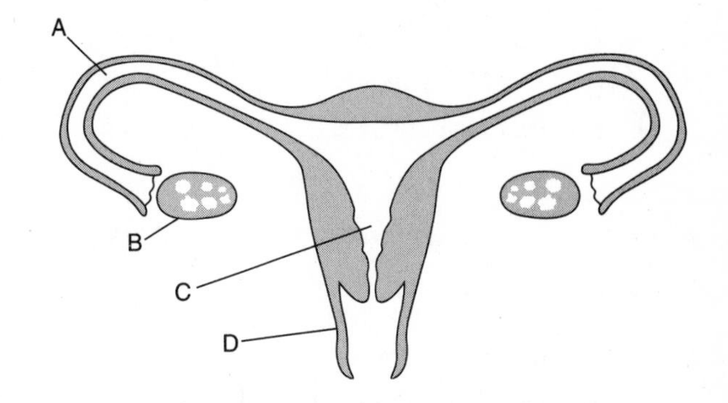

Which structure is correctly paired with its function?
A. *A*—releases estrogen and progesterone
B. *B*—produces and releases the egg
C. *C*—provides the usual site for fertilization
D. *D*—nourishes a developing embryo

64. Which structure is the membrane that serves as the protective, fluid-filled sac in which an embryo is suspended? A. pituitary B. placenta C. corpus luteum D. amnion

65. The technique of uniting a sperm cell and an egg cell outside the female's body is called A. *in vitro* fertilization B. internal fertilization C. gametogenesis D. artificial ovulation

66. Which of the following hormones is *not* involved in the regulation of the human menstrual cycle? A. progesterone B. estrogen C. FSH D. testosterone

67. Identical twins develop from A. one egg and two sperm B. two eggs and one sperm C. two eggs and two sperm D. one egg and one sperm

68. Some body structures of a human male are represented in the diagram below. An obstruction in the structures labeled *X* would directly interfere with the A. transfer of sperm to a female B. production of sperm C. production of urine D. transfer of urine to the external environment

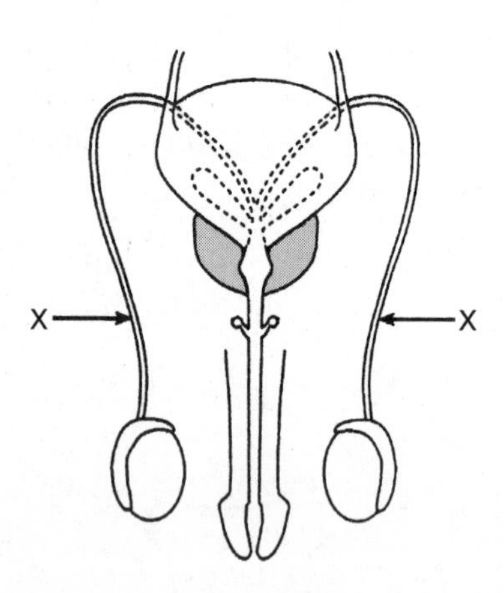

69. Fraternal twins develop from A. one egg and two sperm B. two eggs and one sperm C. two eggs and two sperm D. one egg and one sperm

70. One function of the placenta in a human is to A. surround the embryo and protect it from shock B. allow for mixing of maternal blood with fetal blood C. act as the heart of the fetus, pumping blood until the fetus is born D. permit passage of nutrients and oxygen from the mother to the fetus

Base your answers to questions 71 through 73 on the diagram below, which represents a cross section of a part of the human female reproductive system, and on your knowledge of biology.

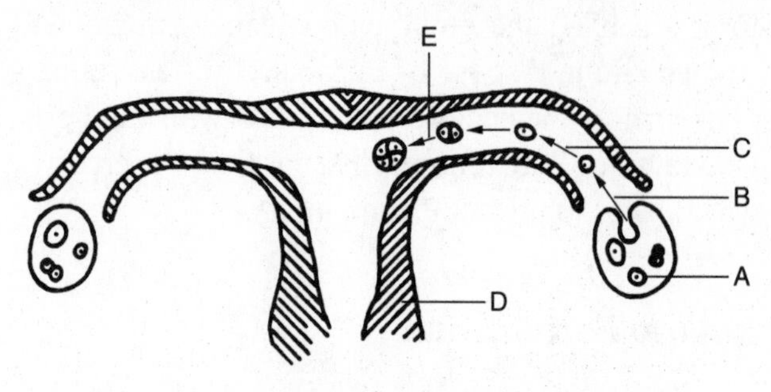

71. Which structure is prepared for the implantation of a fertilized egg as a result of the action of reproductive hormones? A. *E* B. *B* C. *C* D. *D*

72. Within which structure does fertilization normally occur? A. *A* B. *E* C. *C* D. *D*

73. Which step represents the process of ovulation? A. *A* B. *B* C. *C* D. *E*

Base your answers to questions 74 through 76 on the diagram below, which represents a stage in human embryonic development.

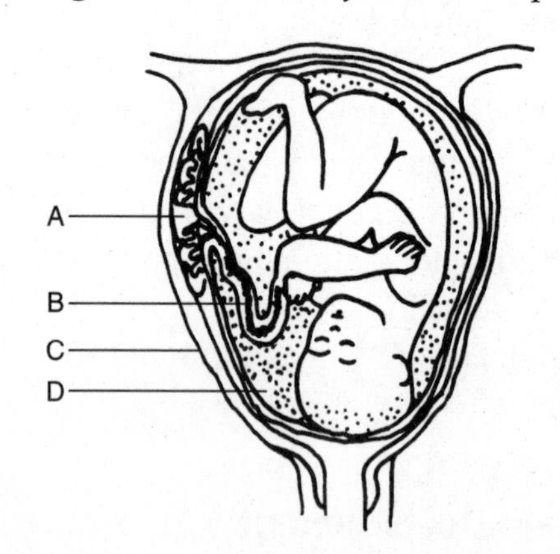

74. The exchange of oxygen, food, and wastes between the mother and the fetus occurs at structure A. *A* B. *B* C. *C* D. *D*

75. What is the function of the fluid labeled *D*? A. nourishment B. protection C. excretion D. respiration

76. The structure labeled *C*, within which embryonic development occurs, is known as the A. oviduct B. birth canal C. uterus D. placenta

For each of the processes described in questions 77 through 79, choose from the list below the correct stage of the human menstrual cycle during which that process occurs.

Human Menstrual Cycle Stage

a. Ovulation b. Follicle stage

c. Menstruation d. Corpus luteum stage

77. The lining of the uterus is shed: A. *a* B. *b* C. *c* D. *d*

78. An egg is released from an ovary: A. *a* B. *b* C. *c* D. *d*

79. An egg matures within an ovary: A. *a* B. *b* C. *c* D. *d*

OPEN RESPONSE

Base your answers to questions 80 through 83 on the following graph, which shows a woman's changing hormone levels for FSH and estrogen over a period of 21 days, and on your knowledge of human reproductive biology.

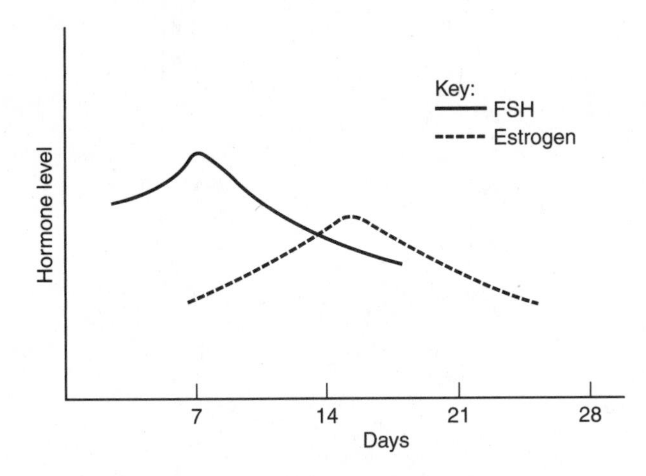

80. Describe the relationship depicted in the graph between FSH and estrogen.

81. Why does the level of estrogen begin rising *after* the FSH level rises?

82. Why does the FSH level begin to fall after the seventh day?

83. On approximately what day are the woman's estrogen levels highest?

84. List the four major hormones that play a role in the menstrual cycle and discuss how they interact during the cycle.

85. Mr. and Mrs. W have been trying to conceive their first child for over one year, with no success. They decide to visit their doctors for medical tests. Blood tests on Mrs. W reveal that her FSH levels are abnormally low. Discuss how this finding might explain the couple's inability to conceive. What medical treatment might help Mrs. W to become pregnant?

86. Briefly discuss the function of the following structures in the development of the human embryo: *placenta, umbilical cord, amnion.*

87. A human is a complex organism that develops from a zygote. Briefly explain some of the steps in this developmental process. In your answer be sure to:

- explain how a zygote is formed;
- compare the chromosomes in a zygote with those of a parent's body cell;
- identify one developmental process involved in the change from a zygote to an embryo;
- identify the body structure in which fetal development usually occurs.

SEXUAL REPRODUCTION IN FLOWERING PLANTS

Flowers are the reproductive organs of the *angiosperms*, or flowering plants.

Structure of Flowers

Flowers may contain the following structures: sepals, petals, stamens, and pistils (Figure 4-13).

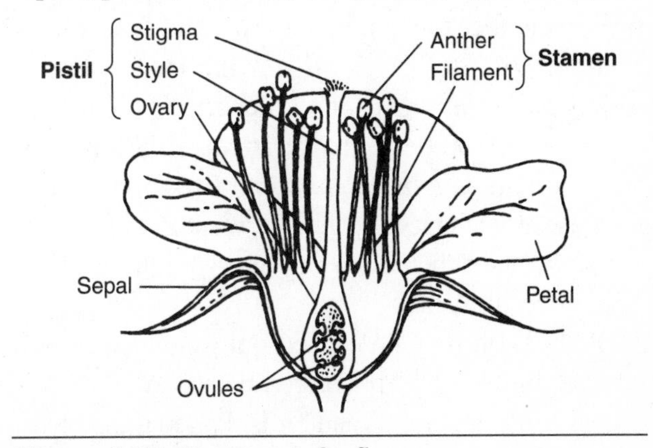

Figure 4-13. Structure of a flower.

Sepals are leaflike structures at the base of a flower that enclose and protect the flower bud. In some species the sepals are green, while in others the sepals are white or brightly colored.

Petals are leaflike structures inside the sepals that surround the reproductive organs of the flower. Petals may be brightly colored or white and often have a sweet fragrance.

Stamens are the male reproductive organs of a flower. Each stamen consists of an oval-shaped *anther* supported by a stalk, or *filament*. *Pollen grains*, which contain several haploid sperm nuclei, are produced by meiosis of the diploid cells in the anther. The thick wall that encloses the pollen grain prevents the contents from drying out. This is an adaptation for life on land.

Pistils are the female reproductive organs of a flower. A pistil consists of a stigma, style, and ovary. The *stigma*, which is a knoblike, sticky structure, is adapted for receiving pollen grains. The stigma is supported by the *style*, a slender stalk that connects the stigma to the *ovary*, which is at the base of the pistil. In the ovary, haploid egg cells are produced by meiosis in structures called *ovules*.

The flowers of some species contain both stamens and pistils. In other species, some flowers contain only stamens, while others contain only pistils. The flowers of some species have both sepals and petals, while the flowers of other species lack one or the other.

Pollination and Fertilization

The transfer of pollen grains from an anther to a stigma is called **pollination**. The transfer of pollen from an anther to a stigma of the same flower or to a stigma of another flower on the same plant is called *self-pollination*. The transfer of pollen from an anther of one flower to the stigma of a flower on another plant is *cross-pollination*. Cross-pollination increases the chances of genetic variation in the offspring because the pollen and the egg cells are from two different plants.

Pollination may be carried out by wind, insects, bats, or birds. Brightly colored petals and the scent of nectar attract insects and birds. Pollen grains from a flower adhere to their bodies and are carried to another flower, where they rub off on the sticky surface of a stigma.

When a pollen grain reaches a stigma, it *germinates*, or sprouts (Figure 4-14). A pollen tube grows from the pollen grain down through the stigma and style to an ovule within the ovary. The growth of the pollen tube is controlled by the tube nucleus. Two sperm nuclei and the tube nucleus pass down through the pollen tube. The sperm nuclei enter an ovule, where one sperm nucleus fertilizes the egg nucleus to form a diploid ($2n$) zygote. The other sperm nucleus fuses with two *polar nuclei* in the ovule to form a triploid ($3n$) *endosperm nucleus*, which divides to form a food storage tissue, called *endosperm*. The zygote undergoes repeated mitotic divisions to form a multicellular plant embryo. After fertilization, the ovule ripens to form a *seed*, while the ovary develops into a *fruit*. The seeds of flowering plants are found inside the fruits.

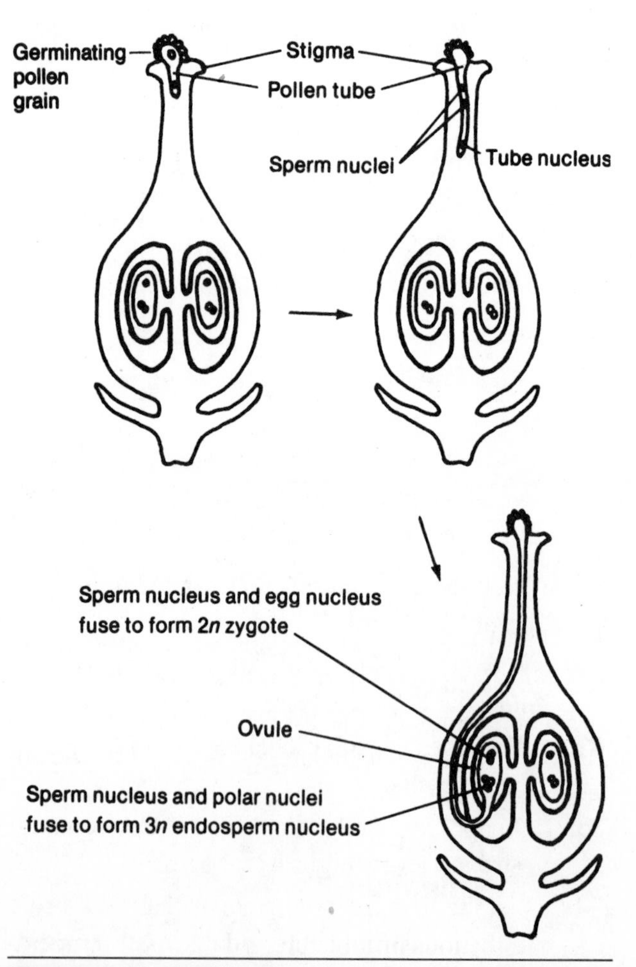

Figure 4-14. Fertilization in flowering plants.

Structure of a Seed

A seed consists of a seed coat and a plant embryo with one or two cotyledons (Figure 4-15). The *seed coat*, which develops from the outer coverings of the ovule, surrounds and protects the embryo. The plant embryo consists of the epicotyl, hypocotyl, and cotyledon. The *epicotyl* is the upper portion of the embryo; it develops into the leaves and upper portion of the stem. The *hypocotyl* is the lower portion of the embryo; it develops into the roots and, in some species, the lower portion of the stem. The *cotyledons* contain endosperm, the stored food that provides nutrients for the developing plant.

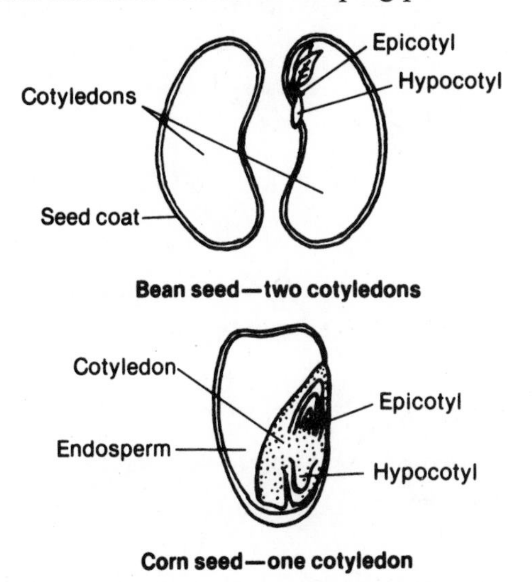

Figure 4-15. Structure of a seed.

Fruits

The fruits of flowering plants are structures that are specialized for seed dispersal. Fruits carry the seeds away from the parent plant, which helps to prevent overcrowding. The fruits of dandelions and maples, for example, are dispersed by wind; coconuts are dispersed by water; and cockleburs are fruits that become attached to the fur of animals and are carried away as they move. Fleshy fruits are eaten by animals, and their seeds are later deposited with the animal's wastes.

Seed Germination

When conditions of moisture, oxygen, and temperature are favorable, seeds *germinate*. The embryo plant develops leaves and roots, and begins to produce its own food by photosynthesis. The development of a mature plant from an embryo involves cell division, cell differentiation, and growth.

Plant Growth

In flowering plants, only certain regions, called *meristems*, are able to undergo cell division. There are two types of meristems: cell division in the tips of roots and stems (in apical meristems) results in an increase in length; cell division between the xylem and phloem (in lateral meristems, or *cambiums*) results in an increase in the diameter of roots and stems. The cells of the meristem regions divide and then undergo elongation and differentiation, forming the different kinds of plant tissues.

QUESTIONS

MULTIPLE CHOICE

88. Which reproductive structures are produced within the ovaries of plants? A. pollen grains B. sperm nuclei C. egg nuclei D. pollen tubes

89. In a flowering plant, the ovule develops within a part of the A. style B. anther C. pistil D. stigma

90. Which embryonic structure supplies nutrients to a germinating bean plant? A. pollen tube B. hypocotyl C. epicotyl D. cotyledon

91. Heavy use of insecticides in springtime may lead to a decrease in apple production in the fall, which is most probably due to interference with the process of A. pollination B. cleavage C. absorption D. transpiration

92. In a bean seed, the part of the embryo that develops into the leaves and upper portion of the stem is known as the A. seed coat B. epicotyl C. hypocotyl D. cotyledon

93. A condition necessary for the germination of most seeds is favorable A. light B. chlorophyll concentration C. temperature D. nitrate concentration

94. In flowering plants, the entire female reproductive organ is called the A. filament B. anther C. style D. pistil

95. In flowering plants, pollen grains are formed in the A. style B. anther C. sepal D. stigma

96. The seeds of a flowering plant develop from the mature A. fruits B. cotyledons C. ovules D. endosperm

97. The endosperm of a bean seed is contained within its A. cotyledons B. ovules C. stamen D. petals

98. The fruits of a flowering plant develop from the ripened A. seeds B. ovules C. ovaries D. pollen tubes

99. Which of the following is *not* part of a plant embryo? A. epicotyl B. seed coat C. hypocotyl D. cotyledon

100. Which portion of a bean seed would contain the greatest percentage of starch? A. seed coat B. epicotyl C. cotyledon D. hypocotyl

Base your answers to questions 101 through 103 on the diagram below, which shows a cross section of a flower, and on your knowledge of biology.

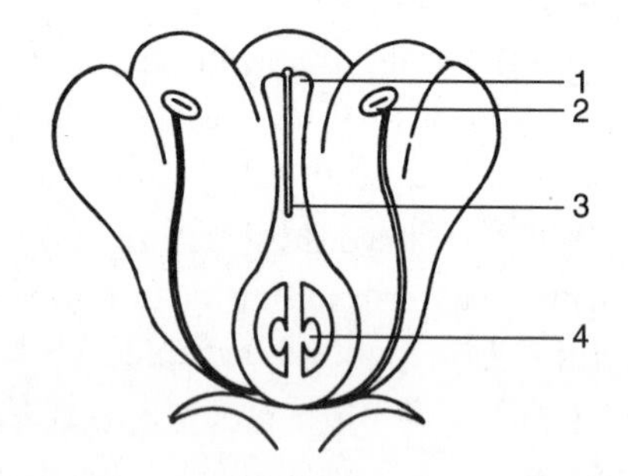

101. In this diagram, the stigma and the anther are A. 1 and 2 B. 1 and 4 C. 2 and 4 D. 2 and 3

102. Which process has occurred in this flower? A. pollen germination B. seed formation C. zygote formation D. fruit production

103. In which part would fertilization occur? A. 1 B. 2 C. 3 D. 4

Base your answers to questions 104 and 105 on the diagram below, which shows the internal structure of half of a bean seed, and on your knowledge of biology.

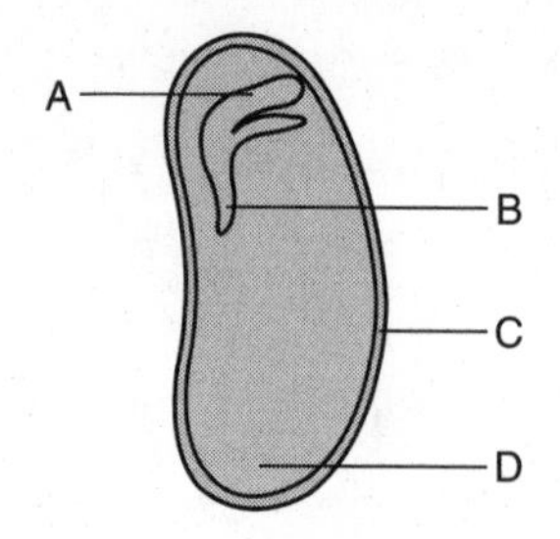

104. In which structure would most of the stored food for the embryo be found? A. *A* B. *B* C. *C* D. *D*

105. The epicotyl and the hypocotyl are represented by A. *A* and *C* B. *B* and *D* C. *A* and *B* D. *C* and *D*

OPEN RESPONSE

106. Explain why cross-pollination increases the chances of genetic variation in the offspring of flowering plants.

107. Compare sexual reproduction in mammals and flowering plants. In what way is the process similar in these two types of organisms?

READING COMPREHENSION

Base your answers to questions 108 through 112 on the information below and on your knowledge of biology. Source: *Science News* (June 4, 2005): Vol. 167, No. 23, p. 366.

Menstrual Cycle Changes the Brain

Hormone fluctuations over the course of a woman's menstrual cycle change the abundance of a particular type of receptor on the surface of nerve cells, say scientists.

The finding may explain why some women with neurological disorders experience flare-ups of their conditions at the same time most months.

Previous research showed that nearly 80 percent of epileptic women have more seizures than usual during the phase of the menstrual cycle when their blood concentration of progesterone declines and that of estrogen increases. Other studies showed that women with a condition called premenstrual dysphoric disorder experience severe anxiety and depression during the same phase.

Istvan Mody and his colleagues at the University of California, Los Angeles examined brain slices from female mice. In search of changes in cells during the animals' 6-day menstrual cycles, the researchers looked at nerve-cell receptors for gamma aminobutyric acid, a chemical that inhibits neurons from firing.

The scientists found that the prevalence of a receptor subtype called delta was high when progesterone concentrations were up and estrogen concentrations were down. The same subtype was less prominent during the rest of the cycle, when the relative hormone concentrations were reversed. Nerve cells with more delta receptors were less likely to fire when stimulated with electricity than were cells with fewer delta receptors.

In a separate experiment, live mice in the high-progesterone phase of their cycle were less likely to have a seizure when given convulsion-inducing drugs than were mice in their high-estrogen phase. The high-progesterone mice also displayed less anxious behaviors than high-estrogen mice did.

Mody and his colleagues say that these results, published in the June *Nature Neuroscience*, suggest that a shortage of delta receptors may increase nerve cell activity during the low-progesterone phase of the menstrual cycle, in turn increasing anxiety and seizure susceptibility.

108. At what point in their menstrual cycle do most epileptic women suffer from more seizures?

109. What function does the chemical gamma aminobutyric acid (GABA) have in the body?

110. Using the information in this article, explain the relationship between the following factors: high progesterone levels; low estrogen levels; number of delta receptors on nerve cells; and likelihood of nerve cells to fire.

111. What conclusion did the researchers draw about the menstrual cycle and seizures, based on their studies of female mice?

112. If delta is the cell membrane receptor for GABA, what must be true about the three-dimensional shapes of delta and GABA?

CHAPTER 5 Genetics and Heredity

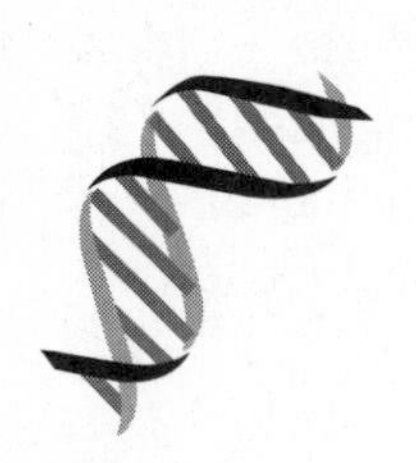

Standards 3.4, 3.5, 3.6 (Genetics)

BROAD CONCEPT: Genes allow for the storage and transmission of genetic information. They are a set of instructions encoded in the nucleotide sequence of each organism. Genes code for the specific sequences of amino acids that comprise the proteins that are characteristic of that organism.

FOUNDATIONS OF GENETICS

Genetics is the branch of biology that deals with patterns of **inheritance**, or heredity. *Heredity* is the biological process by which parents pass on genetic information to their offspring through their gametes. The science of genetics originated with the work of an Austrian monk, *Gregor Mendel*, who performed a series of experiments on pea plants between 1856 and 1868.

Principles of Mendelian Genetics

In his breeding experiments, Mendel (who, like everyone else at that time, had no knowledge of genes or chromosomes) made careful observations of the inheritance patterns of specific contrasting traits found in pea plants. Through a mathematical analysis of the traits found in the large numbers of offspring from his experimental crosses, Mendel developed his principles of *dominance*, *segregation*, and *independent assortment*. Mendel also concluded that the traits he observed were controlled by pairs of inherited "factors," with one member of each pair coming from each parent organism. Thus, in organisms that reproduce sexually, half of the offspring's genetic material is contributed by the female parent and half by the male parent. As a result, the offspring has traits from both parents, and is never identical to either one of them.

Gene–Chromosome Theory

The importance of Mendel's work was not recognized until the early 1900s, when the development of better microscopes enabled biologists to observe chromosome behavior during meiotic cell division. Biologists then linked the separation of homologous chromosome pairs during meiosis and their **recombination** at fertilization with the inheritance of Mendel's factors. Breeding experiments carried out by T. H. Morgan with the fruit fly, *Drosophila*, provided supporting evidence for Mendel's principles of inheritance.

Mendel's inherited, or **hereditary**, factors—now known as **genes**—are arranged in a linear fashion on the chromosomes. Each gene has a position, or *locus* (plural, *loci*), on the chromosome. The two alternate genes that control each trait are called **alleles**, and they are located in the same position on homologous chromosomes. This *gene–chromosome theory* explains the hereditary patterns observed by Mendel. Research done by geneticist Barbara McClintock in the 1940s indicated that some genes are capable of changing their position on the chromosome and are expressed in multiple ways depending on their location. We will discuss this in more detail in the next chapter.

Gene Expression

Every organism has at least two alleles that govern every *trait*, such as eye color or height. As mentioned, these alleles are passed on—one from the mother and one from the father—to the offspring. The genes encode information that is expressed as the traits of the organism, a phenomenon called **gene expression**. A single gene (that is, one set of alleles) may control one or several traits. Alternatively, some traits are determined by more than one gene (that is, by more than one set of alleles). This second type of gene expression is referred to as *polygenic expression*.

Although all the body cells in an organism contain the same genetic instructions, the cells may differ considerably from one another in structure and function. The reason is that, in any given cell, only some of the genes are expressed, while all other genes are inactivated. For example, in liver cells, it is mainly the genes that pertain to liver functions that are active, while the other genes are inactive. The same is true of all other cells in a body. You can think of the genes on a cell's chromosomes as recipes in a cookbook: the book may contain hundreds of recipes, but if you are making a chocolate cake, you will read only the instructions for making that item. Likewise, the cell reads only the instructions for making its specific products.

Genes that are "on" (called *exons*) are expressed, while those that are "off" (called *introns*) are not expressed. There are many mechanisms that can switch genes on and off, including intracellular chemicals, enzymes, regulatory proteins, and the cell's environment. In addition, a particular gene may alternately be expressed or inactivated, depending on the cell's needs at the time.

SOME MAJOR CONCEPTS IN GENETICS

Dominance

In his experiments, Mendel crossed plants that were *pure* for contrasting traits. For example, he crossed pure tall plants with pure short plants. All the offspring of such crosses showed only one of the two contrasting traits. In the cross of tall plants and short plants, all the offspring were tall. In this type of inheritance, the allele that is expressed in the offspring is said to be *dominant*; the allele that is present but not expressed is said to be *recessive*. This pattern illustrates Mendel's Principle of Dominance.

By convention, the dominant allele is represented by a capital letter, while the recessive allele is represented by the lowercase form of the same letter. For example, the allele for tallness, which is dominant, is shown as *T*, while the allele for shortness, which is recessive, is shown as *t*.

If, in an organism, the two genes of a pair of alleles are the same, for example, *TT* or *tt*, the organism is said to be *homozygous*, or pure, for that trait. The genetic makeup of the organism, which is its **genotype**, is either homozygous dominant (*TT*) or homozygous recessive (*tt*). If the two genes of a pair of alleles are different, for example, *Tt*, the organism is said to be *heterozygous*, or *hybrid*, for that trait.

The physical appearance of an organism that results from its genetic makeup is called its **phenotype**. For example, a pea plant that is heterozygous for height has the genotype *Tt* and the phenotype of being tall. When an organism that is homozygous for the dominant trait is crossed with an organism that is homozygous for the recessive trait ($TT \times tt$), the phenotype of the offspring is like that of the dominant parent. Thus, the heterozygous offspring (*Tt*) is tall. Phenotype also refers to traits that result from gene expression that are not necessarily observable from the outside. For instance, if an organism is able to synthesize a particular protein because it has the gene for that protein, this is also known as a phenotype.

In studies involving genetic crosses, the organisms that are used to begin the studies are called the *parent generation*. The offspring produced by crossing members of the parent generation are called the *first filial*, or F_1, *generation*. The offspring of a cross between members of the F_1 generation make up the *second filial*, or F_2, *generation*.

QUESTIONS

MULTIPLE CHOICE

1. When a strain of fruit flies homozygous for light body color is crossed with a strain of fruit flies homozygous for dark body color, all the offspring have light body color.

This illustrates Mendel's principle of A. segregation B. dominance C. incomplete dominance D. independent assortment

2. Two genes located in corresponding positions on a pair of homologous chromosomes and associated with the same characteristic are known as A. gametes B. zygotes C. chromatids D. alleles

3. For a given trait, the two genes of an allelic pair are not alike. An individual possessing this gene combination is said to be A. homozygous for that trait B. heterozygous for that trait C. recessive for that trait D. pure for that trait

4. In pea plants, flowers located along the stem (*axial*) are dominant to flowers located at the end of the stem (*terminal*). Let *A* represent the allele for axial flowers and *a* represent the allele for terminal flowers. When plants with axial flowers are crossed with plants having terminal flowers, all of the offspring have axial flowers. The genotypes of the parent plants are most likely A. $aa \times aa$ B. $Aa \times Aa$ C. $aa \times Aa$ D. $AA \times aa$

5. Curly hair in humans, white fur in guinea pigs, and needlelike spines in cacti all partly describe each organism's A. alleles B. autosomes C. chromosomes D. phenotype

6. The appearance of a recessive trait in offspring of animals most probably indicates that A. both parents carried at least one recessive gene for that trait B. one parent was homozygous dominant and the other parent was homozygous recessive for that trait C. neither parent carried a recessive gene for that trait D. one parent was homozygous dominant and the other parent was hybrid for that trait

7. Which statement describes how two organisms may show the same trait yet have different genotypes for that phenotype? A. One is homozygous dominant and the other is heterozygous. B. Both are heterozygous for the dominant trait. C. One is homozygous dominant and the other is homozygous recessive. D. Both are homozygous for the dominant trait.

8. In cabbage butterflies, white color (*W*) is dominant and yellow color (*w*) is recessive. If a pure white cabbage butterfly mates with a yellow cabbage butterfly, all the resulting (F_1) butterflies are heterozygous white. Which cross represents the genotypes of the parent generation? A. $Ww \times ww$ B. $WW \times Ww$ C. $WW \times ww$ D. $Ww \times Ww$

9. Most of the hereditary information that determines the traits of an organism is located in A. only those cells of an individual produced by meiosis B. the nuclei of body cells of an individual C. certain genes in the vacuoles of body cells D. the numerous ribosomes in certain cells

10. The characteristics of a developing fetus are most influenced by A. gene combinations and their expression in the embryo B. hormone production by the father C. circulating levels of white blood cells in the placenta D. milk production in the mother

OPEN RESPONSE

11. Explain how two organisms can have the same phenotype but different genotypes.

12. To illustrate your answer to question 11, pick a trait and use a letter to represent it. Write the genotypes of the parents and F_1 generations for each organism.

13. Why do the offspring of sexually reproducing organisms resemble both parents? Why are they not identical to either one of the parents?

14. Explain why the body cells of an organism can differ in structure and function, even though they all contain the same genetic information.

PATTERNS OF INHERITANCE

Segregation and Recombination

When gametes are formed during meiosis, the two chromosomes of each homologous pair separate, or *segregate*, randomly. Since only one member of each homologous pair is distributed to a gamete, each gamete contains only one allele for each trait. After the gametes fuse during fertilization, the resulting (zygote) cell contains pairs of homologous chromosomes, but new combinations of alleles may be present. This process is described by Mendel's Principle of Segregation and is a major source of variation among sexually reproducing organisms

Figure 5-1 illustrates segregation and recombination in a cross between two individuals that are heterozygous for tallness. In a large number of such crosses, with a large number of offspring, two types of numerical ratios can be observed. In terms of genotype, the ratio is 1 homozygous dominant (*TT*): 2 heterozygous (*Tt*) : 1 homozygous recessive (*tt*). In terms of phenotype, the ratio is 3 tall : 1 short. These genotype and phenotype ratios are typical for all crosses between organisms that are hybrid for one trait.

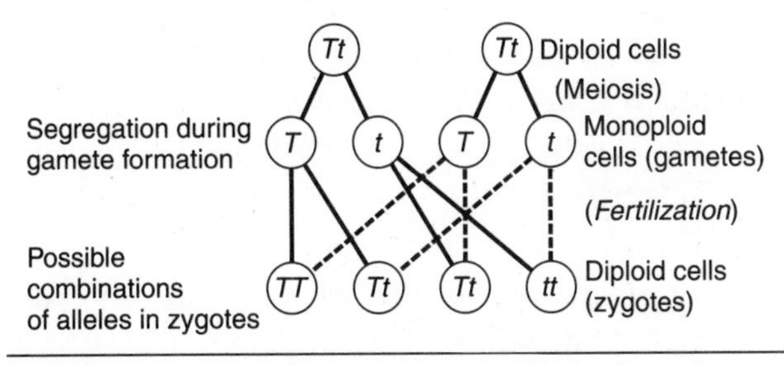

Figure 5-1. Segregation and recombination of alleles.

The Testcross

One way to determine the genotype of an organism that shows the dominant phenotype, is to perform a *testcross*. Recall that an organism with a dominant phenotype, such as tall in pea plants, can be either homozygous (*TT*) or heterozygous (*Tt*) for that trait. You cannot determine the genotype just by observing the organism's phenotype. In a *testcross*, the organism in question is crossed with a homozygous recessive organism (Figure 5-2). If the test organism is homozygous dominant, all the offspring will be heterozygous and show the dominant phenotype. However, if any offspring show the recessive phenotype, the test organism would have to be heterozygous (*Tt*).

For example, let us assume that the unknown genotype is *TT*. When this *TT* (homozygous tall) plant is crossed with a pure recessive *tt* (short) one, all the offspring will have the genotype *Tt* and will appear tall. On the other hand, if the unknown genotype is *Tt* (heterozygous tall) and is crossed with a pure recessive short *tt*, fifty percent of the offspring will be tall (*Tt*) and fifty percent will be short (*tt*). The appearance of just one offspring with the recessive phenotype will indicate that the unknown genotype is heterozygous, because the offspring must have inherited one recessive allele from *each* parent. (Now that techniques for DNA analysis can determine genotypes more quickly, the testcross is not used as often as it was before.)

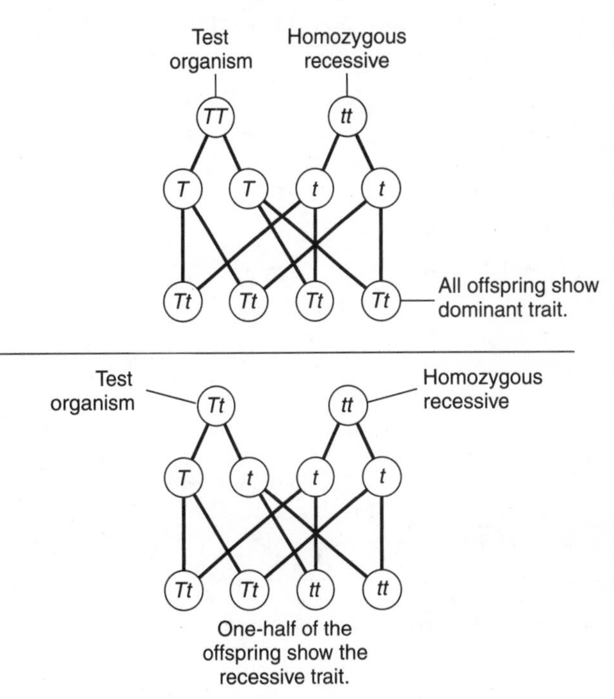

Figure 5-2. Use of a testcross to determine an organism's genotype.

The Punnett Square

The possible offspring of a genetic cross are often shown with a diagram called a *Punnett square*. We can use a Punnett square to show the possible offspring of a cross between a heterozygous tall pea plant (*Tt*) and a homozygous short pea plant (*tt*).

The first step in using a Punnett square is to determine the possible genotypes of the gametes of each parent. In this example, the heterozygous tall plant (*Tt*) produces two types of gametes: half will contain the dominant gene for height, *T*, and half will contain the recessive gene, *t*. The gametes of the homozygous short plant (*tt*) will each contain the recessive gene for height, *t*.

As shown in Figure 5-3, the letters that represent the trait carried by the gametes of one parent are written next to the boxes on the left side of the square; the letters for the gametes of the other parent are written above the boxes on top of the square. The letters are combined to show offspring genotypes as follows: letters on top of the square are written in the boxes below them, and letters on the side are written in the boxes to the right of them. The dominant gene, when present, is written first. The pairs of letters in the four boxes represent the possible combinations of genes in the offspring of the cross. Of the possible offspring of this cross, half would be heterozygous tall (*Tt*) and half would be homozygous (recessive) short (*tt*).

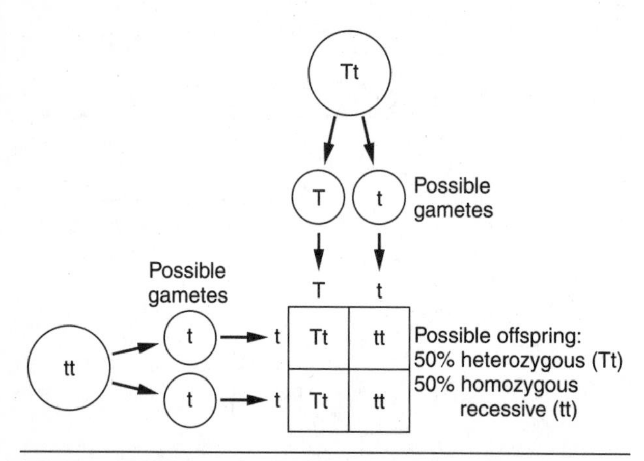

Figure 5-3. Use of a Punnett square to determine possible genotypes of offspring.

Independent Assortment

When Mendel began his experiments on pea plants, he observed only the inheritance of one trait at a time. Soon, he began to look at the inheritance of multiple traits at the same time. For example, he observed the inheritance of plant height and seed (pea) color together. If a pea plant that is homozygous for tall and for yellow seed color (yellow is dominant to green in peas) is crossed with a plant that is homozygous for short and for green seed color, all the offspring will be tall and have yellow seeds, since both these traits are dominant. However, the offspring (F_1) will all be hybrid (heterozygous) for both traits. As such, they are known as *dihybrids* (meaning hybrid for two traits).

The cross can be understood as follows: tall, yellow (*TTYY*) is crossed with short, green (*ttyy*). The gametes for the first parent will have one allele for tall and one for yellow (*TY*). The gametes for the second parent will have one allele for short and one for green (*ty*). When the nuclei of these gametes fuse during fertilization, the zygote will have the following genotype: *TtYy*. The phenotype will be tall and yellow. Note that the offspring is hybrid, or heterozygous, for *both* traits. What would happen if two of these dihybrids were crossed with each other? Remember that during gamete formation (meiosis) the homozygous alleles for each trait become separated from one another (segregation). Thus, the pair of alleles controlling height are separated as are the pair of alleles controlling seed color. An individual with the genotype *TtYy* can form *four* different kinds of gametes: *TY*, *Ty*, *tY*, and *ty*. Thus each gamete has one allele for height and one allele for seed color. If we cross two of these individuals using a Punnett square, we can analyze the types of offspring that will result in the F_2 generation. This cross is known as a *dihybrid cross*.

When we analyze the results of the cross in the Punnett square, we find that there are nine offspring that are tall and yellow, three offspring that are tall and green, three offspring that are short and yellow, and one offspring that is short and green. Thus the ratio of phenotypes in a dihybrid cross is 9:3:3:1. When Mendel observed these results, he stated that different individual traits are inherited separately from one another. This observation is known as his Principle of Independent Assortment (Figure 5-4).

Incomplete Dominance

There are some situations that do not conform to the Mendelian principles just described. Consider, for example, the showy garden plants known as four o'clocks (so named because the flowers open in the late afternoon). If a homozygous red-flowered plant is crossed with a homozygous white-flowered plant, all the offspring develop flowers that are pink. It appears that the two traits of the parents are "blended" in the hybrid offspring. How can this phenomenon be explained? Geneticists state that, in order for sufficient red pigment to be produced in the flower, the plant must have two copies of the gene that codes for

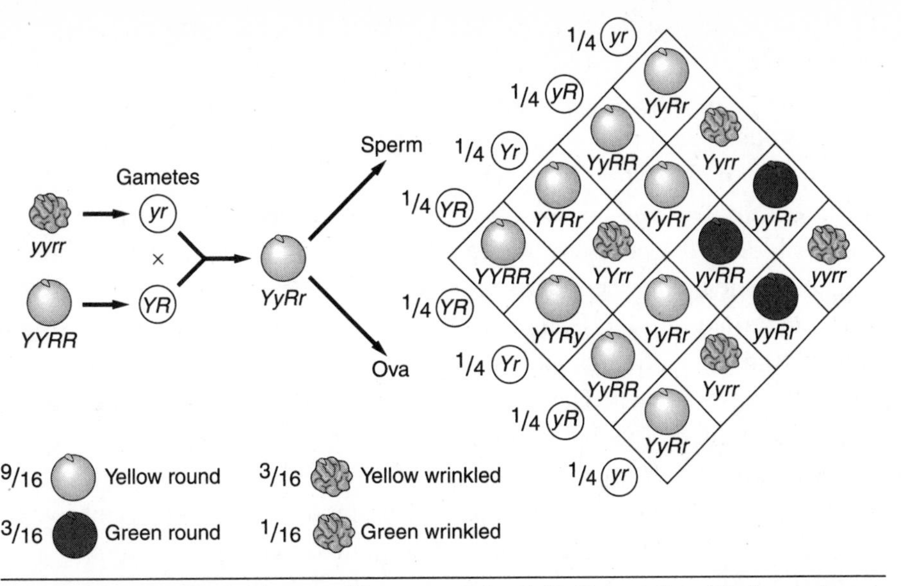

Figure 5-4. A dihybrid cross illustrates Mendel's Principle of Independent Assortment.

the production of red pigment in each of its cells. If the plant has only one of these red genes, less red pigment is synthesized and the flower appears pink. In this case, the red gene is not totally dominant over the white gene, so there is *blended inheritance*.

The homozygous red parent can be represented with the genotype *RR*, while the homozygous white parent can be represented either as *WW* or *R'R'*. (Note that we do not use a lower case letter because it would indicate that the gene is recessive.) A pink individual would be represented by *RW* or *RR'*. Unlike cases in which dominance is complete, the genotype of a hybrid *can* be determined by its phenotype. In other words, we would know by the appearance of a pink-flowered four o'clock plant that it is heterozygous (Figure 5-5). Additionally, we would know that a red-flowered plant is homozygous red and that a white-flowered plant is homozygous white. If two pink individuals were to be crossed, the F_2 generation would show the homozygous and heterozygous traits in the following ratio: 1 *RR* : 2 *RW* : 2 *WW.* The Punnett square for this cross would be:

	R	*W*
R	*RR* (red)	*RW* (pink)
W	*RW* (pink)	*WW* (white)

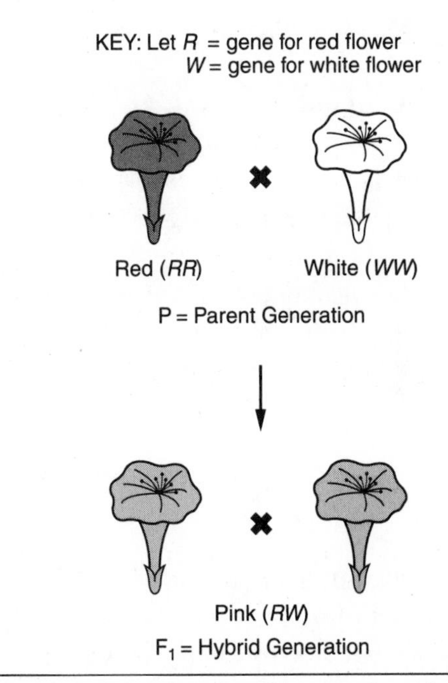

Figure 5-5. An example of incomplete dominance in four o'clock plants: The traits are "blended" in the hybrid offspring, so they all have pink flowers.

Codominance

In some cases, the alleles that control a particular trait are different, but neither allele is dominant over the other. As a result, both alleles are fully expressed in the offspring, so the phenotype shows *codominance*. The inheritance of human blood groups is a good example of this. The red blood cells of humans may contain a protein on their membrane or they may lack this protein entirely. The presence or absence of this protein is controlled by three different alleles. The allele represented as I^A is responsible for the production of a protein called antigen A. The allele I^B controls the production of a protein called antigen B. The allele *i* does not code for the production of any protein; and it is a recessive gene as well. The alleles I^A and I^B are codominant. Neither one is dominant over the other, and when I^A and I^B are present in an individual they are both equally expressed. This explains how three alleles can result in *four* different blood groups in humans. An individual with blood type A must have at least one allele for antigen A. This person could be either I^AI^A or I^Ai. Similarly, a person with blood type B must have at least one allele for antigen B and could have the genotype I^BI^B or I^Bi. Note that the alleles I^A and I^B are dominant over the allele *i*. If a person has blood type O (the O actually represents a zero, meaning there is no antigen present), this individual has neither the gene for antigen A nor the gene for antigen B. In other words, this person has the genotype *ii*. What about a person who has blood type AB? Clearly, the individual must have *both* alleles and both are expressed. This genotype is represented as I^AI^B.

Linkage

Mendel's observation of the independent inheritance of different traits was the basis for his principle of independent assortment. When the events of meiosis were discovered, it became clear that traits are inherited independently of one another only when their genes are on nonhomologous chromosomes. However, when the genes for two different traits are located on the same pair of homologous chromosomes, they tend to be inherited together. Such genes are said to be *linked*. In fact, all the genes located on the same chromosome tend to be inherited together unless some event separates them. The patterns of inheritance and phenotype ratios for linked traits are different from those of nonlinked traits (the kinds observed by Mendel).

Crossing-Over

During *synapsis* in the first meiotic division, when the replicated homologous chromosomes pair up next to one another, the chromatids of a pair of homologous chromosomes often twist around each other, break, exchange segments, and rejoin (Figure 5-6). This exchange of segments, called *crossing-over*, results in a rearrangement of linked genes and produces variations in offspring. Crossing-over is an important source of genetic variation in sexual reproduction.

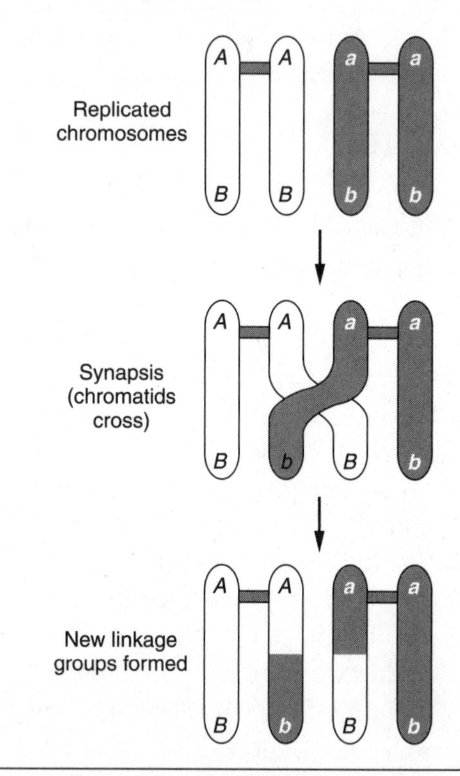

Figure 5-6. Crossing-over of chromatids.

It is important to note that crossing-over is a frequently observed event during meiosis. When chromosomes exchange genes, normally it is the homologous alleles that are exchanged. In addition, the further apart two genes are on the chromosome, the more likely they are to be separated from one another. Genes that lie very close together on the same chromosome are rarely separated. Geneticists have used cross-over frequency to map the location of genes on the chromosomes.

QUESTIONS

MULTIPLE CHOICE

15. Polydactyly is a characteristic in which a person has six fingers per hand. Polydactyly is dominant over the trait for five fingers. If a man who is heterozygous for this trait marries a woman with the normal number of fingers, what are the chances that their child would be polydactyl? A. 0% B. 50% C. 75% D. 100%

16. A cross between two pea plants that are hybrid for a single trait produces 60 offspring. Approximately how many of the offspring would be expected to exhibit the recessive trait? A. 15 B. 45 C. 30 D. 60

17. Which principle states that during meiosis chromosomes are distributed to gametes in a random fashion? A. dominance B. linkage C. segregation D. mutation

18. In guinea pigs, black coat color is dominant over white coat color. The offspring of a mating between two heterozygous black guinea pigs would probably show a phenotype ratio of A. two black to two white B. one black to three white C. three black to one white D. four black to zero white

19. The offspring of a mating between two heterozygous black guinea pigs would probably show a genotype ratio of A. 1 *BB* : 2 *Bb* : 1 *bb* B. 3 *Bb* : 1 *bb* C. 2 *BB* : 2 *bb* D. 2 *BB* : 1 *Bb* : 1 *bb*

20. If a breeder wanted to discover whether a black guinea pig was homozygous (*BB*) or heterozygous (*Bb*) for coat color, the animal in question would have to be crossed with an individual that has the genotype A. *BB* B. *bb* C. *Bb* D. *BbBb*

21. Mendel's principle of independent assortment applies to traits whose genes are found on A. homologous chromosomes B. sex chromosomes C. the same chromosome D. nonhomologous chromosomes

22. The process in which the chromatids of pairs of homologous chromosomes exchange segments is called A. linkage B. crossing-over C. independent assortment D. intermediate inheritance

23. In horses, black coat color is dominant over chestnut coat color. Two black horses produce both a black-coated and a chestnut-coated offspring. If coat color is controlled by a single pair of genes, it can be assumed that A. in horses, genes for coat color frequently mutate B. one of the parent horses is homozygous dominant and the other is heterozygous for coat color C. both parent horses are homozygous for coat color D. both parent horses are heterozygous for coat color

OPEN RESPONSE

24. Based on your answer to question 23, explain how two black horses could produce a chestnut-colored offspring.

Base your answers to questions 25 through 27 on the diagram below, which represents a pair of homologous chromosomes at the beginning of meiosis. The letters A, B, C, a, b, *and* c *represent pairs of alleles located on the chromosomes.*

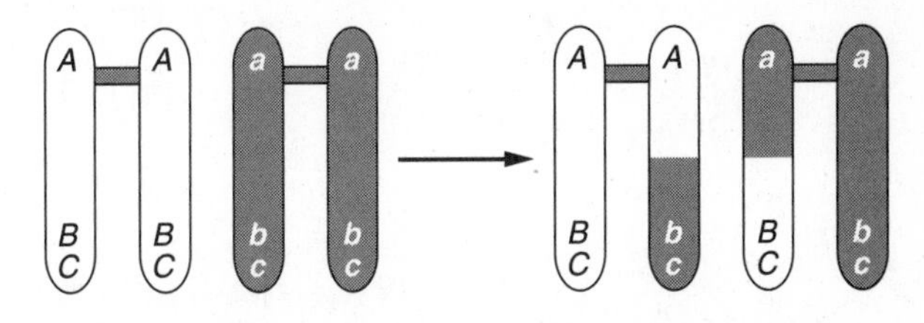

25. Compare the sets of chromosomes on the left with those on the right. Explain what has happened.

26. What process (not shown) is responsible for the observed results?

27. How does this process lead to variations among offspring?

28. When is a testcross used? Explain how it works.

29. Explain the following statement: Traits are inherited independently of one another only if their genes are on non-homologous chromosomes. You may use diagrams to support your explanation.

SEX CHROMOSOMES AND INHERITANCE

Sex Determination

The diploid cells of many organisms contain two types of chromosomes: *autosomes* (which do not determine sex) and *sex chromosomes*. There is generally one pair of sex chromosomes, and all the other chromosomes are autosomes. In human body cells there are 22 pairs of autosomes and one pair of sex chromosomes. The sex chromosomes are called the X and Y chromosomes. Normal females have two X chromosomes, and normal males have one X and one Y chromosome.

During meiotic cell division, the sex chromosomes, like all other chromosome pairs, are separated (Figure 5-7). The resulting gametes contain only one sex chromosome. Since females have two X chromosomes, each female gamete receives an X chromosome. Since the genotype of males is XY, sperm cells may receive either an X or a Y chromosome. The sex of the offspring is determined at fertilization and depends on whether the egg is fertilized by a sperm with an X or a sperm with a Y chromosome. If the sperm has an X chromosome, the resulting zygote will be female (XX). If the sperm has a Y chromosome, the resulting zygote will be male (XY).

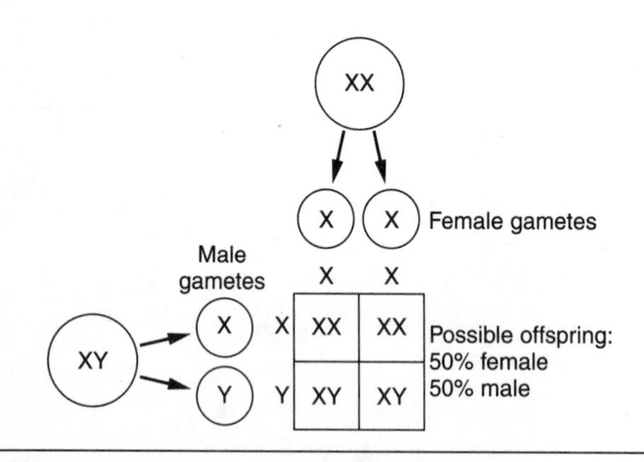

Figure 5-7. Sex determination of offspring.

Sex-linked Traits

T. H. Morgan, in his experiments with fruit flies, found that some rare, abnormal recessive traits appear with greater frequency in males than in females. From his observations, Morgan concluded that the genes for these traits are present on the X chromosome and that there are no corresponding alleles for these traits on the Y chromosome. Genes found on the X chromosome are called *sex-linked genes*. Recessive sex-linked traits appear more frequently in males than in females because in females there is usually a normal, dominant allele on the other X chromosome, so that the phenotype is normal. In males, there is no second allele, so the presence of one recessive gene produces a recessive phenotype.

Both *hemophilia* and *color blindness* are sex-linked disorders; they occur more frequently in males than in females. Hemophilia is a condition in which the blood does not clot properly; color blindness is an inability to distinguish certain colors. The genes for normal blood clotting and normal color vision are dominant; the genes for hemophilia and color blindness are recessive. For a female to show either of these disorders, she must have recessive genes (alleles) on both of her X chromosomes. Females with one normal, dominant gene and one recessive gene for these disorders are called *carriers*. They can pass the disorder to their offspring but do not themselves show symptoms of the disorder. Figure 5-8 shows the possible genotypes of children of a normal male and a female carrier of color blindness.

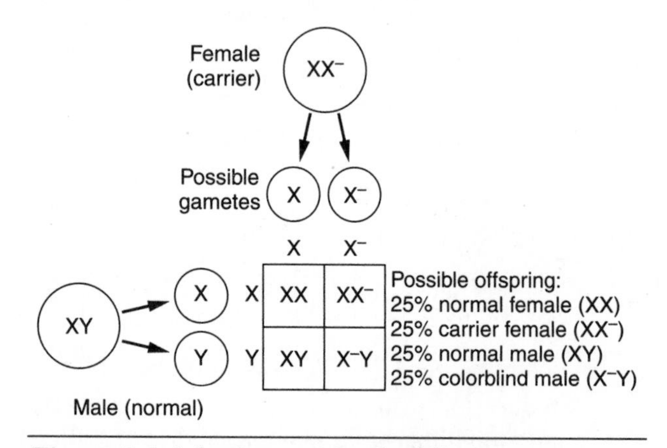

Figure 5-8. Inheritance of color blindness.

QUESTIONS

MULTIPLE CHOICE

30. If a color-blind man marries a woman who is a carrier for color blindness, it is most probable that A. all of their sons will have normal color vision B. half of their sons will be color-blind C. all of their sons will be color-blind D. none of their children will have normal color vision

31. A color-blind man marries a woman with normal vision. Her mother was color-blind. They have one child. What is the chance that this child will be color-blind? A. 0% B. 25% C. 50% D. 100%

32. A color-blind woman marries a man who has normal color vision. What are their chances of having a color-blind daughter? A. 0% B. 25% C. 75% D. 100%

33. Which parental pair could produce a color-blind female? A. homozygous normal-vision mother and color-blind father B. color-blind mother and normal-vision father C. heterozygous normal-vision mother and normal-vision father D. heterozygous normal-vision mother and color-blind father

34. Which statement correctly describes the normal number and type of chromosomes present in human body cells of a particular sex? A. Males have 22 pairs of autosomes and 1 pair of XX sex chromosomes. B. Females have 23 pairs of autosomes. C. Males have 22 pairs of autosomes and 1 pair of XY sex chromosomes. D. Males have 23 pairs of autosomes.

35. Based on the pattern of inheritance known as sex linkage, if a male is a hemophiliac, how many genes for this trait are present on the sex chromosomes in each of his diploid cells? A. 1 B. 2 C. 3 D. 4

36. Traits controlled by genes on the X chromosome are said to be A. sex-linked B. mutagenic C. incompletely dominant D. homozygous

OPEN RESPONSE

37. Use a diagram to show why, for each pregnancy, the chances of giving birth to either a boy or a girl is 50-50. Explain the results shown in your diagram.

38. Explain why hemophilia occurs more often in males than in females. Use a diagram to illustrate your answer.

GENETIC MUTATIONS

Changes in the genetic material are called **mutations**. Mutations in body cells can be passed on to new cells of the individual as a result of mitosis, but they cannot be transmitted to offspring by sexual reproduction. However, mutations in sex cells *can* be transmitted to the next generation. Mutations may involve alterations in chromosomes or alterations in the chemical makeup of genes.

Chromosomal Alterations

Chromosomal alterations involve a change in the structure or number of chromosomes. The effects of chromosomal alterations are often seen in the phenotype of an organism because each chromosome contains many genes.

Nondisjunction. During meiosis, the two chromosomes of each homologous pair separate from each other; each gamete produced by the division receives only one member of each homologous pair. The separation of homologous chromosomes is called *disjunction*. The term *nondisjunction* refers to a type of chromosomal alteration in which one or more pairs of homologous chromosomes fails to separate normally during meiotic cell division (Figure 5-9).

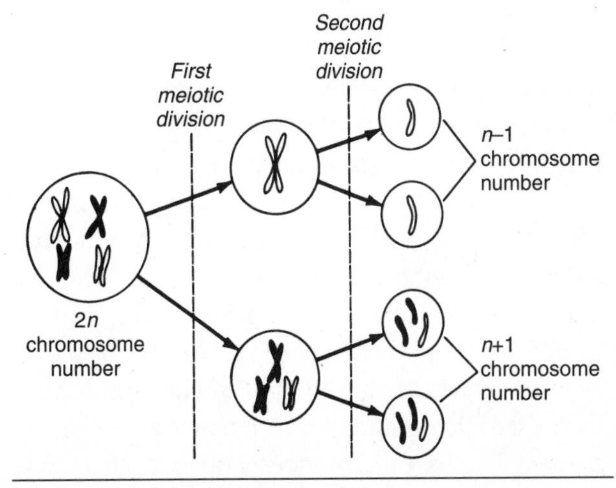

Figure 5-9. Nondisjunction of chromosomes.

As a result of nondisjunction, one of the gametes produced contains both members of the homologous pair, while another gamete contains neither chromosome. Nondisjunction results in the production of some gametes with more chromosomes than normal and some gametes with fewer chromosomes than normal. If one of these abnormal gametes is involved in fertilization, the resulting zygote will have either more than or less than the normal ($2n$) number of chromosomes.

Down syndrome in humans is caused by the presence of an extra chromosome number 21. Nondisjunction during gamete production in one of the parents produces a gamete with an extra chromosome 21. As a result of fertilization, this extra chromosome is transmitted to the offspring, who will have mild to severe mental retardation.

Polyploidy. Occasionally during gamete formation, a complete set of chromosomes fails to undergo disjunction, and a gamete is produced that contains the diploid ($2n$) chromosome number. If a diploid gamete unites with a normal (n) gamete during fertilization, the resulting zygote will have a $3n$ (triploid) chromosome number. If two $2n$ gametes fuse, a $4n$ zygote results. The inheritance of one or more complete extra sets of chromosomes is called *polyploidy.* This condition is common in plants but rare in animals. In plants, polyploid individuals are usually larger or more vigorous than the normal, diploid varieties. Certain strains of wheat, potatoes, alfalfa, apples, tobacco, and zinnias are polyploid. Some polyploid plants produce seedless fruit and are sterile.

Changes in Chromosome Structure. Changes in the makeup of chromosomes may result from random breakage and recombination of chromosome parts. *Translocation* occurs when a segment of one chromosome breaks off and reattaches to a nonhomologous chromosome. *Addition* occurs when a segment breaks off one chromosome and reattaches to the homologous chromosome. *Inversion* occurs when a segment breaks off and reattaches in reverse on the same chromosome. *Deletion* occurs when a segment breaks off and does not reattach to any other chromosome.

Gene Mutations

A random change in the chemical makeup of the DNA (genetic material) is a *gene mutation.* The effects of some gene mutations, such as albinism, are noticeable, but other gene mutations may not produce noticeable effects.

Inheritable gene mutations tend to be harmful to the individual. For example, sickle-cell anemia and Tay-Sachs disease are caused by gene mutations. Fortunately, most gene mutations are recessive and are hidden by the normal, dominant allele. However, if both parents carry the same recessive mutant gene, there is a chance that their offspring will be homozygous recessive and show the harmful trait.

Occasionally, random gene mutations produce changes that make an individual better adapted to the environment. Over time, such helpful mutant genes tend to increase in frequency within a population.

Mutagenic Agents

Although mutations occur spontaneously, the rate of mutation can be increased by exposure to certain chemicals and to forms of **radiation** that act as *mutagenic agents*. For example, forms of mutagenic radiation include x-rays, ultraviolet rays, radioactive substances, and cosmic rays. Mutagenic chemicals include formaldehyde, benzene, and asbestos fibers.

QUESTIONS

MULTIPLE CHOICE

39. Which phrase best describes most mutations? A. dominant and disadvantageous to the organism B. recessive and disadvantageous to the organism C. recessive and advantageous to the organism D. dominant and advantageous to the organism

40. The failure of a pair of homologous chromosomes to separate during meiotic cell division is called A. nondisjunction B. translocation C. addition D. deletion

41. The condition in which a gamete contains the $2n$ or $3n$ number of chromosomes is called A. translocation B. a gene mutation C. polydactyly D. polyploidy

42. The presence of only one X chromosome in each body cell of a human female produces a condition known as Turner syndrome. This condition most probably results from the process called A. polyploidy B. crossing-over C. nondisjunction D. hybridization

43. A random change in the chemical structure of DNA produces A. polyploidy B. a translocation C. nondisjunction D. a gene mutation

44. Down syndrome in humans is characterized by the presence of an extra chromosome 21 in all cells of the body. The number of chromosomes present in the body cells of individuals with this condition is A. $n + 1$ B. $3n$ C. $2n + 1$ D. $4n$

45. The graph below shows the relationship between maternal age and the number of children born with Down syndrome per 1000 births.

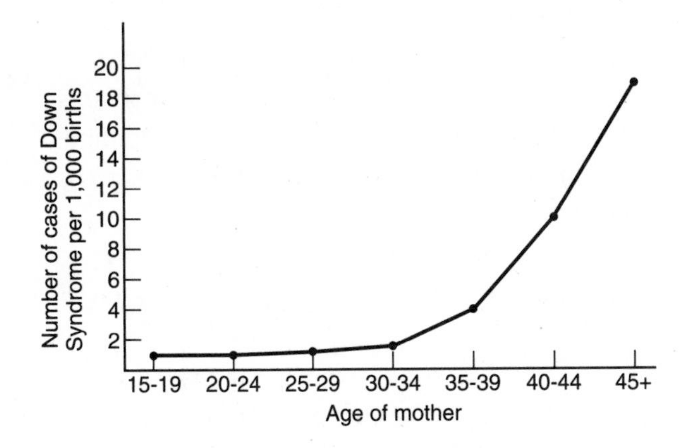

According to the graph, the incidence of Down syndrome A. generally decreases as maternal age increases B. is about nine times greater at age 45 than at age 30 C. stabilized at 2 per 1000 births after age 35 D. is greater at age 15 than at age 35

46. Ultraviolet rays, x-rays, and certain other forms of radiation can increase the rate of gene mutation. These forms of radiation are said to act as A. mutagenic agents B. catalysts C. enzymes D. indicators

47. The large size and exceptional vigor of certain varieties of wheat, apples, and zinnias are due to the possession of extra sets of chromosomes, which result from A. incomplete dominance B. gene mutations C. nondisjunction of complete sets of chromosomes D. nondisjunction of chromosome number 21 only

48. A type of chromosomal alteration in which a segment of chromosome breaks off and does not reattach to any chromosome is called A. addition B. inversion C. deletion D. translocation

49. Changes in the genetic code of a human can be transmitted to offspring if they occur in A. cancer cells B. gametes C. cell membranes D. antibodies

OPEN RESPONSE

Base your answers to questions 50 through 53 on the following information about an experiment and on your knowledge of biology.

Two groups of 100 lima beans each were used. Group *A* was exposed to natural light for a period of 24 hours and then planted. Group *B* was exposed to microwave energy for 24 hours and then planted under the same conditions as Group *A*. When the seeds germinated, the plants were observed for growth over a period of two weeks. The results are summarized in the table below.

	Number of Plants	
Group	***Normal Growth***	***Stunted and/or Pale***
A	83	17
B	54	46

50. What hypothesis was most likely being tested in this experiment?

51. Describe the results of the experiment.

52. Based on the data, propose a conclusion for the experiment.

53. What are some of the implications of the data?

54. Explain how it is possible for an individual to inherit an extra chromosome.

55. Identify one human genetic disorder that is caused by the inheritance of an abnormal number of chromosomes. What effect does this condition have on the offspring?

HEREDITY AND THE ENVIRONMENT

The development and expression of inherited traits can be influenced by environmental factors such as nutrients, temperature, sunlight, and so on. The relationship between gene action and environmental influence can be seen in the following examples.

Temperature affects fur color in the Himalayan rabbit. Under normal circumstances, these rabbits are white with black ears, nose, tail, and feet. (The black fur helps the rabbit absorb more heat in its extremities.) However, when some of the white fur on a Himalayan rabbit's back is shaved off and the area kept covered with an ice pack, the new hairs grow in black. The artificial change in temperature produces a change in fur color.

Experiments have shown that the production of chlorophyll requires exposure to sunlight. When parts of a leaf are covered with dark paper, chlorophyll production stops in the area that is covered. Only the exposed part produces chlorophyll, is green, and performs photosynthesis.

Stress and nutrition can affect gene expression. For example, someone who has a tall genotype may not develop a tall phenotype if his or her growth is stunted by malnutrition.

PLANT AND ANIMAL BREEDING

Using the principles of genetics, plant and animal breeders have been able to produce, improve, and maintain new varieties of plants and animals. Methods of **selective breeding** used by such people include artificial selection, inbreeding, and hybridization.

In *artificial selection*, individuals with the most desirable traits (for example, sheep with thick, soft wool) are crossed or allowed to mate in the hopes that their offspring will show the desired traits.

The offspring of selected organisms may be mated with one another to produce more individuals with the desirable traits. This technique, called *inbreeding*, involves the mating of closely related organisms. (Of course, the risk of inbreeding is that harmful recessive genes are more likely to be inherited and cause disorders in the offspring.)

Two varieties of a species may have different desirable traits. In a technique called *hybridization*, breeders cross two such varieties in the hope of producing hybrid offspring that show the desirable traits of both varieties. For example, if one variety of rose has very large petals and another variety has a very sweet scent, their hybrid might show both desirable traits.

QUESTIONS

MULTIPLE CHOICE

56. If bean plant seedlings are germinated in the dark, the seedlings will lack green color. The best explanation for this condition is that A. bean plants are heterotrophic organisms B. bean seedlings lack nitrogen compounds in their cotyledons C. the absence of an environmental factor limits the expression of a genotype D. bean plants cannot break down carbon dioxide to produce oxygen in the dark

57. In many humans, exposing the skin to sunlight over prolonged periods of time results in the production of more pigment by the skin cells (tanning). This change in skin color provides evidence that A. ultraviolet light can cause mutations B. gene action can be influenced by the environment C. the inheritance of skin color is an acquired characteristic D. albinism is a recessive characteristic

58. Identical twins were separated at birth and brought together after 13 years. They varied in height by 5 centimeters and in weight by 10 kilograms. The most probable explanation

for these differences is that A. their environments affected the expression of their traits B. their cells did not divide by mitotic cell division C. they developed from two different zygotes D. they differed in their genotypes

59. A normal bean seedling that had the ability to produce chlorophyll did not produce any chlorophyll when grown in soil that was totally deficient in magnesium salts. Which statement concerning this plant's inability to produce chlorophyll is true? A. The lack of magnesium prevented the plant's roots from absorbing water. B. The production of chlorophyll was controlled solely by heredity. C. The lack of magnesium caused a mutation of the gene that controlled chlorophyll production. D. The production of chlorophyll was influenced by environmental conditions.

60. To ensure the maintenance of a desirable trait in a particular variety of plant, a farmer would use A. binary fission B. mutagenic agents C. artificial selection D. natural selection

61. The mating of very closely related organisms in order to produce the most desirable traits is known as A. inbreeding B. hybridization C. karyotyping D. crossing-over

62. Plant and animal breeders usually sell or get rid of undesirable specimens and use only the desirable ones for breeding. This practice is referred to as A. vegetative propagation B. artificial selection C. natural breeding D. random mating

63. A single gene mutation results from A. a change in a base sequence in DNA B. recombination of alleles C. the failure of chromosomes to separate D. blocked nerve messages

64. The following table shows relationships between genes, the environment, and coloration of tomato plants. Which statement best explains the final appearance of these tomato plants? A. The expression of gene *A* is not affected by light. B. The expression of gene *B* varies with the presence of light. C. The expression of gene *A* varies with the environment. D. Gene *B* is expressed only in darkness.

Inherited Gene	*Environmental Condition*	*Final Appearance*
A	Light	Green
B	Light	White
A	Dark	White
B	Dark	White

65. Some mammals have genes for fur color that produce pigment only when the outside temperature is above a certain level. This pigment production is an example of how the environment of an organism can A. destroy certain genes B. cause new mutations to occur C. stop the process of evolution D. influence the expression of certain genes

OPEN RESPONSE

66. Identify three environmental factors that can influence phenotype. Give an example of each.

67. Describe some steps a breeder would take to produce an organism that has desirable traits.

HUMAN HEREDITY

The principles of genetics apply to all organisms. However, specific studies of human genetics are limited because humans are not suitable subjects for experimentation: human generation time is too long; there are only a small number of offspring per generation in a human family; and it is unethical to perform such experiments on humans. Knowledge of human heredity has been gathered indirectly through studies of human pedigree charts and information obtained in the course of genetics counseling.

Human Pedigree Charts

The patterns of inheritance of certain traits can be traced in families for a number of generations.

These patterns can be illustrated in *pedigree charts* that show the presence or absence of certain genetic traits in each generation. The use of a pedigree chart may also make it possible to identify carriers of recessive genes.

Human Genetic Disorders

Some diseases caused by genetic abnormalities are sickle-cell anemia, Tay-Sachs disease, phenylketonuria, and Huntington's disease. These disorders are caused by gene mutations.

Sickle-cell anemia is a blood disorder found most commonly in individuals of African descent. The disorder is caused by a gene mutation that results in the production of abnormal hemoglobin molecules and red blood cells. The abnormal hemoglobin and sickle-shaped cells do not carry oxygen efficiently, resulting in anemia. The sickle-shaped red cells also tend to obstruct blood vessels, causing severe pain. Sickle-cell anemia occurs in individuals who are homozygous for the allele. Both homozygous and heterozygous individuals can be identified by blood tests.

Tay-Sachs disease is a recessive genetic disorder in which nerve tissue in the brain deteriorates because of an accumulation of fatty material. The disorder is a result of the body's inability to synthesize a particular enzyme. Tay-Sachs disease, which is fatal, occurs in individuals who are homozygous for the trait and is most common among Jewish people of Central European descent.

Phenylketonuria (PKU) is a disorder in which the body cannot synthesize an enzyme necessary for the normal metabolism of the amino acid phenylalanine. The disease, which occurs in homozygous recessive individuals, is characterized by the development of mental retardation. Analysis of the urine of newborn infants can detect PKU. If PKU is present, mental retardation can be prevented by maintaining a diet free of phenylalanine.

Huntington's disease is a genetic disorder that causes an adult's nervous system to break down gradually. The disease is caused by a dominant gene; thus, if one parent has the disease, it is likely that half of his or her children will also inherit it and develop the condition.

Detection of Genetic Disorders

Some human genetic disorders can be detected either before or soon after birth by the use of one or more of the following techniques.

Advances in genetic research have resulted in the development of simple blood and urine tests that can determine if an individual has certain genetic disorders. Carriers of sickle-cell anemia and Tay-Sachs disease can be identified by these screening techniques.

Karyotyping is a technique in which a greatly enlarged photograph of the chromosomes of a cell is prepared. The homologous pairs of chromosomes are matched together, and the chromosomes are examined to see if there are any abnormalities in number or structure (Figure 5-10).

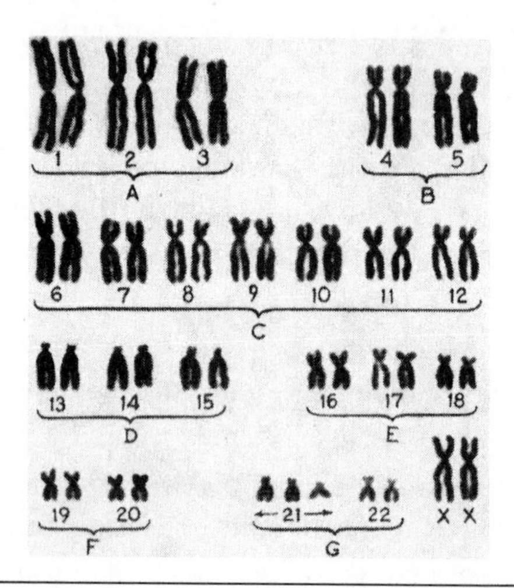

Figure 5-10. The karyotype of a person with Down syndrome shows that the disorder is caused by the inheritance of three copies of chromosome 21.

Amniocentesis is a technique in which a small sample of amniotic fluid is withdrawn from the amniotic sac of a pregnant woman. The fluid contains fetal cells, which can be used for karyotyping or for chemical analysis. Amniocentesis is used in the identification of sickle-cell anemia, Tay-Sachs disease, and Down syndrome in fetuses.

With the completion of the **Human Genome Project**, scientists have now mapped every gene present on human chromosomes. Geneticists can identify *markers* present in the DNA that can cause diseases or whose presence can indicate that the individual is likely to develop a particular genetic illness.

Genetic Counseling

The various techniques described above are used by *genetics counselors* to inform concerned parents about the possible occurrence of genetic defects in their children. For couples whose families show the presence of a particular genetic disorder, a pedigree chart may be developed to predict the probability of their child inheriting the disorder. Amniocentesis, followed by karyotyping and chemical tests, may be performed once pregnancy is established.

QUESTIONS

MULTIPLE CHOICE

68. An inherited metabolic disorder known as phenylketonuria (PKU) is characterized by mental retardation. This condition results from the inability to synthesize a single A. enzyme B. hormone C. vitamin D. carbohydrate

69. Which statement best describes amniocentesis? A. Blood cells of an adult are checked for anemia. B. Saliva of a child is analyzed for the amino acids. C. Urine of a newborn baby is analyzed for the amino acid phenylalanine. D. Fluid surrounding a fetus is removed for chemical and genetic analysis.

70. Which is a genetic disorder in which abnormal hemoglobin leads to fragile red blood cells and obstructed blood vessels? A. phenylketonuria B. sickle-cell anemia C. leukemia D. Down syndrome

71. Human disorders such as PKU and sickle-cell anemia, which are defects in the synthesis of individual proteins, are most likely the result of A. gene mutations B. nondisjunction C. crossing-over D. polyploidy

72. Which technique can be used to examine the chromosomes of a fetus for possible genetic defects? A. pedigree analysis B. analysis of fetal urine C. karyotyping D. blood cell tests

OPEN RESPONSE

73. Give three reasons why a direct study of the inheritance of human traits is difficult to carry out.

74. Briefly describe the two ways that information about patterns of human heredity is usually obtained.

READING COMPREHENSION

Base your answers to questions 75 through 78 on the information below and on your knowledge of biology. Source: *Science News* (July 9, 2005): Vol. 168, No. 2, p. 29.

Stem Cell Shift May Lead to Infections [and] Leukemia

Researchers have long wondered why elderly people suffer more infections and have a greater chance of developing myeloid leukemia, a type of blood cancer, than younger people do. Now, research in mice suggests that the aging of blood-producing stem cells could be responsible for both conditions.

With age, the body of a person or other animal loses its capacity to sustain its tissues and organs. "Since we know the cells mediating this maintenance are stem cells, it doesn't take a great leap of faith to think that stem cells are at the heart of that failure," says Derrick Rossi of Stanford University.

To examine whether the aging of stem cells contributes to infections and leukemia, Rossi and his colleagues irradiated young and old mice to kill off their blood-making stem cells. The scientist then transplanted such stem cells from young donor mice into elderly irradiated animals and from old donors into young irradiated animals.

After several weeks, the researchers found that young animals' stem cells transplanted into the old mice produced the different types of blood cells in ratios much like those in young mice that haven't been irradiated. However, the young animals that received old animals' stem cells had significantly fewer new lymphoid blood cells—which make cells that battle infections—than normal young animals do.

After examining gene activity in the stem cells transplanted from old animals, Rossi's team found a boost in activity among genes responsible for creating myeloid cells. These create red blood cells and some other blood components. Many myeloid-production genes have been associated with myeloid leukemia in people.

The scientists conclude in the June 28 *Proceedings of the National Academy of Sciences* that a shift from lymphoid-cell production to myeloid-cell production could be responsible for the increases both in infections and in risk of leukemia that come with old age.

75. According to the article, what do researchers think is the reason that elderly people develop infections and blood cancer more often than young people do?

76. Briefly describe the experiment the research team conducted on mice to test their hypothesis.

77. Describe the change in gene activity that occurs in the blood-making stem cells of older animals.

78. What conclusions about blood cells and illness in elderly people did the scientists make as a result of their experiments on mice?

CHAPTER 6 Molecular Genetics

Standards 3.1, 3.2, 3.3 (Genetics)

BROAD CONCEPT: Genes allow for the storage and transmission of genetic information. They are a set of instructions encoded in the nucleotide sequence of each organism. Genes code for the specific sequences of amino acids that comprise the proteins that are characteristic of that organism.

STRUCTURE AND FUNCTION OF DNA

Biochemists have learned that the DNA of chromosomes is the genetic material that is passed from generation to generation. Genes are the sections of DNA (deoxyribonucleic acid) molecules that encode information for the synthesis of a particular protein or even a portion of a protein. DNA controls cellular activities by directing the production of such proteins as enzymes, hormones, and structural proteins.

The DNA Molecule

DNA molecules are immense; each one is made up of thousands of repeating units called **nucleotides**. Large molecules that are made up of repeating units of smaller molecules, or *monomers*, are known as **polymers**; thus DNA is a polymer. A DNA nucleotide is composed of three parts: a *phosphate group*; a molecule of the 5-carbon sugar *deoxyribose*; and a *nitrogenous base* (Figure 6-1).

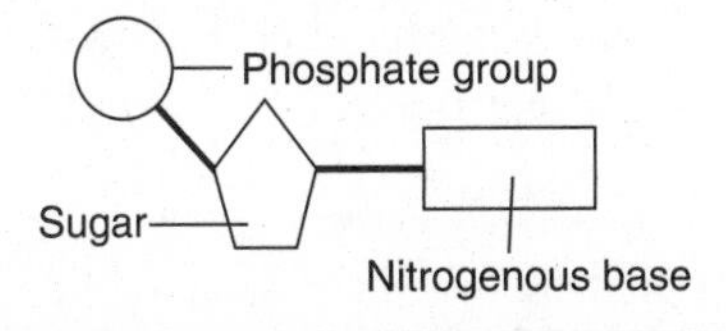

Figure 6-1. Structure of a DNA nucleotide unit.

There are four different nitrogenous bases in the DNA nucleotides: *adenine*, *cytosine*, *guanine*, and *thymine.* Therefore, there are four different kinds of nucleotides, depending on which nitrogenous base is present.

Watson-Crick Model. In the model of DNA developed by scientists James Watson and Francis Crick, the DNA molecule consists of two connected chains of nucleotides forming a ladderlike structure (Figure 6-2). The sides of the "ladder" are composed of alternating phosphate and deoxyribose (sugar) molecules. Each rung of the ladder consists of a pair of nitrogenous bases connected by hydrogen bonds. The two chains of the DNA molecule are twisted to form a spiral, or *double helix.*

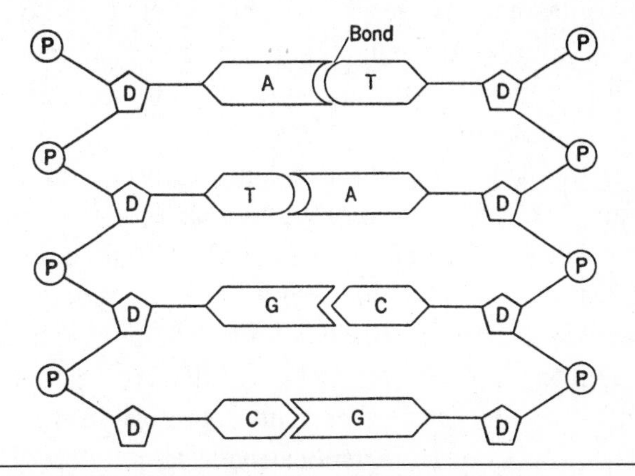

Figure 6-2. Structure of the DNA molecule.

Due to their unique chemical structures, the four nitrogenous bases of DNA nucleotides bond together in only one way: Adenine (A) always pairs with thymine (T); and cytosine (C) always pairs with guanine (G). Because the bases pair together only in this way, the two strands of a DNA molecule are always *complementary*. Where there is an adenine on one strand, there is a thymine on the other strand in the corresponding position; and where there is a cytosine on one strand, there is a guanine on the other strand in the corresponding position. So, if you know the order of bases on one strand, then you can predict the order on the complementary strand. This structural feature of DNA is critical to the function of the molecule as well as to its ability to make copies of itself.

DNA Replication

DNA, unlike any other chemical compound, can make exact copies of itself—that is, the DNA molecule can **replicate**. This process, called *DNA replication*, is a necessary part of the chromosome replication that occurs before mitosis and meiosis. In other words, when the chromosomes replicate prior to cell division, it is actually the DNA molecules within the chromosomes that replicate. Each chromosome consists of a DNA molecule and some associated structural proteins known as *histones*.

In replication, the double-stranded DNA helix unwinds; the two strands then separate by a breaking of the hydrogen bonds between the nitrogenous base pairs. Free nucleotides from the cytoplasm then enter the nucleus, where they bond to their complementary bases on the DNA strands (Figure 6-3). Replication produces two identical DNA molecules that are exact copies of the original molecule. Thus, the genetic information encoded in the original DNA molecule is perfectly reproduced in two (half-new) DNA molecules. These two DNA molecules can be passed on to new cells during cell division to ensure that each daughter cell receives a copy of the same genetic information that was present in the original cell. The process of DNA replication is actually carried out by a team of several important, specific enzymes, which play different roles in the process. For example, the enzyme *helicase* is involved in the unwinding and separation of the two DNA strands, while DNA *polymerase* enzymes are involved in the construction of the new complementary strands, by using the original strands as patterns.

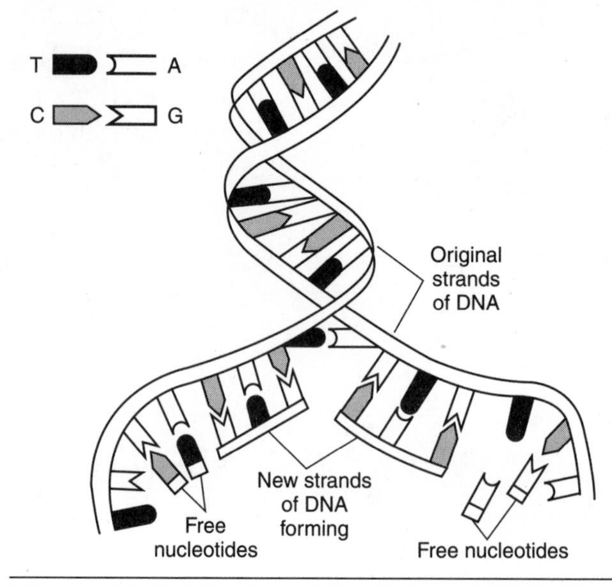

Figure 6-3. Replication of the DNA strands.

Gene Control of Cellular Activities

The unique qualities of an organism are determined by the DNA of its genes. Again, a gene is a specific section of a DNA molecule that codes for a protein, or polypeptide. Many genes control enzyme synthesis, and the enzymes control cell activities. For example, a dominant gene enables people to produce the enzyme lactase, which digests milk sugar (lactose). People who lack an active copy of this gene cannot digest milk sugar and, thus, are lactose intolerant.

The hereditary information is in the sequence of the nucleotides in DNA molecules. The DNA nucleotide sequence determines the sequence of amino acids in enzymes and other proteins. The genetic control of protein synthesis involves RNA as well as DNA.

RNA Molecules

RNA (ribonucleic acid) molecules are similar to DNA in that they also are nucleic acids made up of nucleotide units. However, in RNA nucleotides, the 5-carbon sugar *ribose* is substituted for deoxyribose, and the nitrogenous base *uracil* (U) is substituted for thymine. RNA molecules consist of one strand of nucleotides, whereas DNA molecules have two strands. There are three kinds of RNA molecules in cells: *messenger RNA* (mRNA), *transfer RNA* (tRNA), and *ribosomal RNA* (rRNA).

Messenger RNA is synthesized in the cell nucleus. Portions of a DNA molecule (that is, the *genes*) unwind, and the two strands separate. The RNA nucleotides pair with complementary bases on a DNA strand, thus forming a strand of messenger RNA that is complementary to the DNA strand. The DNA strand serves as a pattern, or **template**, for the synthesis of the messenger RNA. In this way, the hereditary information in the nucleotide sequence of DNA is copied in complementary form into the nucleotide sequence of messenger RNA. The process by which an mRNA copy of the genetic code in DNA is made is referred to as **transcription** (Figure 6-4).

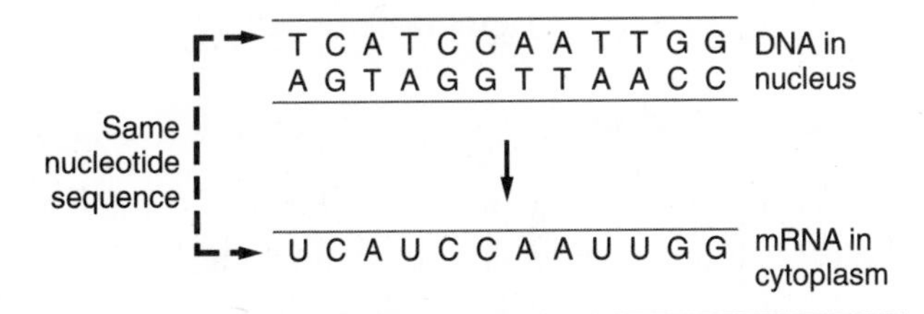

Figure 6-4. Transcription: The DNA sequence is copied into messenger RNA. Note that the RNA base uracil (U) substitutes for the DNA base thymine (T).

The sequence of nucleotides in messenger RNA contains the genetic code, which determines the amino acid sequence of proteins. After much research, it was determined that the genetic code for each amino acid is a specific sequence of *three* nucleotides. The three-nucleotide sequence in messenger RNA that specifies a particular amino acid is called a **codon**. A particular molecule of mRNA may have tens of thousands of codons that specify both the type and sequence of the amino acids in the protein to be synthesized. In addition, different molecules of mRNA have different lengths and base sequences depending on the part of the DNA template from which they were made.

Transfer RNA molecules are found in the cytoplasm. Their function is to carry amino acid molecules to the *ribosomes*, the sites of protein synthesis. Ribosomes are made up of ribosomal RNA and proteins. There are 20 different kinds of amino acids in cells, and there is a different form of transfer RNA for each codon. Each kind of transfer RNA has a three-nucleotide sequence, called an *anticodon*, which is complementary to a codon on the messenger RNA. For example, if an mRNA codon has the sequence ACG, the corresponding tRNA anticodon would be UGC. Note again that the base uracil (U) replaces thymine (T).

Protein Synthesis

The process of **protein synthesis** is a highly complex but extremely orderly process that occurs in cells. It involves DNA, mRNA, tRNA, amino acids, and a host of enzymes all participating together. It begins with the synthesis of a messenger RNA molecule, which moves out through pores in the nuclear membrane into the cytoplasm. Once in the cytoplasm, the strand of messenger RNA becomes attached to the ribosomes (Figure 6-5). Specific amino acids are carried to the ribosomes and messenger RNA by the transfer RNA molecules. It is important to note that each type of tRNA molecule only carries one of the 20 kinds of amino acids. The specificity of the tRNA molecule is determined by its three-base anticodon sequence. The anticodons of the transfer RNAs align with the codons of the messenger RNA on the ribosomes.

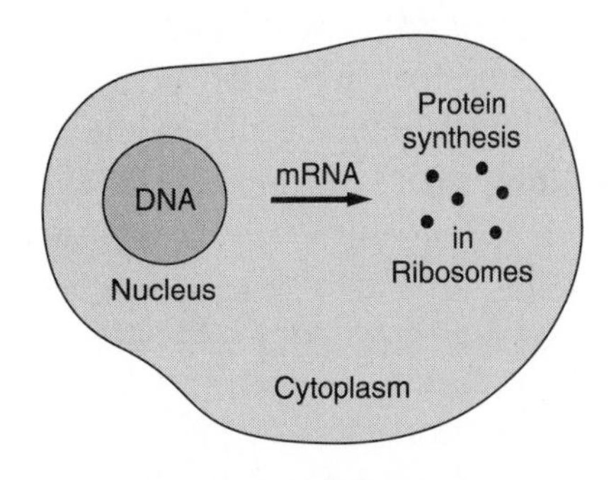

Figure 6-5. The messenger RNA molecule goes out to the ribosomes in the cytoplasm.

The amino acids carried by the transfer RNAs bond together in a sequence determined by the base sequence of the messenger RNA. The resulting chain of amino acids is a polypeptide (Figure 6-6). Some proteins consist of a single polypeptide chain, while others include two or more. The process by which the genetic code is used to synthesize specific proteins or polypeptides with the involvement of mRNA and tRNA is referred to as **translation**.

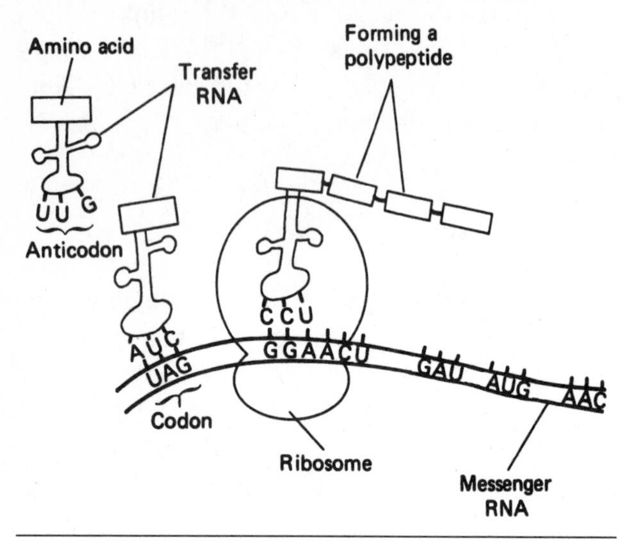

Figure 6-6. Translation: The protein (polypeptide) is synthesized at the ribosome.

To sum up, the process of *replication* results in an exact genetic copy of an original DNA molecule. This allows the genetic information to be passed on from generation to generation. The process of *transcription* results in a molecule of mRNA being synthesized using a specific portion of a DNA molecule now known to be a gene. The mRNA is complementary to the original DNA strand upon which it was produced. The process of *translation* results in a polypeptide being synthesized when tRNA molecules carry specific amino acid molecules to the ribosomes and position them according to the sequence of codons on the mRNA molecule attached to the ribosome. Protein synthesis involves joining the amino acids together with peptide bonds to form the polypeptide chains.

One Gene–One Polypeptide Hypothesis

According to the *one gene–one polypeptide hypothesis*, each gene controls the synthesis of a single polypeptide. A modern definition of the gene is that it is the sequence of nucleotides in a DNA molecule needed to synthesize one polypeptide. However, this hypothesis is now seen as overly simplified. It is now known that genes are not necessarily fixed in one place on the chromosomes. Rather, they can move to different locations, or *loci*, on the chromosomes and form new genetic codes when they are positioned next to different segments of DNA. In this way, a limited number of genes can produce a greater variety of polypeptides than would be possible if the genes remained in one place.

The geneticist Barbara McClintock first proposed the idea of movable, or "jumping," genes in the early 1940s, but other scientists largely dismissed her work at the time. Now known to exist, these movable genes, or transposable genetic elements, are called *transposons*. The recently completed Human Genome Project revealed that we have far fewer genes than had been expected. Now, the large variety of human traits can be better understood because we know about transposons. By "jumping" around the chromosomes, the same gene can be expressed in several different ways.

Gene Mutations

Any change in the sequence of nucleotides in a DNA molecule is a *gene mutation*. If the mutation occurs in the DNA of the sex cells (gametes), it may be inherited. Mutations that occur in body (somatic) cells are *not* passed on to offspring. Only mutations that occur in the gametes (sperm cells or egg cells) can be passed on to the next generation. Gene mutations called *point mutations* may involve the addition or deletion of bases, or the substitution of one base for another. Addition or deletion of bases in a DNA sequence is a very serious mutation because it causes a shift in the "reading frame." Recall that genetic information is encoded as three-base nucleotide sequences called *codons*, such as CUA or AGT (Figure 6-7). If a mutation occurs that somehow adds or deletes a nucleotide base from the sequence, all the information from the point of that mutation onward would be misread, resulting in errors in the amino acids being placed into the polypeptide. This would result in an incorrect or nonfunctional protein. In fact, this kind of mutation would most likely be lethal to an organism.

UUU Phe UUC Phe UUA Leu UUG Leu	UCU Ser UCC Ser UCA Ser UCG Ser	UAU Tyr UAC Tyr UAA *Stop* UAG *Stop*	UGU Cys UGC Cys UGA *Stop* UGG Trp
CUU Leu CUC Leu CUA Leu CUG Leu	CCU Pro CCC Pro CCA Pro CCG Pro	CAU His CAC His CAA Gln CAG Gln	CGU Arg CGC Arg CGA Arg CGG Arg
AUU Ile AUC Ile AUA Ile AUG Met	ACU Thr ACC Thr ACA Thr ACG Thr	AAU Asn AAC Asn AAA Lys AAG Lys	AGU Ser AGC Ser AGA Arg AGG Arg
GUU Val GUC Val GUA Val GUG Val	GCU Ala GCC Ala GCA Ala GCG Ala	GAU Asp GAC Asp GAA Glu GAG Glu	GGU Gly GGC Gly GGA Gly GGG Gly

Figure 6-7. The mRNA amino acid triplet codes. Note that most amino acids are represented by more than one codon.

Consider, for example, the following simplified codon sequence: AAA-UUU-CCC-GGG. As indicated here, there are four codons, each representing four different amino acids. Now, suppose a mutation occurred in the sequence that removed (deleted) one of the adenine nucleotides. The codons would now be read as follows: AAU-UUC-CCG-GG-. Clearly, the mutation has resulted in a change that affects all the codons, starting from the point of the mutation. All the amino acids from the point of the adenine deletion onward would be the wrong ones. The resulting protein would be totally incorrect and would not function for its intended purpose. If the protein were a critical enzyme, this mutation could be lethal because the cell would not be able to perform some vital metabolic function.

Substitution occurs when one base in a codon is replaced by a totally different base, such as adenine replaces cytosine. For example, sickle-cell anemia is caused by the substitution of just one incorrect nitrogenous base in the gene that controls hemoglobin synthesis. A codon is changed by the substitution and the wrong amino acid is inserted in place of the correct one. This, in turn, affects the structure and function of the hemoglobin protein. The red blood cells that contain the misformed hemoglobin assume a curved (sickle) shape, which causes them to become lodged in small blood vessels and capillaries. In addition, they are not capable of carrying as much oxygen as the normally shaped red blood cells can.

There are some situations in which a base substitution has no effect on the resulting protein. This is usually the case when the substitution occurs in the third position of the codon. For many of the 20 amino acids, only the first two positions of the codon are necessary to specify that particular amino acid; the third position does not matter. For example, the amino acid proline has four different codons: CCA, CCU, CCC, and CCG. Thus, a change in the third position for these codons will still specify the correct amino acid. Also, since there are 64 possible codons in the genetic code (four different bases taken three at a time is $4^3 = 64$) but only 20 amino acids, all but one amino acid have more than one codon. (There are even three different "stop" codons: UAA, UAG, and UGA.) If a substitution occurred but the change was to another codon that specifies the *same* amino acid, the mutation would have no effect on the resulting protein.

QUESTIONS

MULTIPLE CHOICE

1. Which diagram illustrates the correct structure of a segment of a DNA molecule? A. 1 B. 2 C. 3 D. 4

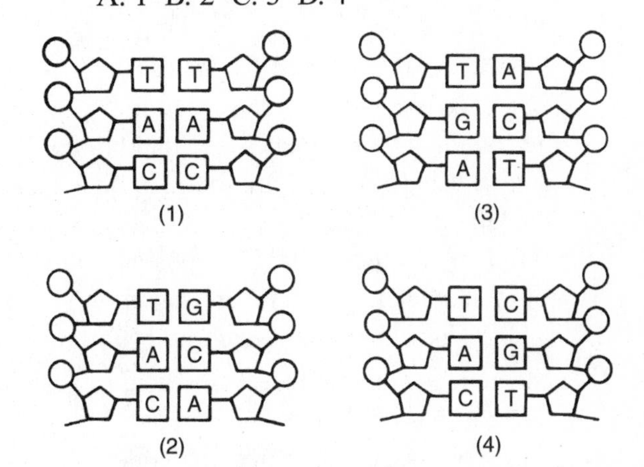

2. DNA and RNA molecules are similar in that they both contain A. nucleotides B. a double helix C. deoxyribose sugars D. thymine

3. Which series is arranged in correct order according to *decreasing* size of structures? A. DNA, nucleus, chromosome, nucleotide, nitrogenous base B. nucleotide, chromosome,

nitrogenous base, nucleus, DNA C. nucleus, chromosome, DNA, nucleotide, nitrogenous base D. chromosome, nucleus, nitrogenous base, nucleotide, DNA

4. Which substances are components of a DNA nucleotide? A. phosphate, deoxyribose, and uracil B. phosphate, ribose, and adenine C. thymine, deoxyribose, and phosphate D. ribose, phosphate, and uracil

5. Which two bases are present in equal amounts in a double-stranded DNA molecule? A. cytosine and thymine B. adenine and thymine C. adenine and uracil D. cytosine and uracil

6. Which terms describe gene activities that ensure the stability of life processes and continuity of hereditary material? A. oxidation and hydrolysis B. enzyme synthesis and DNA replication C. oxygen transport and cyclosis D. pinocytosis and dehydration synthesis

7. In humans, a gene mutation results from a change in the A. sequence of the nitrogenous bases in DNA B. chromosome number in a sperm cell C. chromosome number in an egg cell D. sequence of the sugars and phosphates in DNA.

8. Which set of statements correctly describes the relationship between the terms *chromosomes*, *genes*, and *nuclei*? A. Chromosomes are found on genes. Genes are found in nuclei. B. Chromosomes are found in nuclei. Nuclei are found in genes. C. Genes are found on chromosomes. Chromosomes are found in nuclei. D. Genes are found in nuclei. Nuclei are found in chromosomes.

9. The genetic code for one amino acid molecule consists of A. five sugar molecules B. two phosphates C. three nucleotides D. four hydrogen bonds

10. During the replication of a DNA molecule, separation of the DNA molecule will normally occur when hydrogen bonds are broken between A. thymine and thymine B. guanine and uracil C. adenine and cytosine D. cytosine and guanine

11. In the diagram below, what substance is represented by the letter *X*? A. ribose B. deoxyribose C. phosphate D. adenine

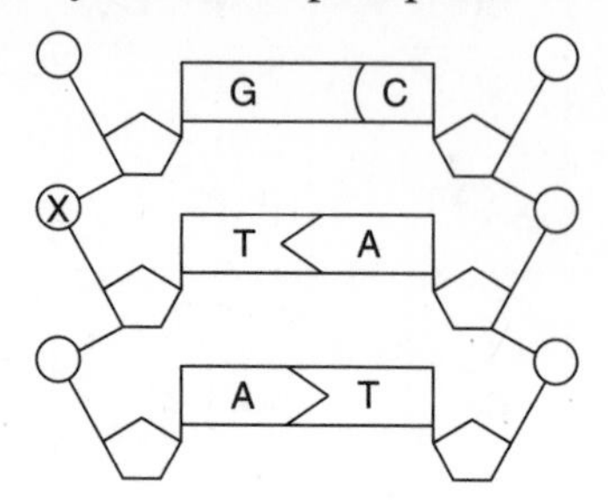

Base your answers to questions 12 through 16 on the following diagram, which represents the process of protein synthesis in a typical cell.

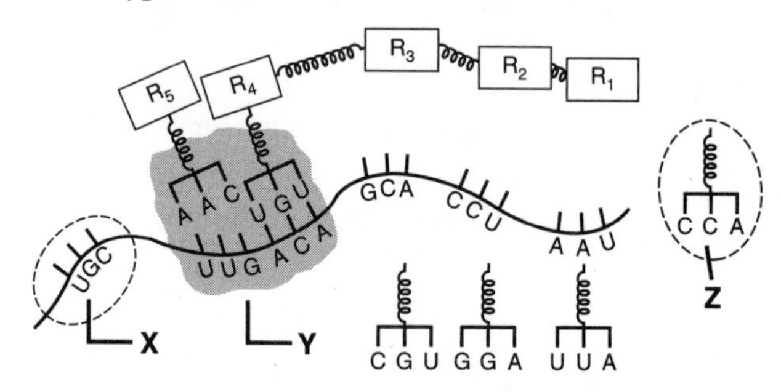

12. The original template for this process is a molecule of A. DNA B. messenger RNA C. transfer RNA D. ribosomal RNA

13. The units labeled R_1, R_2, and R_3 represent A. nucleotides B. RNA molecules C. DNA molecules D. amino acids

14. The organelle labeled *Y*, on which this process occurs, is the A. nucleus B. ribosome C. chloroplast D. mitochondria

15. The circled portion labeled *X* is known as a (an) A. amino acid B. codon C. anticodon D. single nucleotide

16. The circled portion labeled *Z* represents a molecule of A. DNA B. messenger RNA C. transfer RNA D. ribosomal RNA

OPEN RESPONSE

17. Briefly describe the two important functions of DNA.

18. Why is DNA replication critical to the survival of organisms?

Base your answers to questions 19 and 20 on the information and chart below and on your knowledge of biology.

In DNA, a sequence of three bases is a code for the placement of a certain amino acid in a protein chain. The table below shows eight amino acids with their abbreviations and DNA codes.

Amino Acid	*Abbreviation*	*DNA Code*
Phenylalanine	Phe	AAA, AAG
Tryptophan	Try	ACC
Serine	Ser	AGA, AGG, AGT, AGC, TCA, TCG
Valine	Val	CAA, CAG, CAT, CAC
Proline	Pro	GGA, GGG, GGT, GGC
Glutamine	Glu	GTT, GTC
Threonine	Thr	TGA, TGG, TGT, TGC
Asparagine	Asp	TTA, TTG

19. Which amino acid chain would be produced by the following DNA base sequence?
C-A-A-G-T-T-A-A-A-T-T-A-T-T-G-T-G-A

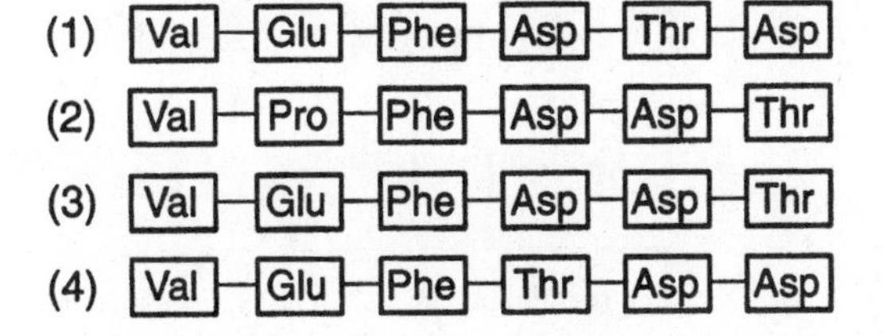

20. Describe how a protein would be changed if a base sequence mutated from GGA to TGA.

BIOTECHNOLOGY

Cloning

The term **cloning** describes the process by which a group of genetically identical offspring is produced from the cells of an organism. The cloning of plants shows great promise for agriculture. Plants with desirable qualities—such as drought and pest resistance—can be produced rapidly from the cells of a single plant. The cloning of animals also has importance for the farming (livestock) industry. It has been achieved in such diverse vertebrates as frogs, mice, sheep, goats, horses, cows, cats, dogs, and monkeys.

Genetic Engineering

Gene splicing, or **genetic engineering**, involves the transfer of genetic material from one organism to another. This **recombination** of genes results in the formation of what is known as *recombinant DNA*. Using gene-splicing techniques, or **biotechnology**, genes from one organism can be inserted into the DNA of another organism. Human genes that control the synthesis of insulin, interferon, and growth hormone have been introduced into bacterial cells, where they function as part of the bacterial DNA. In this way, bacterial cells are used being to synthesize certain substances needed by humans. Genetic engineering may eventually be able to correct some genetic defects and produce commercially desirable plants and animals.

Techniques of Genetic Engineering. The technique of making recombinant DNA (rDNA) molecules involves three important components.

First, a specific enzyme is needed to cut the DNA from the donor genes at a specific site. This enzyme is called a *restriction enzyme*. The enzyme is used to cut out a piece of DNA that contains one or more desired genes from the donor's DNA.

Next, a *vector* is needed to receive the donor DNA. Most frequently, a naturally occurring circular piece of bacterial DNA, called a *plasmid*, is used for this purpose.

Finally, an enzyme is used to "stitch" the donor DNA into the plasmid vector. This enzyme is called *ligase*, and it forms bonds between the donor DNA and the plasmid DNA. The result is that the donor DNA is incorporated into the bacterial plasmid, forming the recombinant DNA (rDNA).

It is important that the donor and the plasmid DNA be cut with the same restriction enzyme. Since each enzyme cuts DNA at only a very specific site, the two different DNAs will have matching cut ends known as *sticky ends*. The nitrogenous bases exposed at these cut sites can then match up according to the base-pairing rules, A to T and G to C (Figure 6-8).

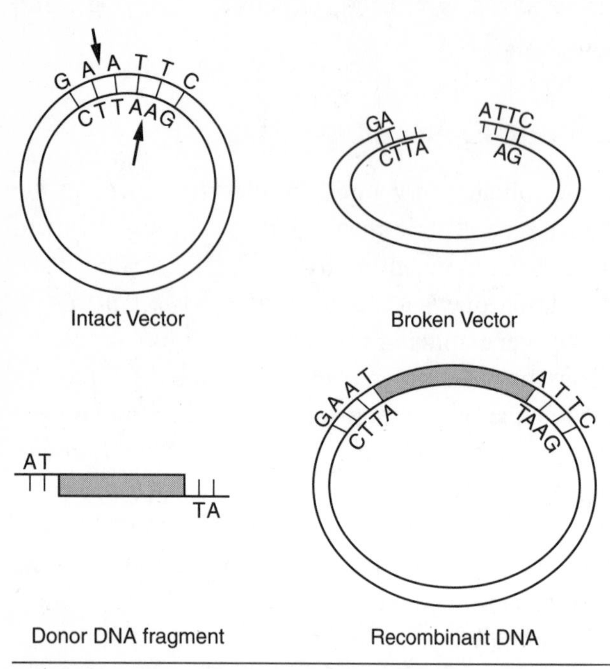

Figure 6-8. Use of a restriction enzyme, plasmid vector, and donor DNA to form recombinant DNA.

The rDNA is then inserted into bacteria. When these bacteria reproduce, they copy the rDNA plasmid along with their own DNA. The plasmid is copied thousands of times, forming a *clone* (a colony having identical genetic material).

In addition to copying the plasmid along with their other DNA, the bacteria *express* the genes that the plasmid carries, including the donor genes. As they reproduce, the bacteria continue to code for production of the desired protein. In this way, the bacteria can produce human proteins because they carry the genes with the instructions. This technique has made it possible to produce many chemicals that are needed by people who cannot produce them, due to genetic disorders. Two human proteins that have been successfully synthesized by rDNA techniques are the hormone insulin and human growth hormone.

Electrophoresis

We have already learned that DNA molecules can be cut with specific enzymes known as *restriction enzymes.* These enzymes cut the DNA molecules at highly specific sites that have a certain sequence of bases, such as AAAGGG. Different restriction enzymes have specific restriction sequences that they recognize. Each time the enzyme encounters its unique restriction site, it cuts the DNA molecule between a phosphate and a sugar subunit in the backbone. If DNA is incubated with a specific restriction enzyme, the molecule will be cut into many fragments of varying sizes, depending on where the restriction site is located in the DNA molecule.

DNA fragments can be separated according to size because they are electrically charged. The phosphate group at the end of a DNA fragment carries a negative charge; thus, it will be attracted to an area with a positive charge. To separate DNA fragments by means of **electrophoresis**, a small chamber or box is connected to an electrical source. DNA samples are loaded into small wells within a medium, called a *gel,* which is prepared so that an electric current can pass through it. When the current is switched on, the DNA fragments begin to move through the gel in response to the electrical field that is created. The fragments move away from the wells (negative end) toward the opposite side (positive end) of the box (Figure 6-9).

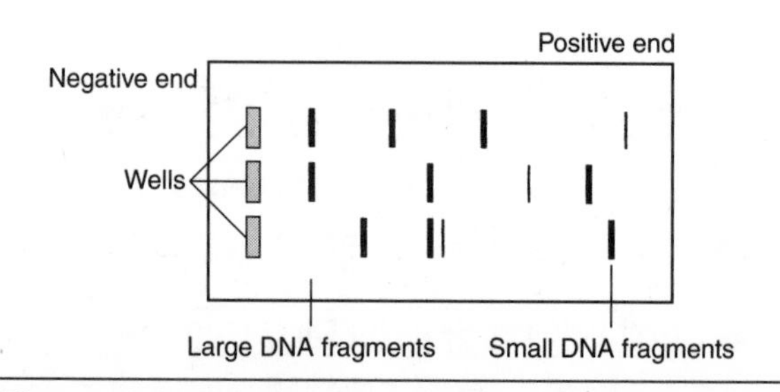

Figure 6-9. Gel electrophoresis: The smaller DNA fragments move farthest from the wells.

The size of a DNA fragment depends on the number of base pairs it contains. The more base pairs, the larger and heavier the fragment. Heavier DNA fragments cannot move as far as smaller DNA fragments can within an electrical field. Thus, the smaller fragments will move the farthest from the wells. The result is a series of bands of DNA running from the wells to the other end of the gel. These DNA fragments are made visible to the human eye by means of a simple staining technique, a special chemical, and/or ultraviolet light.

Electrophoresis is useful in determining the sizes of DNA fragments as well as in comparing DNA from two or more different sources. If two different DNA samples (from different individuals) are cut using the same restriction enzyme and then run on a gel in separate wells, we can determine how closely matched the two samples are. If the DNA samples are very similar, they will produce similar-sized fragments that will line up next to each other on the gel. If the two samples are very different, they will produce many different-sized fragments and relatively few will match up. If DNA samples are run from identical twins, all the fragments should match up, since the DNA in these individuals is identical.

Gel electrophoresis has been extremely helpful in law enforcement and forensics, too. Using this technique, often referred to as "DNA fingerprinting," scientists have been able to match DNA collected at a crime scene with the DNA gathered from a suspect. Biologists also use electrophoresis to determine current and evolutionary relationships between living things.

QUESTIONS

MULTIPLE CHOICE

21. The formation of recombinant DNA results from the A. addition of messenger RNA molecules to an organism B. transfer of genes from one organism to another C. substitution of a ribose sugar for a deoxyribose sugar D. production of a polyploid condition by a mutagenic agent

22. The replication of a double-stranded DNA molecule begins when the strands "unzip" at the A. phosphate bonds B. ribose molecules C. deoxyribose molecules D. hydrogen bonds between the bases

23. Cloning an organism usually produces other organisms that A. contain dangerous mutations B. contain identical genes C. are identical in behavior D. produce completely different enzymes

24. The following diagram represents a section of a molecule that carries genetic information. The pattern of numbers represents A. a sequence of paired bases B. the order of proteins in a gene C. folds of an amino acid D. positions of gene mutations

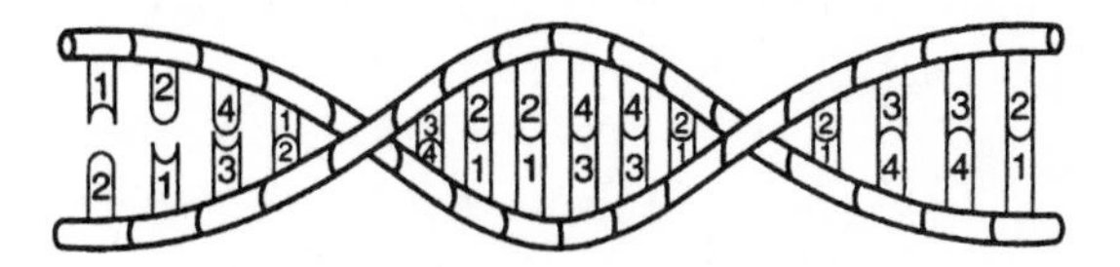

25. Enzymes are used to transfer sections of DNA that code for insulin from the pancreas cells of humans into a certain type of bacterial cell. This bacterial cell will reproduce, giving rise to new cells that can produce A. human insulin B. antibodies against insulin C. enzymes that digest insulin D. a new type of insulin

26. In the human pancreas, acinar cells produce digestive enzymes and beta cells produce insulin. The best explanation for this is that A. a mutation occurs in the beta cells to produce insulin when the sugar level increases in the blood B. different parts of an individual's DNA are used to direct the synthesis of different proteins in different types of cells C. lowered sugar levels cause the production of insulin in acinar cells to help maintain homeostasis D. the genes in acinar cells came from one parent, while the genes in beta cells came from the other parent

27. A gene that codes for resistance to glyphosate, a biodegradable weed killer, has been inserted into certain plants. As a result, these plants will be more likely to A. produce chemicals that kill weeds growing near them B. die when exposed to glyphosate C. convert glyphosate into fertilizer D. survive when glyphosate is applied to them

28. Gel electrophoresis is used to separate DNA fragments on the basis of their A. size B. color C. functions D. chromosomes

29. Which process can be used to rapidly produce a group of genetically identical plants from the cells of a single plant? A. screening B. karyotyping C. gene splicing D. cloning

To answer questions 30 through 33, select from the list below the type of nucleic acid that is best described by the phrase, then match the letter with the description. (Note: *There are only three answer choices for each question.*)

A. DNA
B. Messenger RNA
C. Transfer RNA

30. Genetic material responsible for the traits of an organism, that is passed from parent to offspring A. *A* B. *B* C. *C*

31. Carries genetic information from the cell nucleus out to the ribosomes A. *A* B. *B* C. *C*

32. Contains thymine instead of uracil A. *A* B. *B* C. *C*

33. Carries amino acid molecules to the ribosomes in the cytoplasm A. *A* B. *B* C. *C*

OPEN RESPONSE

34. Explain the role of each of the following items in making recombinant DNA: *restriction enzymes*, *plasmids*, and *ligase*.

35. How are the techniques of genetic engineering making it possible to treat some diseases caused by genetic disorders? Provide an example.

36. The following is a scrambled list of the techniques used in making recombinant DNA. Write these steps in the correct sequence and, for each step, explain why it is placed in that order.

Step
Cut open plasmid with restriction enzyme.
Obtain synthesized protein from the bacteria.
Clone bacterial cells with rDNA plasmids.
Insert donor DNA into the open plasmid.
Cut out donor DNA with restriction enzyme.
Add ligase to bond donor DNA and plasmid.

37. Animal cells utilize many different proteins. Discuss the synthesis of proteins in an animal cell. Your answer must include:

- the identity of the building blocks required to synthesize these proteins;
- the identity of the sites in the cell where the proteins are assembled;
- an explanation of the role of DNA in the process of making proteins in the cell.

READING COMPREHENSION

Base your answers to questions 38 through 41 on the information below and on your knowledge of biology. Source: *Science News* (July 9, 2005): Vol. 168, No. 2, p. 19.

Same Difference: Twins' Gene Regulation Isn't Identical

Although identical twins have identical DNA, they often harbor clear-cut differences: slight variations in appearance or stark distinctions in disease susceptibility, for example. Scientists have suggested that the interplay between nature and nurture could explain such differences, but the mechanism has been poorly understood.

A new study suggests that as identical twins go through life, environmental influences differently affect which genes are turned on and which are switched off.

Called epigenetic modification, such gene activation or silencing typically stems from two types of chemical groups that latch on to chromosomes as charms attach to a bracelet, says Manel Esteller of the Spanish National Cancer Centre in Madrid. Methyl groups that clip on to DNA tend to turn genes off. On the other hand, acetyl groups attaching to histones, the chemical core of chromosomes, usually turn genes on.

Suspecting that such epigenetic differences might account for variations between identical twins, Esteller and his team focused on the two chemical changes. The scientists recruited 80 pairs of identical twins, ranging in age from 3 to 74, from Spain, Denmark, and the United Kingdom.

After extracting DNA from blood, inner-cheek cells, and biopsied muscle, Esteller's team screened the twins' genomes for differences in epigenetic profiles between members of a pair. The researchers also had each twin, or for children a parent, answer a comprehensive questionnaire on the twins' health history and lifestyle, including diet, exercise habits, and alcohol or tobacco use.

In the youngest twins, the scientists found relatively few epigenetic differences. However, the number of differences increased with the age of the twins examined. The number of epigenetic differences in 50-year-old twins was more than triple that in 3-year-old twins. Esteller's group also saw especially large epigenetic differences between twins who had spent most of their lifetimes apart, such as those adopted by different sets of parents at birth, the team reports in an upcoming *Proceedings of the National Academy of Sciences*.

Esteller says that these results suggest that a person's environment—whether he or she is exposed to tobacco smoke, eats particular foods, or suffers an emotionally wrenching event, for example—may affect which genes are turned on or off and so how cells operate. Thus, nurture may have a heavy impact on an individual's nature.

"My belief is that people are 50 percent genetics and 50 percent environment," says Esteller. "It's important to remember that our genes give us features of who we are, but our environment can change how we are."

Arturas Petronis, who studies epigenetics at the Centre for Addiction and Mental Health in Toronto, agrees. He adds that the findings could also have wide-ranging health implications for people who aren't twins.

"About 90 percent of diseases don't follow [simple] rules for inheritance," says Petronis. "By investigating epigenetic changes, to some extent we can understand how environmental factors affect human health."

38. How can environmental influences cause genetic differences between identical twins?

39. Compare the functions of the two DNA modifiers: methyl groups and acetyl groups.

40. Describe the trend that researchers found between the age of the twins and the number of epigenetic differences in their DNA. What could account for this trend?

41. Why does the researcher say his "belief is that people are 50 percent genetics and 50 percent environment"?

CHAPTER 7

Evolution and Classification

Standard 2.3 (Cell Biology)

BROAD CONCEPT: Cells have specific structures and functions that make them distinctive. Processes in a cell can be classified broadly as growth, maintenance, and reproduction.

Standards 5.1, 5.2, 5.3 (Evolution and Biodiversity)

BROAD CONCEPT: Evolution is the result of genetic changes that occur in constantly changing environments. Over many generations, changes in the genetic make-up of populations may affect the biodiversity through speciation and extinction.

Evolution is the process of change over time. The theory of evolution suggests that existing forms of life on Earth have evolved from earlier forms over long periods of time. These earlier forms were usually very different from the related organisms living today. Evolution accounts for the differences in structure, function, and behavior among all life-forms, as well as for the changes that occur within populations over many generations. The process of evolution also helps to explain the great similarities that exist among all life-forms, in spite of their apparent differences. In fact, the theory of evolution is considered to be the great "unifying concept" in the understanding of biology.

EVIDENCE OF EVOLUTION

Observations that support the theory of evolution have been obtained from the study of the geologic record and from studies of comparative anatomy, embryology, cytology (cell structure), and biochemistry. It is important to understand that the word "theory" in science has a very different meaning from the same word in everyday language. A scientific theory is a concept that is supported by much documented evidence and experimental research. Theories are based on the culmination of years of laboratory work, field observation, data gathering, extensive testing, and analysis in order to validate results.

Geologic Record

Geologists estimate the age of Earth to be approximately 4.5 billion years old. This estimate is based on *radioactive dating* (from the decay rates of radioactive isotopes) of the oldest known rocks from Earth's crust. (It is assumed that Earth is at least as old as the oldest rocks and minerals in its crust.)

In studying the geology of the planet, scientists have found many **fossils**, the remains or traces of organisms that no longer exist. From their studies of rocks and fossils, scientists have developed a picture of the changes that have occurred both in Earth itself and in living things on the planet.

Fossils

The earliest known fossils are traces of bacteria-like organisms that are about 3.5 billion years old. (The age of these fossils was determined by radioactive dating of the rocks in which they were found.)

Fossils of relatively intact organisms have been found preserved in ice, tar, and amber (a sticky plant resin that hardens). Mineralized bones, shells, teeth, and other hard parts of ancient organisms are sometimes found intact. (The soft parts generally decay within a short time.)

Some fossils have been formed by *petrification*, a process in which the tissues are gradually replaced by dissolved minerals that produce a stone replica of the original material. Imprints, casts, and molds of organisms or parts of organisms are frequently found in *sedimentary* rock. This type of rock is formed from the deposition of thick layers of soft sediments that eventually harden and turn to rock from the weight of overlying sediments and water. The fossils form when the remains of dead organisms settle to the bottom of a body of water and are quickly covered by sediment. The overlying sediment slows or halts decay. When the layers of sediment harden, traces of the buried organisms are preserved in the rock. This method of fossilization is one way that organisms lacking hard parts have been preserved.

In studying undisturbed sedimentary rock, scientists assume that each layer is older than all the layers, or *strata*, above it. This is called *relative dating*; fossils in the lower strata are older than fossils in the overlying strata. There is often a resemblance between fossils in the upper and lower strata, which suggests a link between the recent and older life-forms (Figure 7-1). The fossil record may also provide evidence of diverging pathways of some organisms from a common ancestor. It is important to note that, even though there is a general trend toward increasing complexity over time, it is not a rule of evolution. There is no predetermined direction that evolution must take, either toward or away from more complex forms of life.

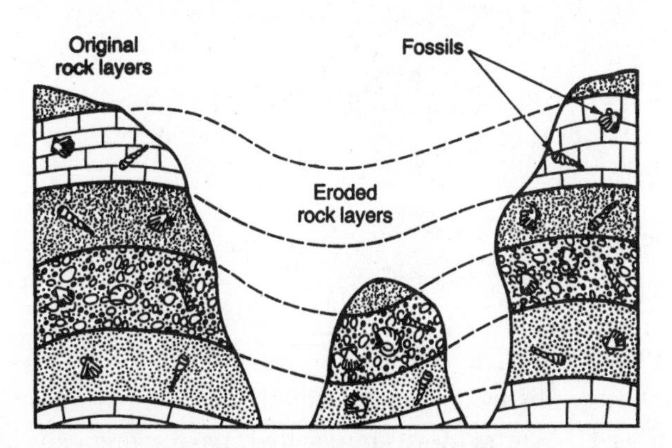

Figure 7-1. A resemblance between fossils in the upper and lower rock strata often indicates an evolutionary link between recent and older life-forms.

Some fossils in older strata are unlike any other kinds of organisms, or *species*, living today. This suggests that many previous species have died out, or gone *extinct*, over time; **extinction** means there are no longer any living members of the species. Other fossils have structures that show ancestral connections to present-day life-forms. There are also fossils that are quite similar to living organisms, suggesting that some species have existed for a long time without much evolutionary change. Fossils of the horseshoe crab look just like those living today; the animal has remained virtually unchanged for over 400 million years.

Comparative Anatomy

Another line of evidence for evolution comes from observations of basic structural, or anatomical, similarities between various organisms. *Homologous structures* are anatomical parts found in different organisms that are similar in origin and structure, although they may differ in function. In other words, the structures arose from a related structure in a common ancestor. For example, the flippers of whales, the wings of bats, the forelimbs of cats, and the arms of humans are homologous structures; they serve different functions, but their basic bone structures are similar because they all evolved from a related structure (Figure 7-2). The presence of such homologous structures suggests that these mammals all evolved from a common (land-dwelling) ancestor.

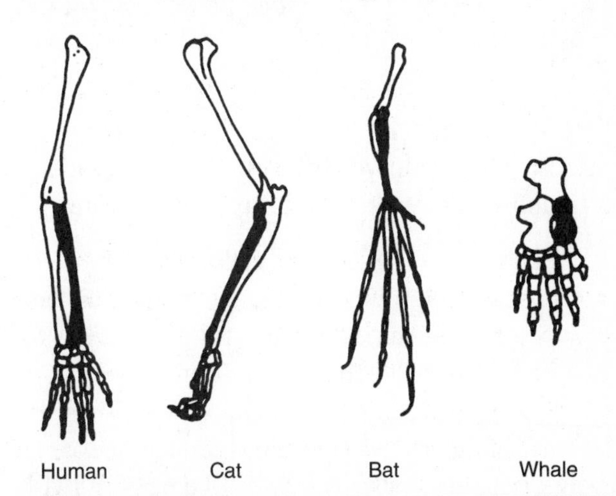

Figure 7-2. The presence of homologous structures in their limb bones suggests that these mammals all evolved from a common ancestor.

Comparative Embryology

Although adult organisms of different species may look very different from one another, a comparison of their early stages of embryonic development may show similarities that suggest a common ancestry. For example, the very early embryos of such vertebrates as fish, reptiles, birds, and mammals show some similarities in structure, such as having a tail. These structures develop in embryos because of common ancestry, but their final fate depends on the particular organism. In other words, as embryonic development continues, the characteristic traits of each species become more apparent (Figure 7-3).

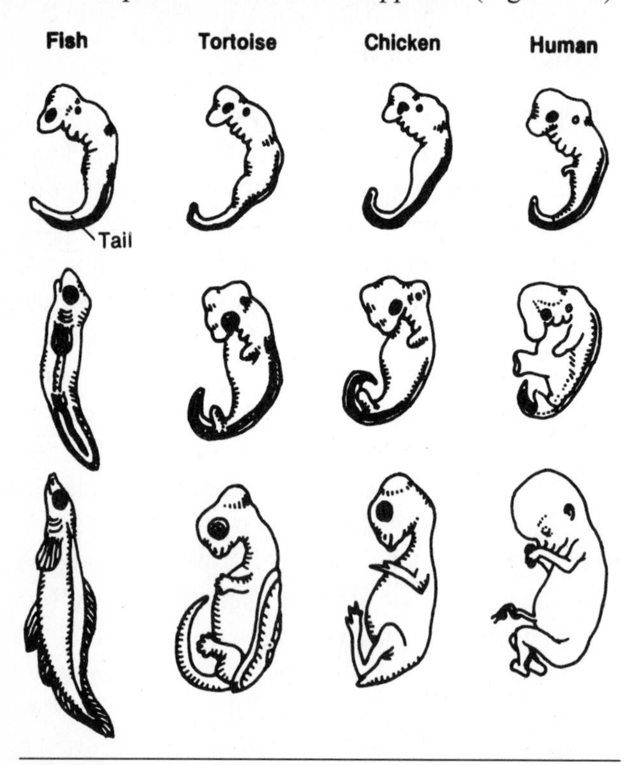

Figure 7-3. Similar features, such as a tail, in the early stages of embryonic development point to a common ancestry for these different vertebrates.

In humans, a passageway called the *Eustachian tube* connects the ear canals and the throat. This tube evolved from the same structure that gives rise to the gills in fish; and gill slits are seen in the very early stages of human embryonic development. Thus, the presence of this ancient structure is strong evidence for evolutionary relationships between humans and fish. Another example is the tail bone, or *coccyx*, in humans. Although we lack an external tail, we still develop a tail-like appendage during early embryonic stages.

Comparative Cytology

As stated in the *cell theory*, all living things are made up of cells. Cell organelles, including the cell membrane, ribosomes, and mitochondria, are structurally and functionally similar in most living organisms. The structure, or *morphology*, of nearly all animal cells is similar, just as the structure of all plant cells is similar. In bacteria, however, there are differences in cell structure. Recall that bacteria are prokaryotic, whereas animals, plants, protists, and fungi are eukaryotic. In addition, the structure of some organelles, such as ribosomes, are somewhat different in bacteria from those of eukaryotic cells. This would suggest a more distant relationship between bacteria and other life-forms. Overall, however, the similarities are greater than the differences.

Comparative Biochemistry

All living things contain similar biochemical compounds. For example, the structure and function of DNA, RNA, and proteins (including enzymes) are similar in all organisms. The closer the relationship between any two organisms, the greater are their biochemical and genetic similarities (Figure 7-4). Today, biochemical relationships can be shown by means of DNA comparisons. If two organisms are closely related, their DNA samples will show very similar nucleotide sequences. Similarly, the proteins (such as enzymes) of the two organisms would be very similar, since a major function of DNA is the synthesis of proteins.

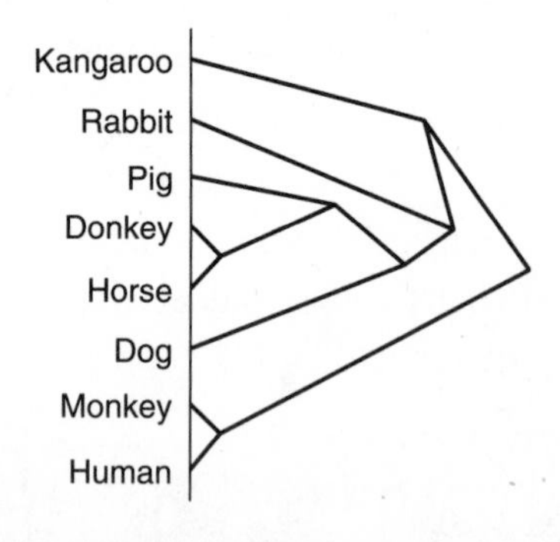

Figure 7-4. The evolutionary relationship between two organisms can be determined by comparing their DNA; the more similar their base sequences, the more recently they evolved from a common ancestor. For example, the donkey and the horse are more closely related than are the pig and the horse.

Vestigial Structures

Some organisms possess body structures that have no known function; yet similar structures in related organisms do have definite functions. These features are called *vestigial* (meaning "lost") *structures* and they indicate strong evolutionary relationships. In humans, a number of such structures exist. For example, the appendix—a small sac found in the intestines—has no function in humans. However, the rabbit has a similar but larger pouch called the *cecum*, which plays a role in the digestion of plant matter. Also, other mammals have muscles that are used to move their ears; similar ear muscles in humans are largely nonfunctional. Finally, muscles in the skin give people the "goose bumps" when they are frightened or cold; in other mammals they function to make their hair or fur stand up when they are threatened or cold. There are vestigial structures in other groups of animals, too. For example, the skeletons of some types of snakes have tiny, vestigial leg bones, indicating that snakes evolved from ancestors that had functional legs (Figure 7-5).

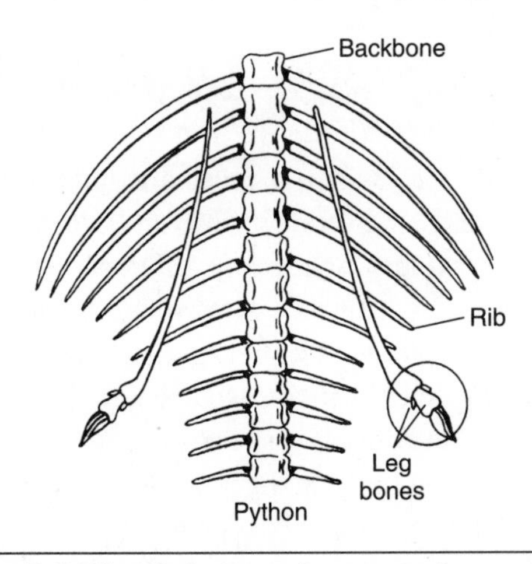

Figure 7-5. The skeletons of some snake species have tiny leg bones that serve no current function. These vestigial structures indicate that snakes evolved from animals that had functional legs.

QUESTIONS

MULTIPLE CHOICE

1. The wings of a bat and the front legs of a horse are examples of structures that are A. heterotrophic B. homozygous C. hermaphroditic D. homologous

2. Which conclusion may be made when comparing fossils in undisturbed strata of sedimentary rock? A. The fossils in the upper strata are younger than those in the lower strata. B. The fossils in the upper strata are older than those in the lower strata. C. The fossils in the upper strata are generally less complex than those in the lower strata. D. There are no fossils in the upper strata that resemble those in the lower strata.

3. The similarity between the blood proteins of all mammals may be taken as evidence of evolutionary relationships based on A. comparative anatomy B. geographic distribution C. comparative embryology D. comparative biochemistry

4. The diagram below represents a section of undisturbed rock and the general location of fossils of several closely related species. According to current theory, which assumption is probably the most correct concerning species *A*, *B*, *C*, and *D*? A. *A* is most likely the ancestor of *B*, *C*, and *D*. B. *B* was extinct when *C* evolved. C. *C* evolved more recently than *A*, *B*, and *D*. D. *D* was probably the ancestor of *A*, *B*, and *C*.

Species C & D
Species C
Species A & B & C
Species A & B
Species A

5. Which assumption is a basis for the use of fossils as evidence for evolution? A. Fossils show a complete record of the evolution of all animals. B. In undisturbed layers of Earth's surface, the oldest fossils are found in the lowest layers. C. Fossils are always found deep in volcanic rocks. D. All fossils were formed at the same time.

6. Many related organisms are found to have the same kinds of enzymes. This suggests that A. enzymes work only on specific substrates B. enzymes act as catalysts in biochemical reactions C. organisms living in the same environment require identical enzymes D. these organisms probably share a common ancestry

7. Which is an example of evidence of evolution based on comparative biochemistry? A. Sheep insulin can be substituted for human insulin. B. The structure of a whale's flipper is similar to that of a human hand. C. Human embryos have a tail during an early stage of their development. D. Both birds and bats have wings.

8. The presence of gill slits in early-stage human embryos is considered to be evidence of the A. likelihood that all vertebrates share a common ancestry B. theory that the first organisms on Earth were heterotrophs C. close relationship between fish and human reproductive patterns D. close relationship between humans and amphibians

9. The DNA in a chimpanzee's cells are shown, through gel electrophoresis, to be similar to the DNA in a human's cells. This is an example of which type of evidence supporting the theory of evolution? A. comparative habitat B. comparative anatomy C. comparative embryology D. comparative biochemistry

OPEN RESPONSE

10. Describe the five main types of evidence used to support the theory of evolution.

11. The diagram below represents a cross section of undisturbed rock layers. A scientist discovers bones of a complex vertebrate species in layers *B* and *C*. In which layer would an earlier, less complex form of this vertebrate most likely first appear? Explain the reasons for your answer.

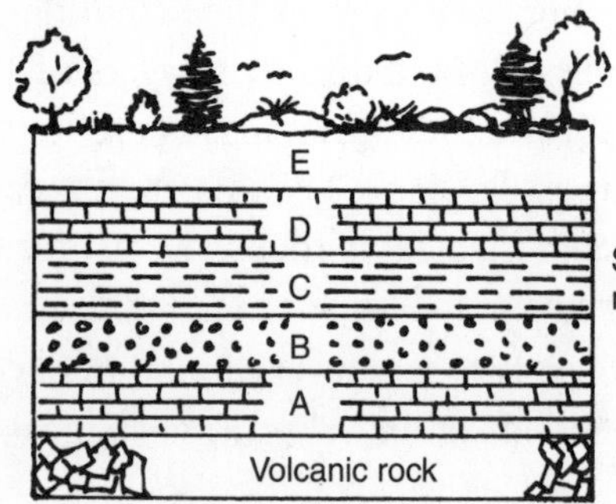

12. *R*, *S*, and *T* are three species of birds. Species *S* and *T* show similar coloration. The enzymes found in species *R* and *T* show similarities. Species *R* and *T* also exhibit many of the same behavioral patterns. Show the relationship between species *R*, *S*, and *T* by placing the letter representing each species at the top of the appropriate branch on the diagram below.

13. Frogs and lizards are very different animals, yet they have many similarities. Provide a brief explanation for why frogs and lizards have many characteristics in common.

EVOLUTION AND THE ORIGIN OF LIFE

Any theory of evolution must attempt to explain the origin and biodiversity of life on Earth. The term **biodiversity** refers to the many different forms of life that exist, each with its own distinct genetic traits. Such biodiversity is essential to a balance in nature. Evolutionary theory must also account for the wide variety of **adaptations**—special characteristics that aid survival—found among both living and extinct species. Different theories may account for different aspects of the evolutionary process. However, taken together, they can explain how life on Earth came to be and how it has progressed from the relatively simple and few to the very complex and diverse.

The Heterotroph Hypothesis

The *heterotroph hypothesis* is one proposed explanation of how life arose and evolved on the primitive Earth. According to this hypothesis, the first life-forms were heterotrophic and therefore had to obtain organic nutrients from their environment. Most evidence suggests that the earliest cellular life-forms were very similar to the bacteria living today, especially those in the Kingdom Archaebacteria.

The Primitive Earth

It is assumed that during the period preceding the development of the first life-forms, the primitive Earth was an extremely hot and inhospitable planet, consisting of inorganic substances in solid, liquid, and gaseous states. Evidence for this idea comes from the conditions known to exist on other planets in our solar system, such as Mars and Venus.

The **atmosphere** of primitive Earth is thought to have had no free oxygen; instead, it consisted of hydrogen (H_2), ammonia (NH_3), methane (CH_4), and water vapor (H_2O). As Earth cooled, much of the water vapor condensed and fell as rain, which carried dissolved atmospheric gases (ammonia, methane, and hydrogen) into the seas that formed. The seas became rich in these dissolved substances and minerals and are often described by biologists as having been a "hot, thin soup."

The primitive Earth provided an energy-rich environment. In addition to the heat, there was electrical energy in the form of lightning, **radiation** (x-rays and ultraviolet rays) from the sun, and radioactivity from rocks.

Synthesis of Organic Compounds

The large amount of available energy was the driving force for synthesis reactions on the primitive Earth. In these reactions, the inorganic raw materials in the seas became chemically bonded to form organic molecules, including simple sugars and amino acids. These organic molecules were the building blocks for the first life-forms.

Scientists Stanley Miller and Harold Urey devised an apparatus in which they simulated the conditions thought to exist in the primitive environment. Their experiments showed that in the presence of heat and electrical energy, dissolved gases could combine to form simple organic compounds.

Formation of Aggregates

In time, simple organic molecules accumulated in the seas. Eventually, they combined in synthesis reactions to form more complex organic molecules. (Such interactions between organic molecules have been demonstrated in laboratories.) Some of the large, complex molecules formed groupings, or clusters, called *aggregates*. These aggregates developed a membrane that enclosed them, thus forming a barrier between themselves and the surrounding water. This made it possible for the substances inside an aggregate to remain separate from those in the surrounding water. It is thought that aggregates absorbed simple organic molecules from the environment for "food." Thus, they carried on a form of heterotrophic nutrition. Over time, the aggregates became more complex and highly organized. Eventually, they developed the ability to reproduce. At that point, when their ability to reproduce had evolved, the aggregates are considered to have been living cells.

Heterotrophs to Autotrophs

It is thought that these early heterotrophic life-forms carried on a form of anaerobic respiration, or *fermentation* (in which glucose is converted to energy and CO_2 without O_2 being present). As a result of very long periods of fermentation, carbon dioxide was added to the atmosphere. Eventually, as a result of evolution, some heterotrophic forms developed the capacity to use carbon dioxide from the atmosphere in the synthesis of organic compounds. These organisms became the first *autotrophs* (meaning "self-feeders"). Some bacteria are autotrophs, but most of the autotrophs alive today are green plants and algae.

Anaerobes to Aerobes

Autotrophic activity (photosynthesis) added oxygen molecules to the atmosphere. Over time, the capacity to use free oxygen in respiration (aerobic respiration) evolved in both autotrophs and heterotrophs.

There are both autotrophs and heterotrophs on Earth today. Some life-forms still carry on anaerobic respiration; but in most life-forms, respiration is aerobic. This is because aerobic respiration releases much more energy from food than does anaerobic respiration.

QUESTIONS

MULTIPLE CHOICE

14. According to the heterotroph hypothesis, the first living things probably were anaerobic because their environment had no available A. food B. energy C. water D. oxygen

15. Which is one basic assumption of the heterotroph hypothesis? A. More complex organisms appeared before less complex organisms. B. Living organisms did not appear until there was oxygen in the atmosphere. C. Large autotrophic organisms appeared before small photosynthesizing organisms. D. Autotrophic activity added oxygen molecules to the environment.

16. The heterotroph hypothesis is an attempt to explain A. how Earth was originally formed B. why simple organisms usually evolve into complex organisms C. why evolution occurs very slowly D. how life originated on Earth

17. The heterotroph hypothesis states that heterotrophic life-forms appeared before autotrophic forms as the first living things. A major assumption for this hypothesis is that A. sufficient heat was not available in the beginning for the food-making process B. the heterotrophic organisms were able to use molecules from the sea as "food" C. lightning and radiation energy were limited to terrestrial areas D. moisture in liquid form was limited to aquatic areas

Base your answer to the following question on the chart below and on your knowledge of biology.

A	B	C
The diversity of multicellular organisms increases.	Simple, single-celled organisms appear.	Multicellular organisms begin to evolve.

18. According to most scientists, which sequence best represents the order of biological evolution on Earth? A. $A \rightarrow B \rightarrow C$ B. $B \rightarrow C \rightarrow A$ C. $B \rightarrow A \rightarrow C$ D. $C \rightarrow A \rightarrow B$

OPEN RESPONSE

19. Identify the source of oxygen in Earth's early atmosphere; explain why this was important later to the evolution of life.

EVOLUTION BY NATURAL SELECTION

Darwin's Theory of Natural Selection

One of the most well-known theories that explains how living things have evolved is the *theory of evolution by means of natural selection*, proposed by Charles Darwin in the mid-1800s. Darwin's theory was based on the presence of variations among members of a population and their interaction with the process he called **natural selection**. The heart of Darwin's theory is that those organisms with the most favorable adaptations to their environment are the most likely to survive, reproduce, and pass those adaptations on to their offspring. Those with less favorable traits will be selected against, and so probably will not survive to reproduce. Darwin's theory of evolution by means of natural selection includes the following four main ideas:

Overpopulation: Within a population, there are more offspring produced in each generation than can possibly survive. In general, species tend to produce more offspring than can be supported by the environment in which they live.

Competition: The natural resources, such as food, water, and space, available to a population are limited. Because there are more organisms produced in each generation than can survive, there must be *competition* among them for the resources needed for survival.

Survival of the fittest: Variations among members of a population make some of them better adapted to the environment, or to changes in it, than others. Such **variability** within populations means that, due to competition, the best-adapted individuals are most likely to survive. Thus the environment is the agent of natural selection, determining which variations are helpful and which are harmful. For example, in an environment that is undergoing a particularly cold period, animals that have thicker fur than most other members of their population are more likely to survive. In this case, their variation—thicker fur—makes them more fit to survive the environmental conditions.

Reproduction: Individuals with useful variations tend to survive and reproduce at a higher rate than other members of their population, thus transmitting these adaptations to their offspring. Likewise, those individuals that do not have such favorable adaptations tend to die off within the population; so the less favorable traits are not passed on to future generations.

The development of new species, a process called *speciation*, occurs as certain variations, or adaptations, accumulate in a population over many generations. Speciation is also a major driving force in increasing biodiversity, that is, as more species evolve, the level of biodiversity in a community goes up as well.

According to Darwin's theory, environmental pressures act as a force for the natural selection of the best-adapted individuals in a population—those with helpful adaptations that enable them to survive and reproduce successfully. However, Darwin's theory did not explain *how* variations arise among the members of a population. (*Note:* In Darwin's time, chromosomes, genes, and DNA had not been discovered yet; thus the scientific study of mutations had not begun.) It is important to understand that *individuals* do not evolve; rather it is the *population* that evolves as the percentage of changes in its gene pool (all of its members' genetic traits) increases over time.

QUESTIONS

MULTIPLE CHOICE

20. Darwin's theory of evolution did *not* contain the concept that A. genetic variations are produced by mutations and sexual recombination B. organisms that are best adapted to their environment survive C. population sizes are limited due to the struggle for survival D. favorable traits are passed from one generation to the next

21. Natural selection is best defined as A. survival of the strongest organisms only B. elimination of the smallest organisms by the largest organisms C. survival of those organisms best adapted to their environment D. reproduction of those organisms that occupy the largest area in an environment

22. Although similar in many respects, two species of organisms exhibit differences that make each one well adapted to the environment in which it lives. The process of change that helps account for these differences is A. evolution by natural selection B. parthenogenesis C. comparative embryology D. inheritance of acquired traits

23. A key idea in Darwin's theory of evolution is that members of a population A. are always identical B. compete for limited resources in the environment C. all get to reproduce and pass on their traits D. are all equally well adapted to the environment

24. Which characteristics of a population would most likely indicate the lowest potential for evolutionary change in that population? A. sexual reproduction and few mutations B. sexual reproduction and many mutations C. asexual reproduction and few mutations D. asexual reproduction and many mutations

25. The theory of biological evolution includes the concept that A. species of organisms found on Earth today have adaptations not always found in earlier species B. fossils are the remains of present-day species and were all formed at the same time C. individuals may acquire physical characteristics after birth and pass these acquired characteristics on to their offspring D. the smallest organisms are always eliminated by the larger organisms within the ecosystem

26. The graph on page 144 shows the populations of two species of ants. Ants of species 2 have a thicker outer covering than the ants of species 1. The outer covering helps prevent excessive evaporation of water. Which statement would best explain the population changes shown in the graph?

A. The food sources for species 1 increased while the food sources for species 2 decreased from January through November. B. Disease killed off species 1 beginning in May. C. The weather was hotter and dryer than normal from April through September. D. Mutations occurred from April through September in both species, resulting in their both becoming better adapted to the environment.

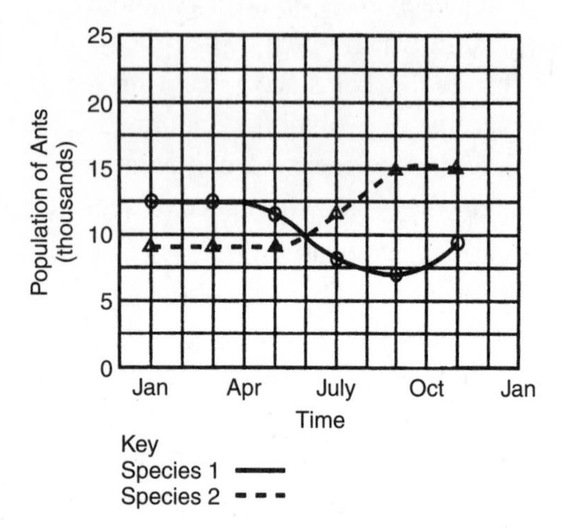

27. When a particular white moth lands on a white birch tree, its light color is beneficial, or *adaptive*, for survival. If the birch trees become covered with black soot, the white color of this particular moth in this environment would most likely A. remain just as adaptive for survival B. become more adaptive for survival C. change to black to be more adaptive for survival D. become less adaptive for survival

28. The fact that a light color is beneficial to a moth that lands on a white tree is most directly related to Darwin's ideas about A. overpopulation B. natural selection C. speciation D. reproduction

29. In order for new species to develop, there *must* be a change in the A. temperature of the environment B. migration patterns within a population C. accumulation of new traits in a population D. rate of succession in the environment

30. The following diagram shows the evolution of some different species of flowers. Which statement about the species is correct? A. Species *A*, *B*, *C*, and *D* came from different ancestors. B. Species *C* evolved from species *B*. C. Species *A*, *B*, and *C* can interbreed successfully. D. Species *A* became extinct.

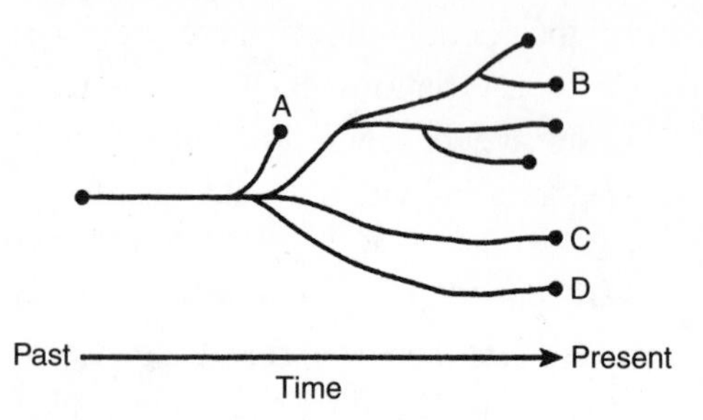

OPEN RESPONSE

31. Identify the four main ideas that make up Darwin's theory of evolution by natural selection.

32. Explain what is meant by "the environment is the agent of natural selection."

33. What is the relationship between speciation and the level of biodiversity in a community?

MODERN EVOLUTIONARY THEORY

The modern theory of evolution includes both Darwin's ideas of variation and natural selection and the genetic basis of variations within populations.

Sources of Genetic Variations

Variations within a population result from two kinds of genetic events. First, recombination of alleles during sexual reproduction is a source of variations. (Recall that sexually reproducing species produce offspring with many variations as a result of the recombination of alleles during gamete formation and fertilization.) Second, random and spontaneous gene and chromosome mutations produce genetic variations. Mutations may arise spontaneously in organisms, or they may be caused by exposure to *mutagenic* (mutation-causing) chemicals or radiation, such as ultraviolet rays and x-rays. These variations provide the raw material for evolution within a population.

Natural Selection and Genetic Variation

Natural selection involves the struggle of organisms to survive and reproduce in a given environment. In order for natural selection to occur, the mutations in genes or chromosomes must have an effect on the *phenotype* (physical appearance, biochemistry, or behavior) of an organism. This is because natural selection works by "choosing" favorable traits that are *expressed*. As a result, individuals having favorable **genetic variations** are more likely to survive, reproduce, and pass those traits on to future generations. If a genetic change results in a *silent mutation*, it means the variation is not expressed in the phenotype. Thus natural selection cannot work on this type of mutation.

Favorable Variations. Favorable characteristics tend to increase in (genetic) frequency within a population. Favorable variations may include physical traits, such as larger muscles and increased speed, or behavioral traits, such as better food-finding or nest-building skills.

If environmental conditions change, traits that formerly had low survival value may come to have greater survival value. Likewise, traits that were favorable may no longer be so *adaptive*. For example, it is adaptive for mammals that live in a cold climate to have a fur covering. If the climate gets even colder, the individuals with thicker fur would have an advantage. If the climate gets warmer, those individuals would be at a disadvantage; the heavy fur would no longer be adaptive. The survival value of traits that had been neither helpful nor harmful may also change. In all of these cases, those traits that prove to be most favorable under the new environmental conditions will increase in frequency within the population (that is, those traits will come to have a higher *gene frequency*).

Unfavorable Variations. Unfavorable characteristics tend to decrease in frequency from generation to generation. Individuals with unfavorable or *non-adaptive* traits may be so severely selected against that, over time, populations that have unfavorable traits may become extinct. Indeed, the fossil record shows that extinction is a fairly common event, having been the fate of about 99 percent of all species that have ever existed on Earth.

Geographic Isolation

Changes in gene frequencies that lead to the development of a new species are more likely to occur in small populations than in large ones. Small groups may be segregated from the main population by a geographic barrier, such as a body of water or a mountain range. As a result of this *geographic isolation*, the small population cannot interbreed with the larger, main population. In time, the isolated population may evolve into a new species.

The following factors may be involved in the evolution of a new species: (a) the gene frequencies in the isolated population may already have been different from the gene frequencies in the main population; (b) different mutations occur in the isolated population and the main population; and (c) different environmental factors exert different selection pressures on each population. Since there is no interbreeding between the two populations, any mutations that occur in one population cannot be transmitted to the other. Over a long period of time, the two populations may become so different that they can no longer interbreed even if direct contact is made. In such a case, two new species have evolved from the one. An example of this *divergent evolution* is seen in the two populations of Grand Canyon squirrels: the Kaibab and the Abert. These two squirrel populations were originally members of one species that became separated, over time, by the formation of the canyon. As a result of natural selection, the divided populations evolved to be the two different species that exist today.

Darwin observed the effect of geographic isolation among several finches he collected on the Galápagos Islands. Darwin hypothesized that the different species he observed had evolved from a single species that had originally migrated to the islands from the mainland of South America. Over time, the different environments on the islands had gradually resulted in the evolution of new, separate species. This process, by which several populations (each a different species adapted to a different environmental condition) evolve from one original population (the parent species), is known as *adaptive radiation* (Figure 7-6).

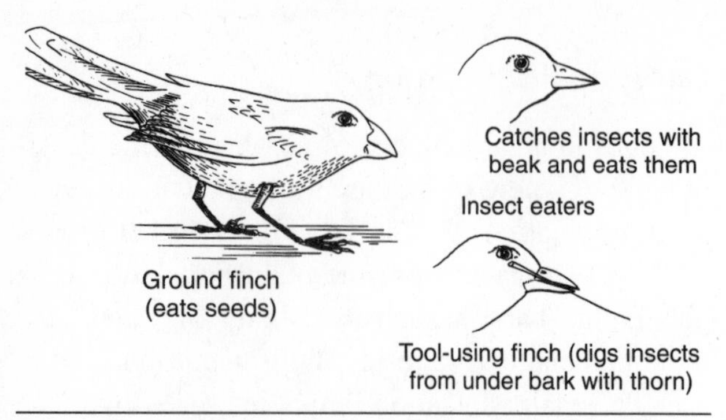

Figure 7-6. These Galápagos finches show a variety of adaptations for getting food in their particular environments.

Reproductive Isolation

Geographic isolation can eventually lead to *reproductive isolation*. The isolated population becomes so different from the main population that members of the two groups can no longer interbreed, even if the geographic barriers were removed. A **species** is defined as a population whose members can successfully interbreed and produce fertile offspring. When two populations can no longer do that, they have become two distinct species; speciation has occurred. One or more *new* species has evolved, increasing the biodiversity in the area. A population of squirrels in a forest can interbreed; and their offspring are also capable of reproducing. The squirrels, therefore, are considered members of the same species. But when a horse and a donkey are mated, they produce a *mule*, which is sterile. The horse and the donkey, although closely related, are not in the same species. However, there are exceptions to this rule. For example, captive lions and tigers can sometimes mate and produce fertile offspring; but the hybrid male offspring are usually sterile.

Time Frame for Evolution

Although scientists generally agree on the basic factors involved in evolutionary change, there is some disagreement about the time frame in which such change occurs.

According to Darwin's original theory, evolutionary change occurs very gradually and continuously over the course of **geologic time** (millions of years). This theory, called *gradualism*, proposes that new species develop as a result of the gradual accumulation of small genetic variations that eventually, together, cause reproductive isolation and lead to speciation.

The more recent theory of *punctuated equilibrium* proposes that most species have long periods (several million years) of relative stability, or stasis, interrupted by geologically brief periods during which major changes occur, possibly leading to the evolution of new species (Figure 7-7). In this way, drastic environmental changes, for example, a global cooling event, could cause species to evolve—or become extinct.

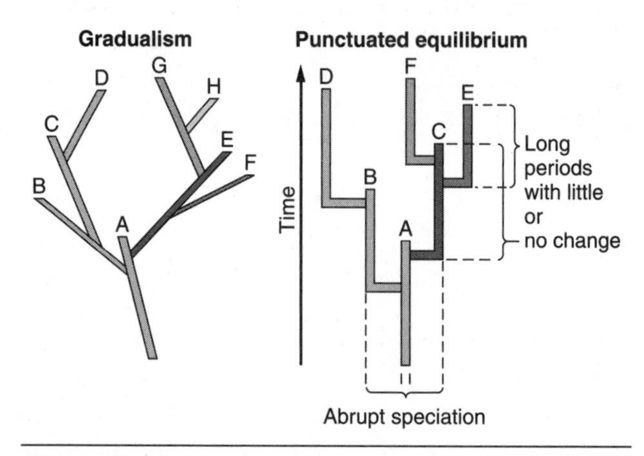

Figure 7-7. Two evolutionary processes: gradualism and punctuated equilibrium.

In the fossil records of some evolutionary branches, or *lineages*, there are transitional forms (such as feathered dinosaurs) that support the theory of gradualism. However, in many evolutionary lineages, there is an apparent lack of transitional forms, which better supports the theory of punctuated equilibrium.

Effect of Humans on Natural Selection

It has been found that some insects have a genetic mutation that makes them resistant to the effects of some insecticides (special **pesticides** developed to kill insect pests). Before the widespread use of insecticides, this trait was of no particular survival value. With the increased use of insecticides, however, this trait developed a very high survival value. Because the insects that are resistant to insecticides have survived and reproduced, the frequency of insecticide *resistance* has increased greatly in insect populations. Resistance to **antibiotics** (drugs that fight bacterial infections) in populations of bacteria has followed the same pattern. The frequency of resistant individuals in bacterial populations has increased with the increasing use of certain antibiotics.

It is important to note that resistance to insecticides and antibiotics did not arise as a result of exposure to these substances. These traits were already present in some members of the organisms' populations, and the insecticides and antibiotics simply acted as the selecting agents. Individuals susceptible to these chemicals died, leaving the resistant ones to survive, reproduce, and pass on the genes for resistance to their offspring. These are examples of *survival of the fittest*.

Humans and Artificial Selection

There is no particular direction in which each species must evolve. Over time, many variations appear within the populations of organisms, similar to the branching of twigs on a tree. Natural selection continuously "prunes" these branches, eliminating those with unfavorable adaptations while letting those with favorable adaptations survive.

In contrast, humans can and do have an effect on the inheritance of traits in some populations of organisms. For example, human actions, such as pesticide use, have led to unexpected changes in the genetic makeup of some insect populations. Modern humans have intentionally altered the traits of many plant and animal species, as well. In the process of domesticating organisms, people have carried out **artificial selection** by intentionally breeding particular plants and animals for desired traits (Figure 7-8). In such cases, it is people, not the environment, who are the selecting agents. Advances in **biotechnology**, or bioengineering techniques, have also had an effect on the genetic traits of some plant and animal populations (Figure 7-8).

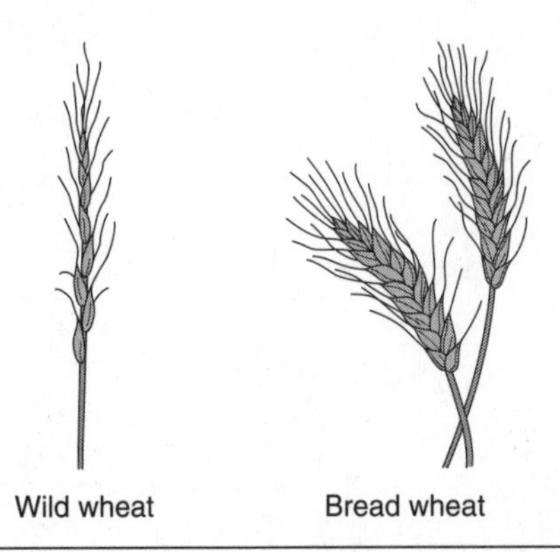

Figure 7-8. As a result of centuries of selective breeding, the small kernels of wild wheat have been transformed into the large kernels of bread wheat—a more useful crop for people.

It is now possible to introduce specific genes into the chromosomes of individual plants and animals in an attempt to "improve" them for human interests. This is an area of biotechnology that has generated much controversy, since it is not known how these genetic changes will affect future generations. Of particular concern is the introduction of genetically altered organisms into the human food supply.

QUESTIONS

MULTIPLE CHOICE

34. A population of mosquitoes is sprayed with a new insecticide. Most of the mosquitoes are killed, but a few survive. In the next generation, the spraying continues, but even more mosquitoes hatch that are immune to the insecticide. How could these results be explained according to the present concept of evolution? A. The insecticide caused a mutation in the mosquitoes. B. The mosquitoes learned how to fight the insecticide. C. A few mosquitoes in the first population were resistant and transmitted this resistance to their offspring. D. The insecticide caused the mosquitoes to develop an immune response, which was inherited.

Refer to the following set of diagrams, which illustrates the forelimb bones of three different mammals, to answer question 35.

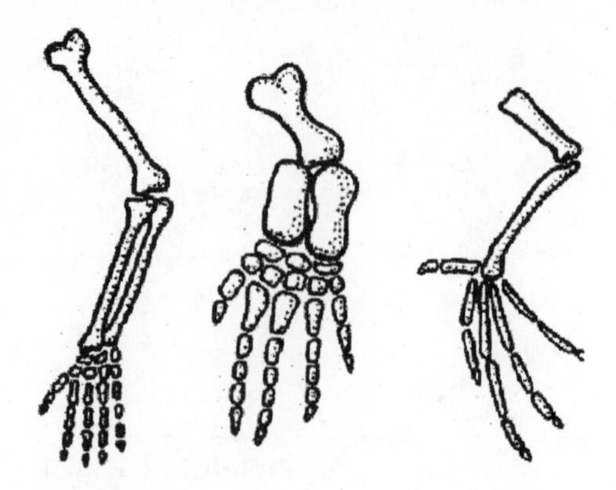

35. Differences in the bone arrangements support the hypothesis that these animals
A. are probably members of the same species
B. have adaptations for different environments
C. most likely have no ancestors in common
D. all contain the same genetic information

36. Two organisms can be considered to be two different species if they A. cannot mate with each other and produce fertile offspring
B. live in two different geographical areas
C. mutate at different rates depending on their environment D. have genes drawn from the same gene pool

37. Certain strains of bacteria that were susceptible to penicillin have now become resistant. The probable explanation for this is that A. the gene mutation rate must have increased naturally
B. the strains have become resistant because they needed to do so for survival C. a mutation that gave some of them resistance was passed on to succeeding generations because it had high survival value D. the penicillin influenced the bacterial pattern of mating

38. The continents of Africa and South America were once a single landmass but have drifted apart over millions of years. The monkeys of both continents, although similar, show several genetic differences from each other. Which factor is probably the most important for causing and maintaining these differences?
A. fossil records B. comparative anatomy
C. use and disuse D. geographic isolation

39. A change in the frequency of any mutant allele in a population most likely depends on the A. size of the organisms possessing the mutant allele B. adaptive value of the trait associated with the mutant allele C. degree of dominance of the mutant allele D. degree of recessiveness of the mutant allele

Base your answer to question 40 on the data in the following paragraph and map.

Thousands of years ago, a large flock of hawks was driven from its normal migratory route by a storm. The birds scattered and found shelter on two distant islands, shown on the map below. The environment of Island A is very similar to the hawk's original nesting region. But, the environment of Island B is very different from that of Island A. The hawks have survived on these two islands to the present day with no interbreeding between the two populations.

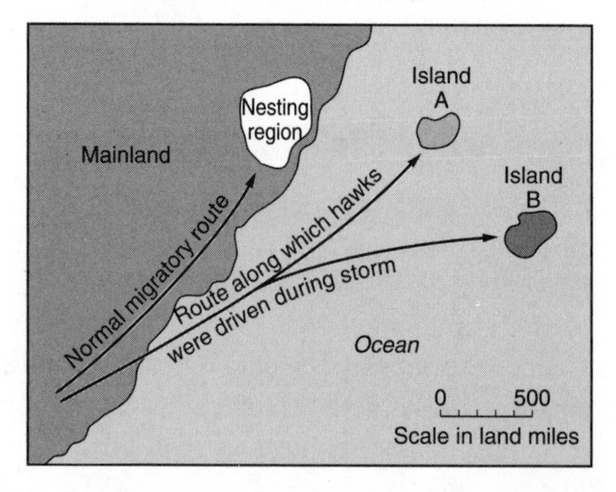

40. Which statement most likely predicts the present-day condition of these two hawk populations? A. The hawks that adapted to Island B have changed more than those on Island A. B. The hawk populations on Islands A and B have undergone identical mutations. C. The hawks that landed on Island A have evolved more than those on Island B. D. The hawks on Island *A* have given rise to many new species of hawks.

41. As a result of sexual reproduction, the potential for evolutionary change in plants and animals is greatly increased because
A. the offspring show more variability than

those from asexual reproduction B. characteristics change less frequently than in asexual reproduction C. environmental changes never affect organisms produced by asexual reproduction D. two parents have fewer offspring than one parent

42. Populations of a species may develop traits that are different from each other if they are geographically isolated for sufficient lengths of time. The most likely explanation for these differences is that A. acquired traits cannot be inherited by the offspring B. the environmental conditions in the two areas are identical C. mutations and selective forces will be different in the two populations D. mutations will be the same in both populations

OPEN RESPONSE

43. State how genetic variations and natural selection in a population can lead to the evolution of a new species.

44. How did Darwin explain the evolution of new finch species from one ancestral finch species? Use the terms *geographic isolation* and *reproductive isolation* in your answer.

Base your answers to questions 45 through 48 on the following information and data table.

A population of snails was living on a sandy beach. The snails' shells appeared in two colors: tan or black. The sand on the beach was a tan color. One day, a volcano in a nearby mountain range erupted, spewing out tons of ash and debris. The ash and debris coated the sand on the beach, blackening it. Biologists had kept careful records of the snail population before and after the volcanic eruption. Their data are presented in the table below.

Time	*Number of Tan Snails*	*Number of Black Snails*
Before volcano erupted	6000	50
After volcano erupted (one year later)	400	3000

45. Explain why the numbers of tan snails and black snails changed.

46. How does this event support the idea of evolution by natural selection?

47. Give one reason why the tan snails might disappear within a few years.

48. Using the data in the preceding table, prepare a bar graph that shows the information on snail populations before and after the volcanic eruption. Include all labels.

49. A species of wildflower grows in a meadow. The flowers are of two color varieties: yellow and purple. There are about equal numbers of yellow flowers and purple flowers. A biologist observes that bees frequently visit the yellow flowers but seldom go to the purple ones.

    Use the above data and your knowledge of biology to write a brief experimental procedure that addresses the following:

    - a question prompted by the information given;
    - a hypothesis that addresses your question;
    - a brief experimental procedure that could be used to test your hypothesis;
    - a description of the main selecting force on the flowers in this meadow;
    - a prediction of what may happen to this population of wildflowers in 50 years.

50. Describe how the continued widespread use of antibiotics may result in the evolution of more resistant strains of bacteria. How does the antibiotic act as a selecting agent? How does this illustrate the concept of natural selection?

CLASSIFICATION

Hundreds of years ago, scientists classified organisms as belonging to either the plant kindom or the animal kingdom. Today, biologists recognize six kingdoms into which all living and extinct organisms can be classified; these are the Archaebacteria, Eubacteria, Protista, Fungi, Plantae, and Animalia kingdoms. Some scientists also classify organisms within three major groups called *domains*, which are based on similarities at the cellular level; these are the Bacteria, Archaea, and Eukaryotes.

The Kingdoms of Life

The **classification** of organisms, called *taxonomy*, is based on several important factors, including cell structure, mode of nutrition, biochemistry, body structure, reproduction and development, and evolutionary history. Organisms that are placed in the same kingdom share several very general characteristics with each other. The classification of organisms becomes more and more specific within each kingdom as life-forms that are more closely related are placed into increasingly smaller classification groups, or *taxa* (singular, *taxon*). Going from the largest to the smallest taxa, the seven main levels of classification are kingdom, phylum, class, order, family, genus, and species. The species level is the smallest and most narrowly defined taxon, with just one type of organism included (Figure 7-9). Scientists use a naming system called *binomial nomenclature* to identify each organism by its own genus and species names. These names are usually derived from Greek and Latin words that describe the organism, such as *Homo sapiens* for humans.

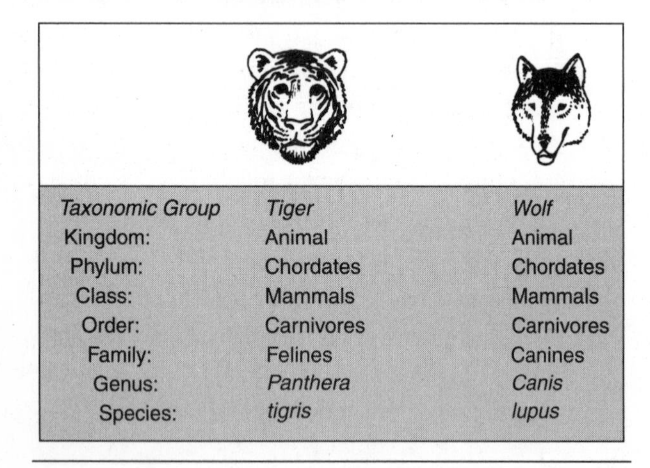

Taxonomic Group	*Tiger*	*Wolf*
Kingdom:	Animal	Animal
Phylum:	Chordates	Chordates
Class:	Mammals	Mammals
Order:	Carnivores	Carnivores
Family:	Felines	Canines
Genus:	*Panthera*	*Canis*
Species:	*tigris*	*lupus*

Figure 7-9. Taxonomic classification of two mammals, from kingdom (the largest group) to species (the smallest group).

Kingdom Archaebacteria

Previously placed in one large kingdom along with the Eubacteria, the *Archaebacteria* are now considered sufficiently distinct to be placed in their own kingdom. The Archaebacteria, which are unicellular prokaryotic organisms, have several unique characteristics. They differ from the Eubacteria in the following important ways: the biochemistry of their cell membranes and cell walls is completely different; their enzymes, genetics, RNA sequences, and reproduction are unique; and they live in extremely harsh environments (salty, acidic, or hot) that other bacteria could not tolerate. The Archaebacteria are thought to be among the first organisms to have evolved on Earth and hence are a very ancient group.

Kingdom Eubacteria

The *Eubacteria* are considered to be the "true" bacteria (also called *monerans*) and they are found nearly everywhere on Earth. These bacteria are also unicellular prokaryotic organisms; but they have a wide variety of metabolic adaptations. Some are heterotrophic (they must consume food), while others are autotrophic (able to synthesize their own food). Some of the autotrophic bacteria make food by a process called *chemosynthesis*, in which chemicals are broken down to provide the energy they need. Bacterial cells are usually rod, ball, or spiral shaped, depending on the species (Figure 7-10). Most Eubacteria have cell walls that contain a rigid substance called *peptidoglycan*. The DNA replication and transcription processes of Eubacteria are different from those of the Archaebacteria; and they reproduce asexually by means of binary fission (the parent cell splits into two daughter cells). Many Eubacteria are useful to humans; for example, bacteria belonging to the genus *Lactobacillus* convert milk into cheese and yogurt. However, many Eubacteria are harmful; they cause a variety of human diseases such as strep throat, tuberculosis, and leprosy. In recent years, bacteria have been widely used in biotechnology as hosts for transplanted genes from other organisms.

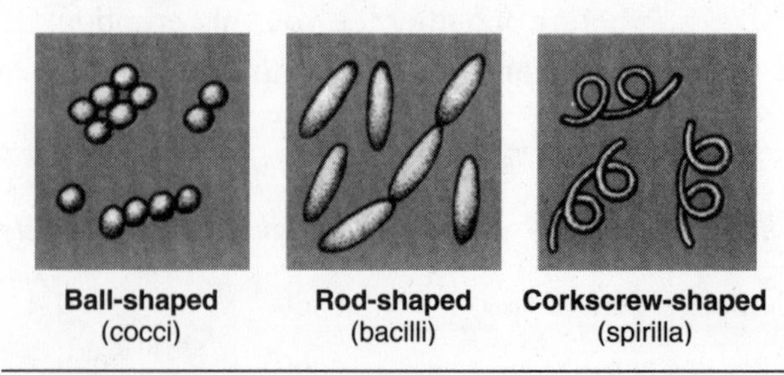

Figure 7-10. The three main shapes of the Eubacteria.

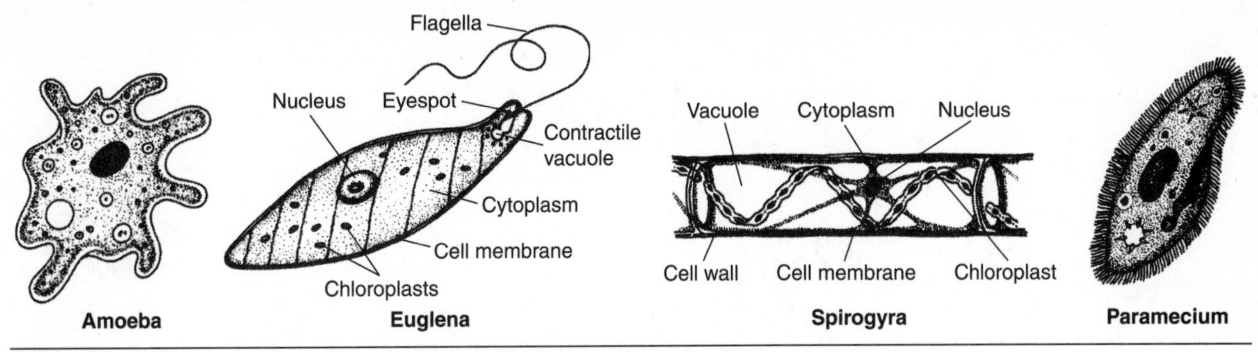

Figure 7-11. Some different types of protists.

Kingdom Protista

The *Protista* kingdom includes both unicellular and multicellular eukaryotic organisms. Recall that all eukaryotic organisms have membrane-bound organelles, such as nuclei, in their cells. Protists have a great diversity of cell types and include the following: plantlike organisms, such as algae, which are autotrophs; animal-like organisms called *protozoa*, such as the amoeba and paramecium, which are heterotrophs; and funguslike organisms, which are the slime molds (Figure 7-11). Euglena is an unusual plantlike protist that can be either autotrophic or heterotrophic, depending on environmental conditions. Protozoans are found in damp or aquatic environments and can be either free-living or parasitic; some have flagella or cilia for locomotion. A number of protozoa are important to humans because they cause diseases; for example, malaria and sleeping sickness are both caused by protozoans. Protists reproduce both sexually and asexually. Scientists think that protists gave rise to all other eukaryotic organisms.

Kingdom Fungi

Members of the *Fungi* kingdom are all multicellular eukaryotic organisms, except for yeasts, which are unicellular. Most fungi have cell walls composed of *chitin*, a nitrogen-containing polysaccharide. All fungi are heterotrophic; they absorb the nutrients they need by first digesting organic matter in the environment. Many fungi are **saprophytes**, organisms that obtain nutrients from dead or decaying plants and animals. As such, the fungi are important as agents of decomposition, breaking down the tissues of dead organisms and recycling the molecules back into the environment. Mushrooms, molds, mildews, and smuts are some of the members of this kingdom (Figure 7-12).

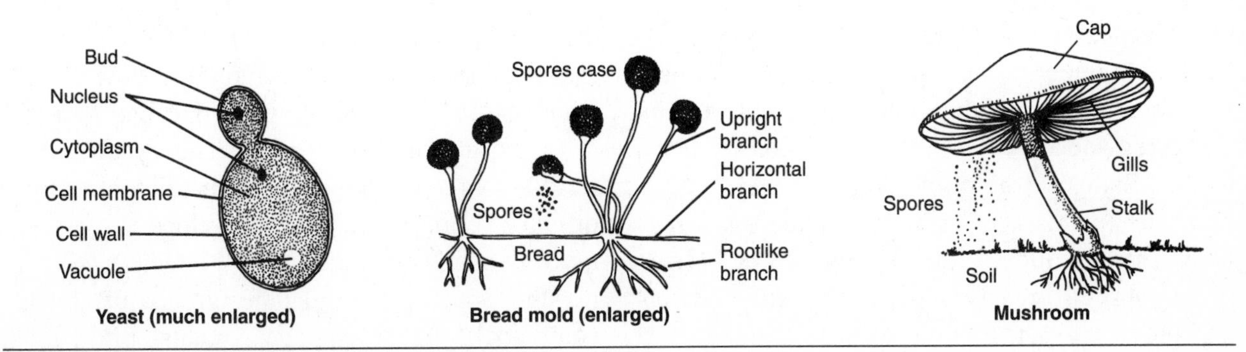

Figure 7-12. Some different types of fungi.

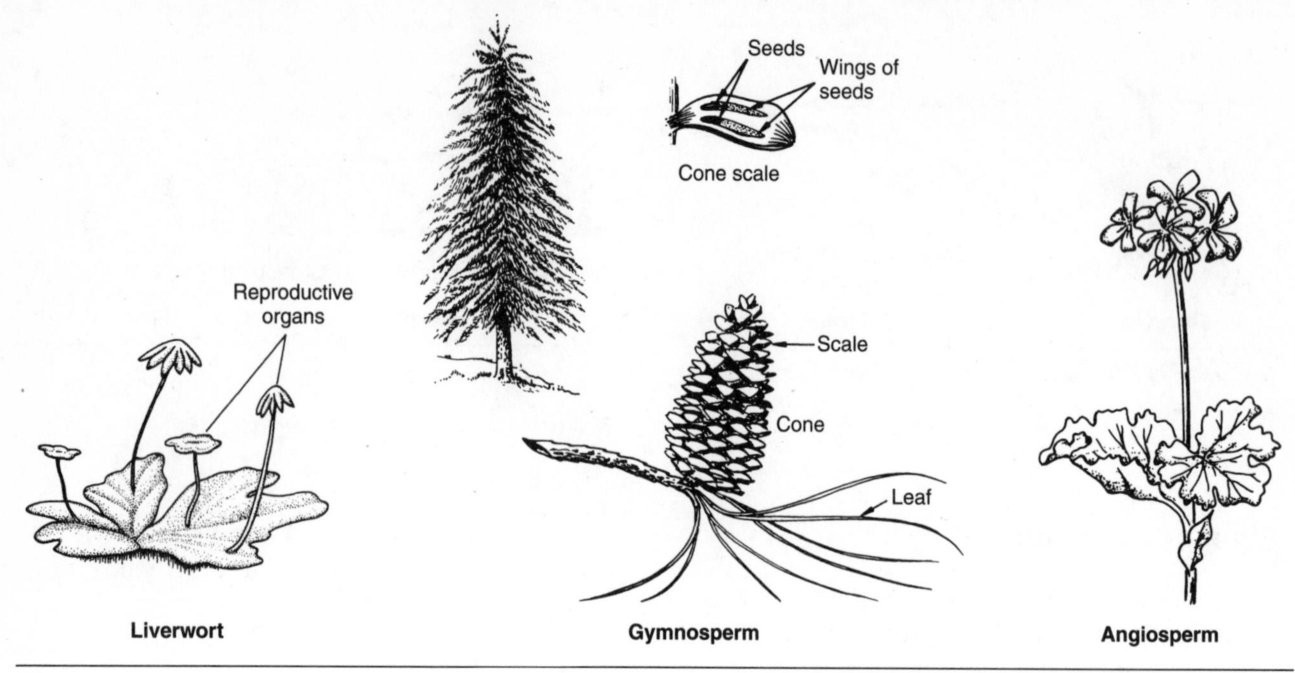

Figure 7-13. Some different types of plants.

Kingdom Plantae

The *Plantae* are all multicellular, eukaryotic autotrophic organisms. All plants carry out photosynthesis; to do so, they contain chlorophyll in chloroplasts. Plants are the producers for most other living things on Earth because, along with the algae, they are able to convert inorganic molecules into the organic molecules of food. The cell walls of plants consist mainly of cellulose (a polysaccharide). Most plants are terrestrial (land dwelling), but some live in water. Plants have a life cycle known as **alternation of generations**, which involves a sexually reproducing generation (gametophyte) that alternates with an asexually reproducing generation (sporophyte). Members of this kingdom range in size from the very tiny duckweed on ponds to the giant redwood trees on the west coast of the United States. In addition, some of the longest-lived organisms on Earth are plants. Examples of plants include mosses, liverworts, ferns, pinecone-bearing gymnosperms, and flower-bearing angiosperms (Figure 7-13).

Kingdom Animalia

The *Animalia* are all multicellular, eukaryotic heterotrophic organisms. Nearly all animals depend on plants, either directly or indirectly, for their food. Unlike plants, animals do not have cell walls around their cell membranes; and the great majority of animals are capable of locomotion. However, like plants, animals vary greatly in size, ranging from microscopic organisms such as rotifers to massive whales. Most animals have a digestive cavity that functions in the digestion and absorption of food. Most can reproduce only sexually, although some animals do reproduce asexually. For example, cnidarians (such as the jellyfish) have a life cycle that includes a polyp generation, which reproduces asexually. In sexual reproduction, the male and female sex cells fuse in the process called *fertilization*, and a zygote is formed. The **zygote** is the fertilized egg cell that contains the diploid number of chromosomes, and which develops into an embryo. Examples include insects (which have more member species than all other groups of animals combined), worms, sponges, corals, birds, fish, frogs, lizards, elephants, and humans (Figure 7-14).

Figure 7-14. Some different types of animals.

QUESTIONS

MULTIPLE CHOICE

51. Which of the following is the correct order of taxa, going from the smallest to the largest groups? A. phylum, kingdom, genus, class B. genus, species, phylum, class C. species, genus, phylum, kingdom D. phylum, class, genus, species

52. The scientific name for the red maple tree is *Acer rubrum*. This name includes its A. class and phylum B. family and species C. genus and species D. genus and order

53. Which kingdom includes organisms that have characteristics of both animals and plants? A. Plantae B. Fungi C. Animalia D. Protista

54. Except for yeasts, the members of Kingdom Fungi are all A. prokaryotic and unicellular B. prokaryotic and multicellular C. eukaryotic and unicellular D. eukaryotic and multicellular

55. Which kingdom contains autotrophic organisms whose life cycles include alternation of generations? A. Fungi B. Plantae C. Animalia D. Protista

56. The following diagrams represent embryos of three different vertebrate species. It is thought that they provide evidence of evolution based on their similar A. sizes B. fossils C. structures D. molecules

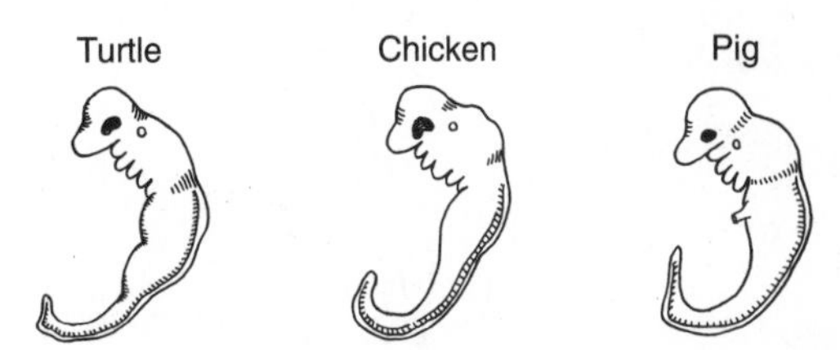

57. In most life-forms alive today, respiration is aerobic rather than anaerobic because aerobic respiration releases more A. oxygen B. carbon dioxide C. energy from food D. glucose molecules

58. Archaebacteria and Eubacteria differ from each other in A. their biochemistry and genetics B. the materials in their cell walls C. the type of environments they inhabit D. all of these characteristics

59. The best title for the chart below would be A. Evolutionary Pathways B. Evidence for Evolution C. Natural Selection D. Mutations in Evolution

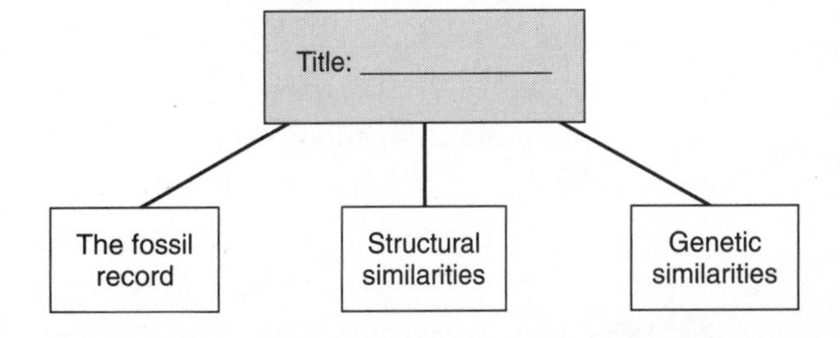

60. At the heart of Darwin's theory is the idea that organisms with the most favorable adaptations to their environment are most likely to

A. be selected against B. migrate the farthest C. survive to reproduce D. grow to the largest sizes

61. All the following groups include organisms that are producers *except* the A. algae B. protists C. fungi D. plants

62. According to the evolutionary "family tree" below, what was the most recent common ancestor of organisms 2 and 4? A. *A* B. *B* C. *C* D. *D*

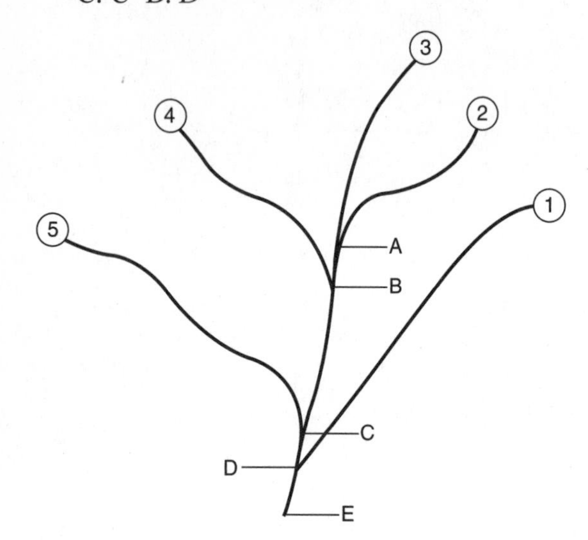

OPEN RESPONSE

63. Why is the evolution of living things considered to be an ongoing process?

64. What determines the "adaptive value" of a trait within a population? Give an example, either real or imagined.

65. Describe how a "family tree" type of diagram can be used to show evolutionary relationships among organisms.

66. Explain how studies of similarities in the biochemistry of proteins can be useful in determining evolutionary relationships among organisms.

READING COMPREHENSION

Base your answers to questions 67 through 70 on the information below and on your knowledge of biology. Source: *Science News* (May 21, 2005): Vol. 167, No. 21, p. 324.

New Mammals: Coincidence [and] Shopping Yield Two Species

After 21 years without a new kind of monkey being reported in Africa, two research teams working independently in different mountain ranges have described the same novel species. And other researchers, after poking through meat for sale in Southeast Asia, report a rodent that they say justifies a new family among mammals, the first in 31 years.

The monkey species, now called the highland mangabey or *Lophocebus kipunji*, has turned up at locations in southern Tanzania 370 kilometers apart. One discovery came from scientists' curiosity about stories around Mount Rungwe of an elusive monkey. In December 2003, Tim Davenport of Mbeya, Tanzania, who works for the New York–based Wildlife Conservation Society, and his team got a good-enough look to recognize it as a new species.

Meanwhile, ornithologists had told Trevor Jones at Udzungwa Mountains National Park in Tanzania that they had spotted sanje mangabeys, an endangered monkey, in a remote forest. But as soon as Jones saw one of the mangabeys there, he says that he knew the brownish color and high crest of hair were all wrong for a sanje mangabey. "I was immediately gobsmacked," he says.

Last October, in Dar es Salaam, one of Jones' colleagues happened to be in the same hotel as Davenport. Jones says that a conversation in the bar, with veiled hints of working on "something a bit special," escalated to revelations that both teams had found the same mangabey species. The Jones team withdrew a paper that it had previously submitted, and the two groups united to describe the new species in the May 20 *Science.*

There's "no question" about it being a new species, comments primate systematist Colin Groves of the Australian National University in Canberra. During the past 2 decades, monkey species not previously described by scientists have turned up only rarely, he says, and the finding of one in Africa surprised him. "The startling new discoveries have mainly been in Asia," he notes.

Loggers are cutting down the forests where the highland mangabey was found near Mount Rungwe, and Davenport calls for immediate protection of the monkey's already fragmented habitat.

The other new species announced this month, an unusual rodent, was reported by a team surveying biodiversity in the forests of Khammouan province in the Lao People's Democratic Republic. The Wildlife Conservation Society–sponsored survey included routine trips through village food markets. Starting in 1996, Robert J. Timmins and others occasionally bought what the local people call the *khanyou.* Dark fur covers a somewhat ratlike body about 25 centimeters long with a furry tail.

The biodiversity surveyors sent market specimens to Paulina Jenkins of the Natural History Museum in London. Details of the bones and teeth, plus DNA analysis, place the animal in the new family Laonastidae, the researchers argue in the latest quarterly issue of *Systematics and Biodiversity*, dated December 2004. They've christened the rodent *Laonastes aenigmamus.*

Dorothée Huchon, a rodent specialist at Tel Aviv University in Israel, says that the DNA evidence hasn't yet convinced her that this should be a new family. However, the rodent group "is full of surprises," she says.

67. What two features about the supposed sanje monkey (seen in the Udzungwa Mountains National Park) aroused Trevor Jones' suspicions that this could be a new species of mangabey?

68. How can the scientists be sure that this monkey is indeed a new species of mangabey?

69. Why is it critical that the highland mangabey's habitat be protected? How is the area now threatened?

70. What three pieces of evidence indicate that the new rodent species *Laonastes aenigmamus* probably should be placed in a new taxonomic family?

CHAPTER 8 Ecology and the Environment

Standards 6.1, 6.2, 6.3, 6.4 (Ecology)

BROAD CONCEPT: Ecology is the interaction among organisms and between organisms and their environment.

Ecology is the study of the relationships between organisms and between organisms and their physical (that is, external) **environment**. No organism exists in nature as an entity apart from its environment. All organisms are affected by their environment and, in turn, all organisms have an effect on their environment.

ECOLOGICAL ORGANIZATION

In ecology, the relationships between organisms and the environment may be considered at various levels. The smallest, least inclusive level in terms of ecological organization is the population; the largest and most inclusive level is the biosphere.

Levels of Organization

All members of a species living in a given location make up a **population**. For example, all the water lilies in a pond make up a population, and all the goldfish in a pond make up a population. Together, all the interacting populations in a given area make up a **community**. For example, all the plants, animals, and microorganisms in a pond make up the pond community.

An **ecosystem** includes all the members of a community along with the physical environment in which they live. The living and nonliving parts of an ecosystem function together as an interdependent and relatively stable **system**. The **biosphere** is the portion of Earth in which all living things exist. Composed of numerous, complex ecosystems, the biosphere includes all the water, soil, and air.

QUESTIONS

MULTIPLE CHOICE

1. All the different species within an ecosystem are collectively referred to as the
A. niche B. community C. consumers
D. population

2. Which term includes the three terms that follow it? A. population: community, ecosystem, organism B. community: ecosystem, organism, population C. ecosystem: organism, population, community D. organism: ecosystem, community, population

3. Which sequence shows increasingly complex levels of ecological organization?
A. biosphere, ecosystem, community
B. biosphere, community, ecosystem
C. community, ecosystem, biosphere
D. ecosystem, biosphere, community

4. The members of the mouse species *Microtus pennsylvanicus* living in a certain location make up a A. community B. succession
C. population D. phylum

5. Which term includes all the regions (on land and in water) in which life exists? A. marine biome B. climax community C. biosphere
D. tundra

OPEN RESPONSE

6. List the four ecological levels of organization in their order of increasing complexity.
7. Define each of the following ecological levels: *ecosystem, population, community, biosphere.*

CHARACTERISTICS OF ECOSYSTEMS

Ecosystems are the structural and functional units studied in ecology. Recall that an ecosystem consists of the living community plus the nonliving parts of the environment. The living organisms in an ecosystem are called the **biotic** factors, while the nonliving parts are known as the **abiotic** factors. There are always interactions between the biotic and abiotic factors in an ecosystem. These interactions are vital for the stability of the ecosystem.

Requirements of Ecosystems

An ecosystem is a self-sustaining unit when the following two conditions are met. First, there must be a constant flow of **energy** into the ecosystem, and there must be organisms within the ecosystem that can use this energy for the synthesis of organic compounds. The primary source of energy for most ecosystems on Earth is sunlight; the organisms that can use this energy for the synthesis of organic compounds are green plants, algae, and other photosynthetic autotrophs. Second, there must be a recycling of materials between the living organisms and the physical, nonliving parts of the ecosystem.

Until recently, it was thought that all life-forms depend, either directly or indirectly, on solar energy to carry out their life activities. However, this is not true. There are some organisms that survive in the deepest parts of the ocean, where no light penetrates. They live around hot-water vents on the seafloor and rely on bacteria that use chemicals—hydrogen sulfide and carbon dioxide—in the seawater to produce energy-rich sugars. These bacteria, which function as the food producers in this ecosystem, carry out *chemosynthesis* rather than photosynthesis. They use the energy of chemical reactions, rather than the energy of the sun, to produce carbohydrates. Survival of the other organisms in that ecosystem, such as worms, clams, shrimp, crabs, and octopuses, depends on the food energy that the bacteria produce.

Abiotic Factors of Ecosystems

As stated above, the components of an ecosystem include both biotic and abiotic factors. The abiotic features of the environment are physical factors that sustain the lives and reproductive cycles of organisms. These factors are: intensity of light; temperature range; amount of water; type of soil; availability of minerals and other inorganic substances; supply of gases, including oxygen, carbon dioxide, and nitrogen; the pH (acidity or alkalinity) of soil or water; and turbulence (from wind and waves).

Abiotic factors vary from one environmental area to another. The abiotic conditions in any particular environment determine the types of plants and animals that can exist there. Thus, abiotic factors are **limiting factors**. For example, the small amount of available water in a desert limits the kinds of plants and animals that can live in that environment.

Biotic Factors of Ecosystems

The biotic factors of an ecosystem are all the living things that directly or indirectly affect each other and the environment. The organisms of an ecosystem interact in many ways. These interactions include nutritional and symbiotic relationships.

Nutritional Relationships. Nutritional relationships involve the transfer of nutrients from one organism to another within the ecosystem.

Autotrophs are organisms that can use energy from the environment to synthesize their own food from inorganic compounds. Most autotrophs are photosynthetic, using energy from sunlight along with carbon dioxide and water from the environment to synthesize organic compounds. The autotrophic chemosynthetic organisms use chemical energy, rather than solar energy, to make their food.

Heterotrophs cannot synthesize their own food and must obtain nutrients from other organisms.

Depending on their source of food, heterotrophs are classified as saprophytes, herbivores, carnivores, or omnivores.

Saprophytes are organisms that obtain nutrients from the decaying remains of dead organisms; saprophytes include numerous species of bacteria and fungi. Animals that feed exclusively on plants are called **herbivores**; examples include deer, cattle, and rabbits. Animals that consume other animals are called **carnivores**. The carnivores include two general groups: **predators**, such as lions, eagles, and sharks, which directly kill and eat their **prey**; and **scavengers**, such as vultures and hyenas, which usually feed on the remains of animals that they have not killed (that is, prey that was already dead). **Omnivores** are animals that consume both plant and animal matter; examples include humans, chimpanzees, and black bears.

Symbiotic Relationships. Different species of organisms sometimes live together in a very close association. Such a close relationship, or **symbiosis**, may or may not be beneficial to both organisms involved. Most types of symbiosis are based on the nutritional needs of one or both organisms.

A type of symbiotic relationship in which one organism benefits while the other is neither helped nor harmed is called *commensalism*. Remoras on sharks, barnacles on whales, and orchids on large, tropical trees all obtain favorable places to live without doing any noticeable harm to the other organism (Figure 8-1). The remoras (a type of fish) eat food scraps left over by the sharks on which they ride; the barnacles gain access to new feeding areas by living attached to whales; and the orchids receive more sunlight by living on the branches of tall trees.

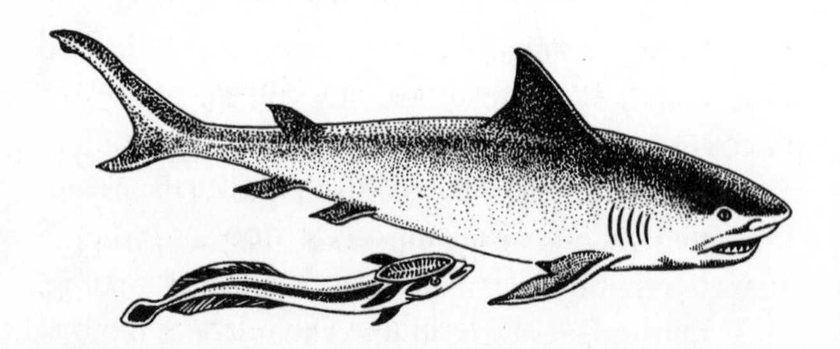

Figure 8-1. The shark and remora have a type of symbiosis known as commensalism.

A symbiotic relationship in which both organisms benefit is called *mutualism*. For example, certain protozoans (unicellular organisms) live within the digestive tracts of termites. Wood eaten by the termites is digested by the protozoans, and both organisms benefit from the nutrients that are released. Without these protozoans, the termites would starve because they cannot produce the enzymes needed to break down the cellulose in wood.

Another example of mutualism is found in lichens, which are made up of both algal and fungal cells. The algal cells carry on photosynthesis, which provides food for the lichen, while the fungal cells provide moisture and minerals, and anchor the lichen to a surface. Lichens are often found attached to the bark of trees, where they have access to sunlight. Nitrogen-fixing bacteria live in the roots of legumes (such as the peanut plant). The relationship between these organisms is mutualistic because the bacteria provide nitrogen compounds for the plant, while the plant provides the bacteria with nutrients and a good place to live (Figure 8-2).

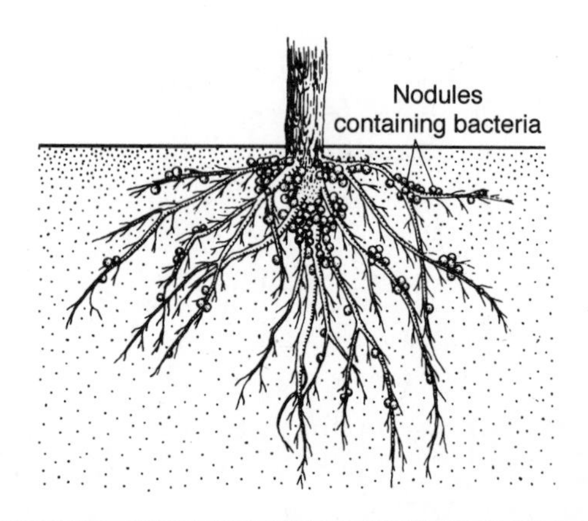

Figure 8-2. Nitrogen-fixing bacteria live in the roots of legumes in a type of symbiosis known as mutualism.

A symbiotic relationship in which one organism, the **parasite**, benefits while the other, the **host**, is harmed is called *parasitism*. Parasites get their nutrients, and a place to live, from the host organism. Many small organisms live as parasites in humans and other animals. Examples include the athlete's foot fungus (on humans) and heartworm (in dogs). Parasites can rob their host organisms of nutrients; they also can cause a variety of illnesses, such as malaria (caused by a protozoan) and elephantiasis (caused by a worm).

QUESTIONS

MULTIPLE CHOICE

8. Different species of animals in a community would most likely be similar in their A. physical structure B. size C. abiotic requirements D. number of offspring produced

9. All the living things that affect each other and their environment are considered to be A. biotic factors B. inorganic substances C. physical conditions D. chemical factors

10. A study was made over a period of years in a certain part of the country. It showed that the area had a low amount of rainfall, a wide seasonal variation in temperature, and short periods of daylight. These environmental factors are A. abiotic factors of little importance to living things B. abiotic factors that limit the type of organisms that live in the area C. biotic factors important to living things in the area D. biotic factors that are affected by the abiotic factors

11. The presence of nitrogen-fixing bacteria in nodules on the roots of legumes (such as the peanut plant) illustrates an association known as A. commensalism B. mutualism C. parasitism D. environmentalism

12. At times, hyenas will feed on the remains of animals that they have not killed themselves. At other times, they will kill other animals for food. Based on their feeding habits, hyenas are best described as both A. herbivores and parasites B. herbivores and predators C. scavengers and parasites D. scavengers and predators

13. Which is an abiotic factor in the environment? A. water B. earthworm C. fungus D. human

14. The organisms that prevent Earth from becoming covered with the remains of dead organisms are known as A. herbivores B. parasites C. autotrophs D. saprophytes

15. A particular species of fish has a very narrow range of tolerance for changes in water temperature and dissolved oxygen content. For this fish, the temperature and oxygen content represent A. autotrophic conditions B. a community C. limiting factors D. symbiosis

16. An example of a parasitic relationship would be A. tapeworms living in the intestines of a dog B. algal and fungal cells living together as a lichen C. barnacles living on a whale D. wood-digesting protozoa living in the gut of a termite

17. Parasitism is a type of nutritional relationship in which A. both organisms benefit B. both organisms are harmed C. neither organism benefits D. one organism benefits and the other is harmed

18. For an ecosystem to be self-sustaining, it must A. contain more animals than plants B. receive a constant flow of energy C. have a daily supply of rainwater D. contain only heterotrophs

19. Heterotrophs include A. autotrophs, saprophytes, and herbivores B. omnivores, carnivores, and autotrophs C. saprophytes, herbivores, and carnivores D. herbivores, autotrophs, and omnivores

20. The primary source of energy for most ecosystems is A. radioactivity B. sunlight C. animal proteins D. carbon dioxide

21. An ecosystem that does *not* depend on sunlight for its energy source is found A. at the tops of mountains B. on tropical islands C. in hot deserts D. near deep-ocean vents

22. A particular species of unicellular organism inhabits the intestines of termites, where they can live protected from predators. The unicellular organisms digest wood that has been ingested by the termites, thus providing nutrients to the termites. The relationship between these two species can be described as A. harmful to both species B. harmful to the host C. beneficial to both species D. beneficial to the parasite only

23. A diagram of several interconnected, nutritional relationships is shown below. Letter *X* most likely represents the A. autotrophs B. carnivores C. decomposers D. parasites

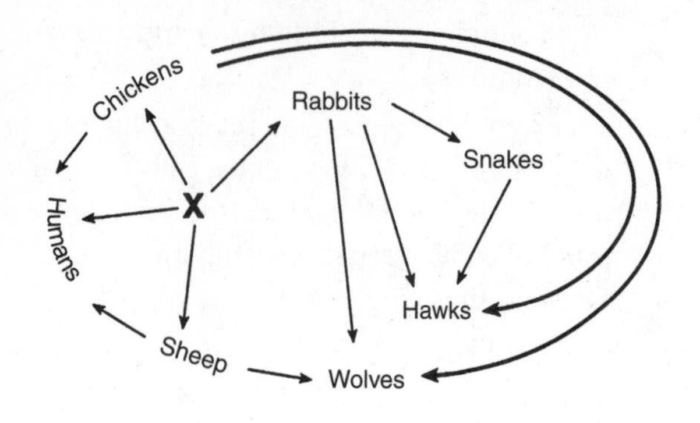

OPEN RESPONSE

24. Identify the two main conditions that must exist for an ecosystem to be self-sustaining.

25. Explain why abiotic factors are considered to be limiting factors; give an example of one.

26. How does a parasite benefit from living in or on its host? In what way does this type of relationship harm the host organism?

ENERGY-FLOW RELATIONSHIPS

For an ecosystem to be self-sustaining, there must be a source of energy and a flow of energy between organisms. The pathways of chemical energy from food through the organisms of an ecosystem are represented by food chains and food webs.

Food Chains

The transfer of energy from autotrophic organisms through a series of heterotrophic organisms with repeated stages of eating and being eaten is described as a **food chain** (Figure 8-3). Green plants obtain energy for their life processes from the radiant energy of sunlight (that is, **solar energy**), which they convert to usable chemical energy (glucose) by

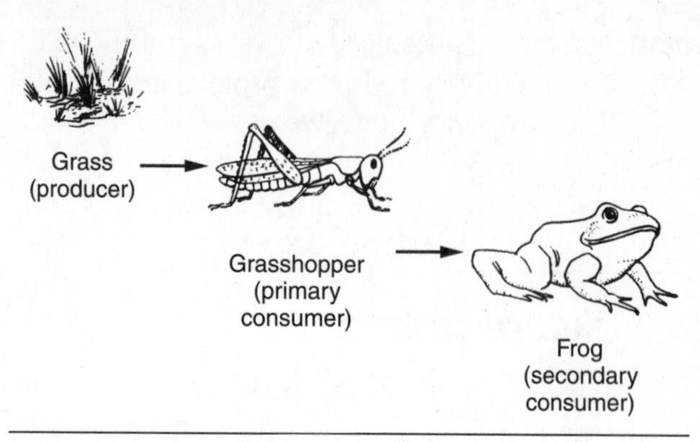

Figure 8-3. A food chain (not drawn to scale).

photosynthesis. For all other organisms in the food chain, energy is obtained from the breakdown of food. The organisms in a food chain are described in terms of the following categories.

Green plants and other autotrophs are the **producers** in the food chain. All the food energy for a community is derived from the organic compounds synthesized by producers (for example, grass in a savannah, algae in the sea, or chemosynthetic bacteria near a deep-ocean vent).

All the heterotrophic organisms in a community are **consumers**. They must obtain energy from the food that they eat. Animals that feed on green plants and algae (that is, directly on producers) are called *primary consumers*, or herbivores. Animals that feed on primary consumers are called *secondary consumers*, or carnivores. Omnivores may be either primary or secondary consumers; that is, omnivores may feed on plants and/or animals. Humans are a good example of an omnivore, since we eat both plant and animal foods.

Saprophytes are **decomposers**, the organisms that break down the remains of dead organisms and organic wastes. Decomposers return substances in the remains and wastes of plants and animals to the environment, where other living organisms can use them again. Most decomposers are either bacteria or fungi. This recycling of materials is critical to the survival of an ecosystem; it ensures that the limited supply of materials can be used over and over again. In addition, the remains, or **residue**, of dead organisms do not accumulate in the ecosystem, since they are broken down to simpler compounds.

Food Webs. In a natural community, most organisms eat more than one species and may be eaten, in turn, by more than one species. Thus, the various food chains in a community are interconnected, forming a **food web** (Figure 8-4). Food webs have the same levels of organisms (producers, consumers, and decomposers) as food chains, but the flow of energy and materials is much more complex.

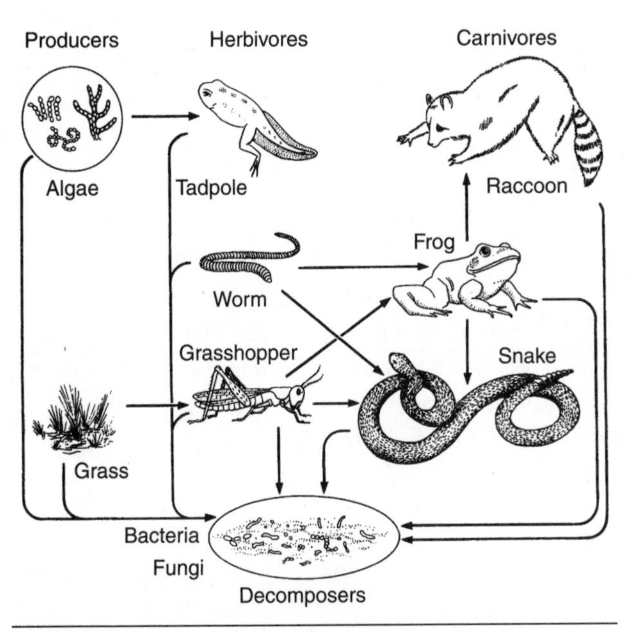

Figure 8-4. A food web (not drawn to scale).

Pyramid of Energy. The greatest amount of energy in a community is present in the organisms that make up the producer level. Only a small portion of this energy is passed on to primary consumers; and only a small portion of that energy in the primary consumers is passed on to secondary consumers; and so on. An **energy pyramid** can be used to illustrate the loss of usable energy at each feeding, or *trophic*, level (Figure 8-5). Producers always occupy the first **trophic level** in an energy pyramid. Consumers can occupy different trophic levels depending on the food chain; that is, they may be a primary consumer (second trophic level) in one food chain, and a secondary consumer (third trophic level) in another one.

At each consumer level in an energy pyramid, only about 10 percent of the ingested nutrients are used to synthesize new tissues, which are the food available for the next feeding level. The remaining energy is used by the consumers for their life functions and is eventually converted to heat, which is lost from the ecosystem. Thus, an ecosystem cannot sustain itself without the constant input of energy from an external source. In most ecosystems, this energy source is the sun. An energy pyramid that has many trophic levels will experience a significant loss of available energy at the top. This means that very few organisms (that is, top predators) can be supported at the highest trophic level.

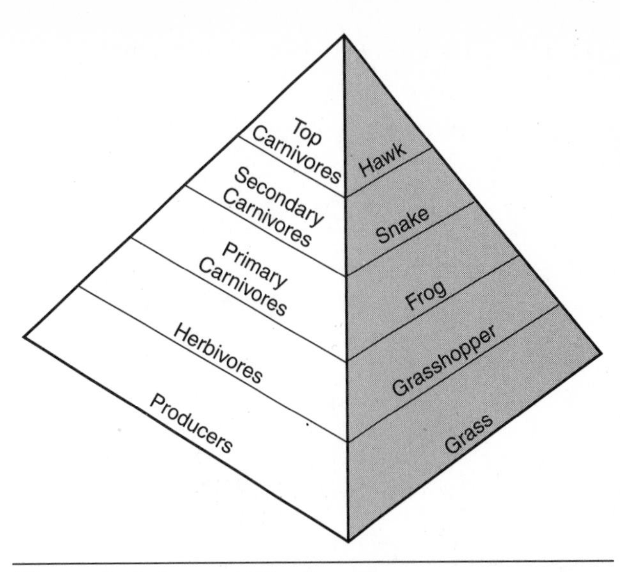

Figure 8-5. A pyramid of energy.

Pyramid of Biomass. In general, the decrease in available energy at each higher feeding level means that less organic matter, or *biomass*, can be supported at each higher level. Thus, the total mass of producers in an ecosystem is greater than the total mass of primary consumers; and the total mass of primary consumers is greater than the total mass of secondary consumers; and so on. A *biomass pyramid* can be used to illustrate this decrease in biomass at each higher feeding level (Figure 8-6 on page 162).

If, for some reason, biomass becomes greater at a higher trophic level than at a lower level, the ecosystem will become unbalanced and unsustainable. The consumers at the higher trophic level will have exceeded the capacity of the ecosystem to support them. Eventually, they will begin to die off as their food supply runs out, and the ecosystem will return to a state of balance.

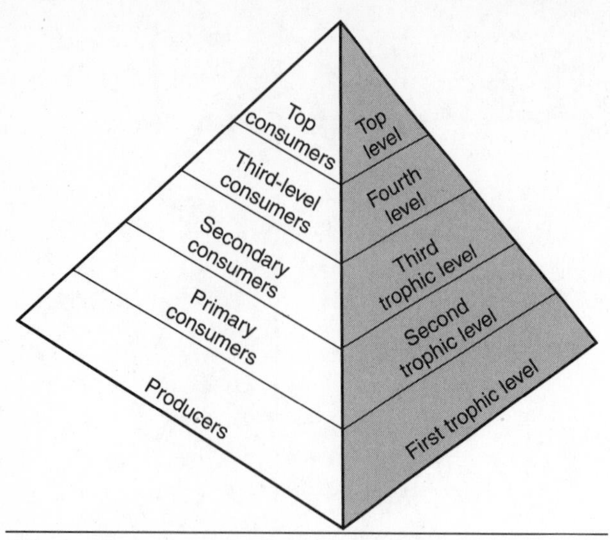

Figure 8-6. A pyramid of biomass.

QUESTIONS

MULTIPLE CHOICE

27. Which occurs within self-sustaining ecosystems? A. The producers have a limited source of energy. B. Consumers eventually outnumber producers. C. Carnivores usually outnumber herbivores. D. Organisms recycle materials with each other and the environment.

28. Which food chain relationship illustrates the nutritional pattern of a primary consumer? A. seeds eaten by a mouse B. an earthworm eaten by a mole C. a mosquito eaten by a bat D. a fungus growing on a dead tree

29. Which term describes both the bird and the cat in the following food chain?

sun → grass → grasshopper → bird → cat

A. herbivores B. saprophytes C. predators D. omnivores

30. Organisms from a particular ecosystem are shown in the box. Which statement concerning an organism in this ecosystem is correct? A. Organism 2 is heterotrophic. B. Organism 3 helps recycle materials. C. Organism 4 obtains all of its nutrients from an abiotic source. D. Organism 5 must obtain its energy from organism 1.

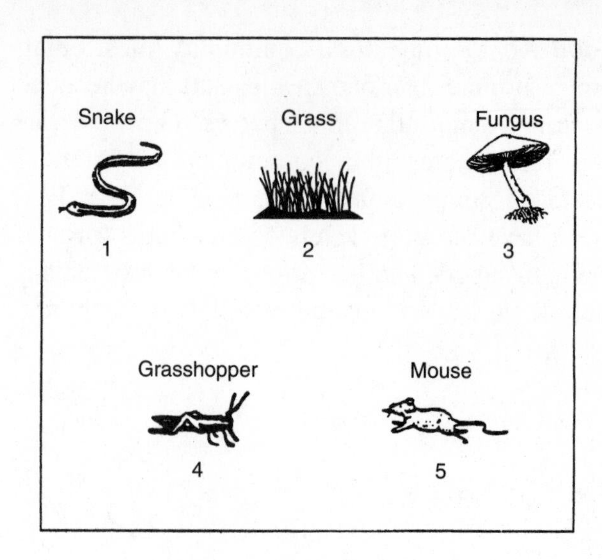

31. The elements stored in living cells of organisms in a community will eventually be returned to the soil for use by other living organisms. The organisms that carry out this process are the A. producers B. herbivores C. carnivores D. decomposers

32. In the food chain below, what is the nutritional role of the rabbit?

lettuce plant → rabbit → coyote

A. parasite B. saprophyte C. primary consumer D. primary producer

33. Which level of the food pyramid shown below represents the largest biomass? A. bass B. minnows C. copepods D. algae

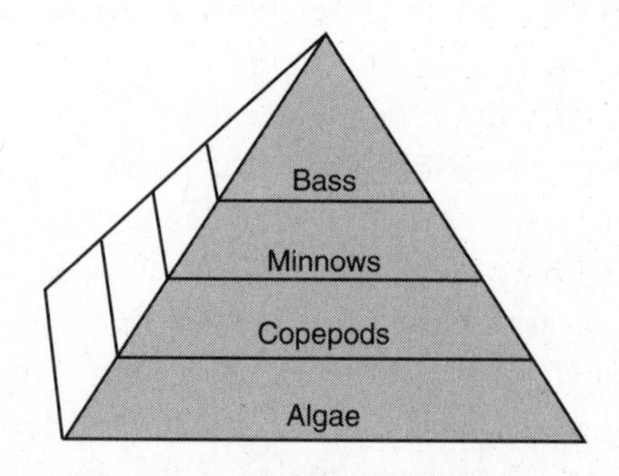

34. Fly larvae consume the body of a dead rabbit. In this process, they function as A. producers B. scavengers C. herbivores D. parasites

35. Which diagram of boxes best represents the usual relationships of biomass in a stable community? A. 1 B. 2 C. 3 D. 4

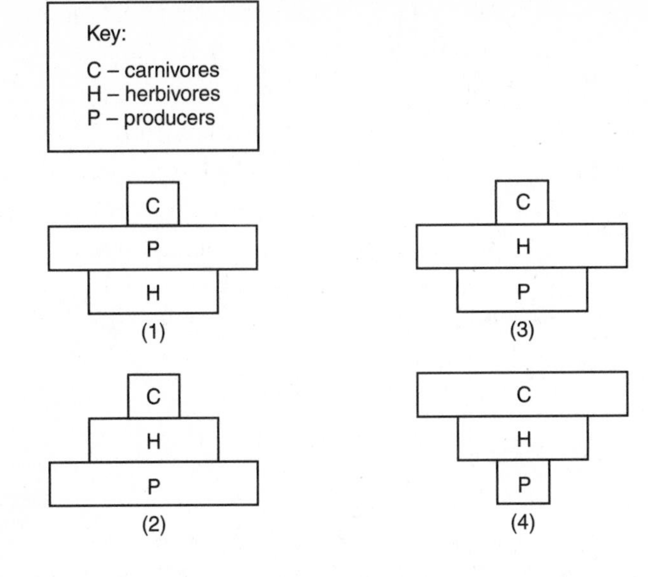

36. Which level in an energy pyramid has the highest amount of available energy? A. highest level consumers B. secondary consumers C. primary consumers D. producers

Base your answers to questions 37 through 40 on the diagram below and on your knowledge of biology. The diagram represents different species of organisms that may interact with each other in and around a pond environment.

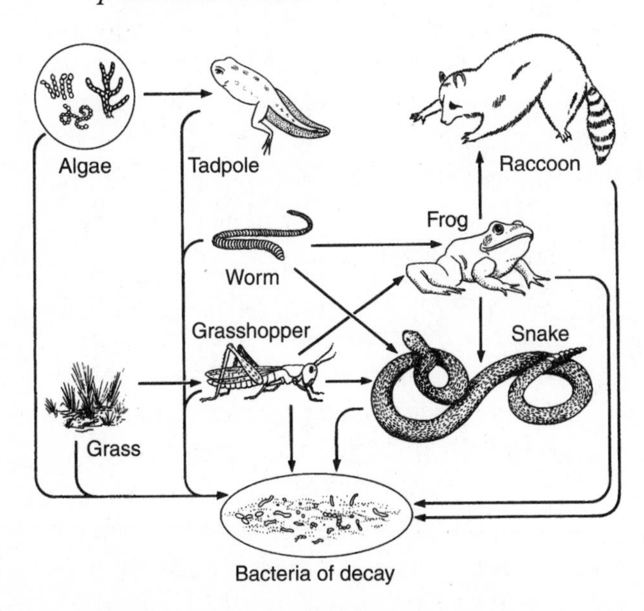

37. The adult frog represents a type of consumer known as a A. producer B. carnivore C. saprophyte D. parasite

38. Which organisms may be classified as herbivores? A. algae, tadpole, raccoon B. worm, snake, bacteria C. tadpole, worm, grasshopper D. grasshopper, bacteria, frog

39. Which statement about the algae and grass is true? A. They are classified as omnivores. B. They are parasites in the animals that eat them. C. They contain the greatest amount of stored energy. D. They decompose nutrients from dead organisms.

40. The interactions among the organisms shown in this diagram illustrate A. a food web B. geographic isolation C. abiotic factors D. organic evolution

Base your answers to questions 41 through 43 on the following food chain and on your knowledge of biology.

rosebush → aphid → ladybug beetle → spider → toad → snake

41. Which organism in the food chain can transform light energy into chemical energy? A. spider B. ladybug beetle C. rosebush D. snake

42. At which stage in the food chain will the population with the smallest number of animals probably be found? A. spider B. aphid C. ladybug beetle D. snake

43. Which organism in this food chain is a primary consumer? A. rosebush B. aphid C. ladybug beetle D. toad

Base your answers to questions 44 through 47 on the diagram below, which represents four possible pathways for the transfer of energy stored by green plants.

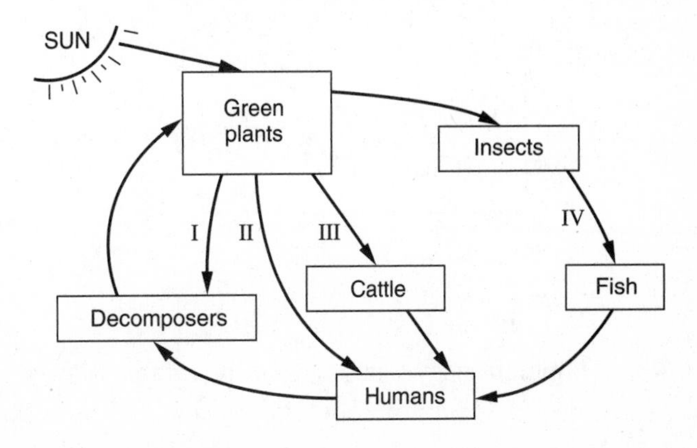

44. The pathway labeled IV represents a (an) A. food chain B. population C. ecosystem D. abiotic factor

45. Through which pathway would the sun's energy be most directly available to humans? A. I B. II C. III D. IV

46. In this diagram, humans are shown to be A. herbivores only B. carnivores only C. omnivores D. parasites

47. The cattle in the diagram represent A. primary consumers B. secondary consumers C. producers D. autotrophs

MULTIPLE CHOICE AND OPEN RESPONSE

Base your answers to questions 48 through 50 on the food web and graph below and on your knowledge of biology. The graph represents the interaction of two different populations, A *and* B, *in the food web.*

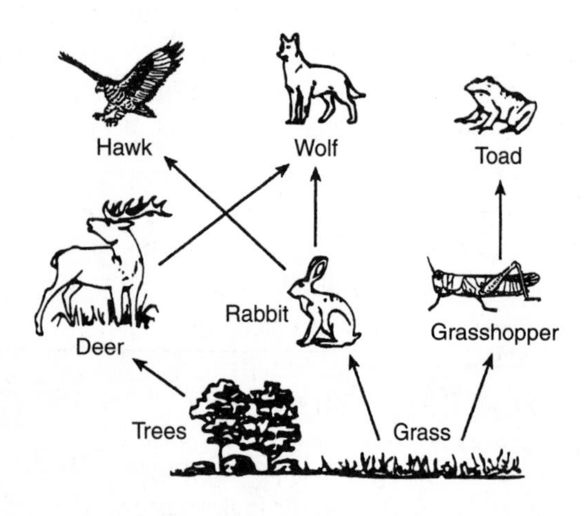

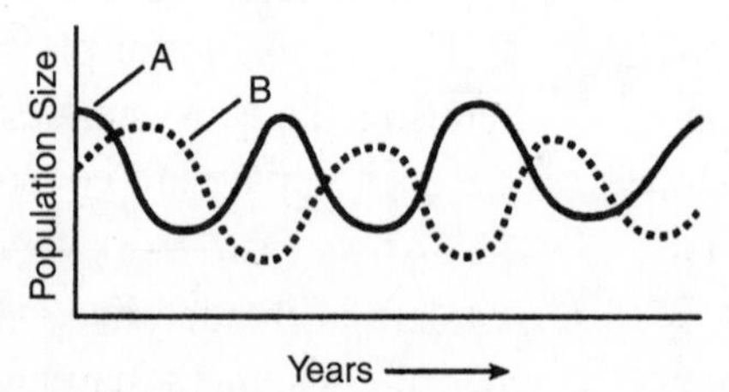

48. Population *A* is made up of living animals. The members of population *B* feed on these living animals. The members of population *B* are most likely A. scavengers B. autotrophs C. predators D. parasites

49. Identify one specific heterotroph from the food web that could be a member of population *A*.

50. An energy pyramid is shown below. Which organism shown in the food web would mostly likely be found at level *X*? A. wolf B. grass C. deer D. toad

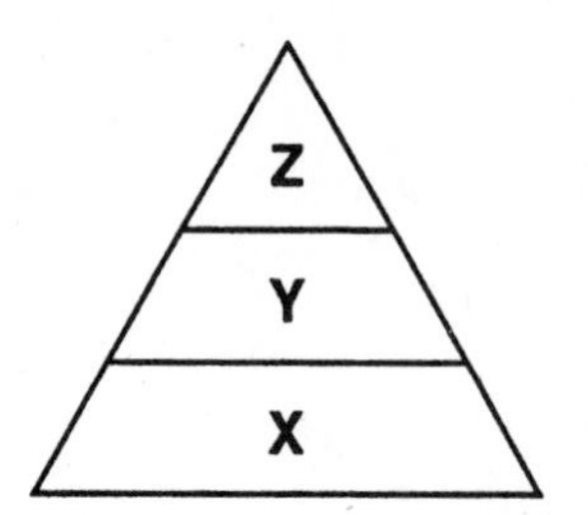

51. Draw, using specific organisms in one community as examples, a pyramid of energy that has three trophic levels.

52. Distinguish between a pyramid of energy and a pyramid of biomass. Explain the relationship between them.

53. Explain why there is a "loss" of energy as one goes step-by-step up an energy pyramid. Discuss where the seemingly "lost" energy actually goes.

54. Explain why an ecosystem could not sustain itself without the constant input of energy from an outside source.

55. Consider the following food pyramid: corn → mice → snakes → hawks. If the total amount of energy captured by the corn is 1,000,000 Calories per day, and only about 10 percent of this energy is passed on at each higher trophic level, calculate:
   - how much energy (in Calories) would be available per day at each higher level (for mice, snakes, and hawks);
   - how many hawks this ecosystem could support, if the hawk population needs 500 Calories per bird per day.

POPULATIONS IN ECOSYSTEMS

Population Size and Carrying Capacity

In every ecosystem on Earth, there are limited amounts of available resources. These resources include food, water, energy, minerals, and living space (territory). Even though some of these resources may be recycled through the actions of bacteria and fungi, the pace of recycling may not keep up with the demand for these materials. Thus the amount of resources available limits the number of organisms that an ecosystem can support.

The maximum number of organisms of a particular type that can be supported in an area is known as the **carrying capacity**. In a stable ecosystem, a population will fluctuate slightly (due to seasonal and other factors), as shown in Figure 8-7. If resources are very plentiful, the population can increase in size. However, if the population increases significantly above its carrying capacity, many individuals will die off because there are insufficient resources available to support them. Other factors that play a part in determining a population's size are immigration (when members of a population enter an ecosystem) and emigration (when members of a population leave an ecosystem).

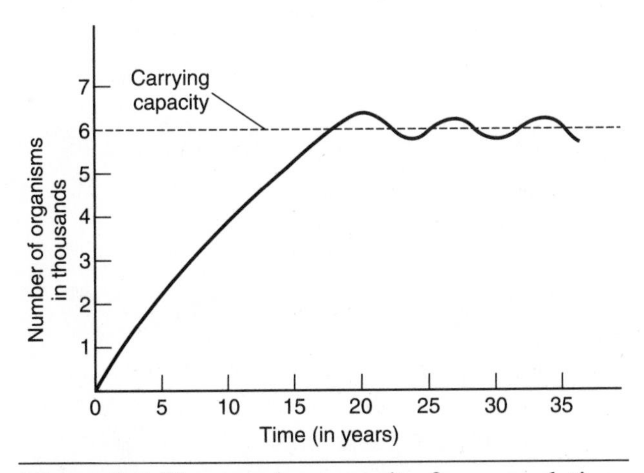

Figure 8-7. The carrying capacity for a population of organisms will fluctuate slightly in a stable ecosystem.

Competition Between Organisms

Different species living in the same environment, or **habitat**, may require some of the same resources for their survival. Since resources (such as food, water, space, light, and minerals) are usually limited, competition occurs among the various species. As stated in Darwin's theory of evolution by means of natural selection, **competition** is the struggle between different organisms for the same limited resources.

The more similar the needs of the species, the more intense the competition. For example, lions, leopards, and hyenas may compete to consume the same type of antelope. In addition, because their requirements are most similar, the strongest competition for resources often occurs among members of the same species. For example, competition for antelope prey (or water, mates, territory, and so on) may be more intense between neighboring prides of lions than between lions and other nearby large predators.

Each species occupies a particular ecological niche in a community. A **niche** is the role that the species fills in its habitat. A species' niche includes the type of food, or **nutrients**, it requires, where and how it lives, where and how it reproduces, and its relationships with other species in the area. For example, a woodpecker lives in the forest because its niche involves finding insects that live inside trees. When two species compete for the same niche, the one that is more successful at utilizing the available resources will out-compete the other, thereby maintaining just one species per niche in the community.

QUESTIONS

MULTIPLE CHOICE

56. Carrying capacity is best thought of as the amount of A. abiotic factors present in an ecosystem B. light available for photosynthesis C. organisms the ecosystem can support D. producers compared to consumers in the ecosystem

57. A stable ecosystem is characterized by A. a greater number of consumers than producers B. population sizes at or near the carrying capacity C. a greater need for energy than is available D. a lack of decomposers to recycle materials

58. In a freshwater pond community, a carp (a type of fish) eats decaying matter from around the bases of underwater plants, while a snail scrapes algae from the leaves and stems of the same plant. They can survive at the same time in the same pond because they occupy A. the same niche but different habitats B. the same habitat but different niches C. the same habitat and the same niche D. different habitats and different niches

59. The role a species plays in a community is called its A. habitat B. biotic factor C. territory D. niche

60. When two different species live in the same environment and use the same limited resources, which interaction will usually occur? A. competition B. cooperation C. commensalism D. mutualism

61. Which of the following is an important *biotic* factor in an ecosystem? A. availability of water B. level of atmospheric oxygen C. activity of decomposers D. amount of soil erosion

Base your answer to question 62 on the information and diagram below and on your knowledge of biology.

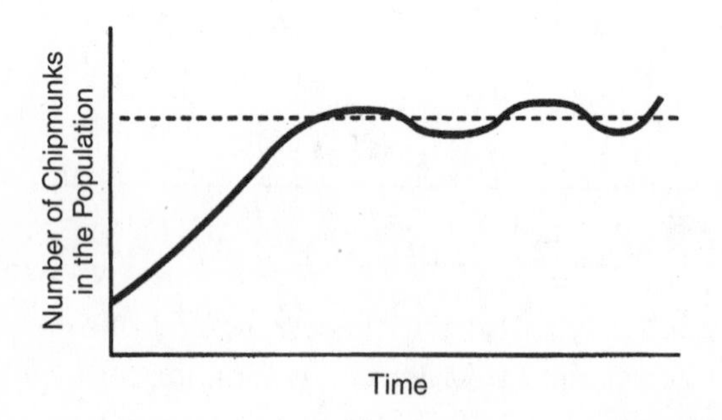

A population of chipmunks migrated to an environment where they had little competition. Their population quickly increased but eventually stabilized, as shown in the graph above.

62. Which statement best explains why the population stabilized? A. Interbreeding between members of the population increased the mutation rate. B. The population size became limited due to factors such as availability of food. C. An increase in the chipmunk population caused an increase in the producer population. D. A predator species came to the area and occupied the same niche as the chipmunks.

63. The size of a mouse population in a natural ecosystem tends to remain relatively constant due to A. the carrying capacity of the environment B. the lack of natural predators C. cycling of energy D. increased numbers of decomposers

64. Purple loosestrife plants are replacing cattail plants in freshwater swamps in New York State. The two species have very similar environmental requirements. This observation best illustrates A. variations within a species B. dynamic equilibrium C. random recombination D. competition between species

OPEN RESPONSE

65. Some bacteria can reproduce once every 20 minutes. As a result, their populations can double several times an hour. Even at this phenomenal rate of reproduction, bacteria do not overrun the planet. Give a brief, valid explanation for this fact.

66. Explain why competition between individuals of the same species is often more intense than competition between members of different species.

CYCLING OF MATERIALS

In a self-sustaining ecosystem, various materials are recycled between the organisms and the abiotic environment. The recycling process allows materials to be used over and over again by living things.

Carbon-Hydrogen-Oxygen Cycle. The elements carbon, hydrogen, and oxygen are recycled through the environment by the processes of respiration and photosynthesis (Figure 8-8). During aerobic cellular respiration, plants and animals use oxygen (O_2) from the air and release carbon dioxide (CO_2) and water (H_2O) via the breakdown of glucose. During photosynthesis, plants use carbon dioxide (CO_2) from the air and water (H_2O) from the environment in the synthesis of glucose ($C_6H_{12}O_6$) and oxygen (O_2) is given off as a by-product.

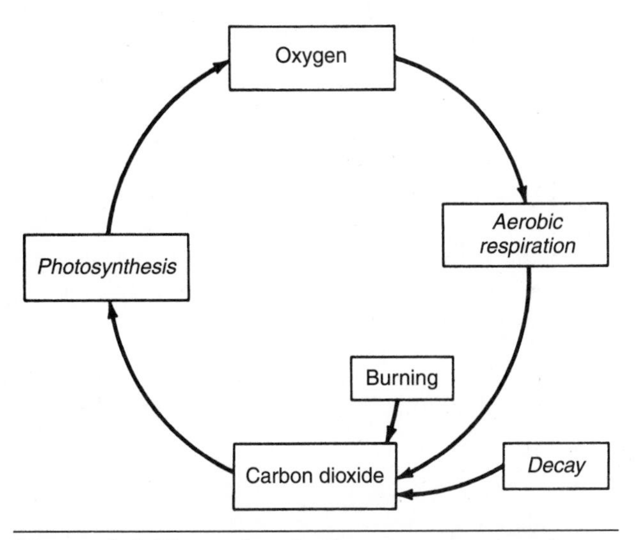

Figure 8-8. The carbon-hydrogen-oxygen cycle.

Water Cycle. In the water cycle, water moves between Earth's surface and the atmosphere (Figure 8-9). The main processes involved in this cycle are *evaporation* and *condensation*. Liquid water on Earth's surface changes to a gas by the process of evaporation and enters the atmosphere in the form of water vapor. As a result of condensation, water vapor is returned to the liquid state (precipitation) and falls to Earth. Some water vapor is added to the atmosphere by aerobic respiration in plants and animals and by *transpiration* (evaporation from the leaves) in plants. Water is also a vital nutrient for all living things, allowing them to carry out essential life processes and chemical reactions.

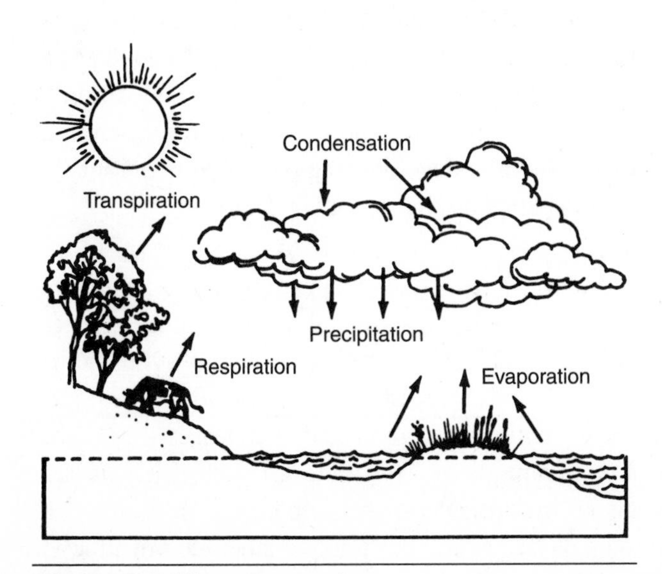

Figure 8-9. The water cycle.

Nitrogen Cycle. The element nitrogen is needed by all living things because it is part of the structure of amino acids and proteins. Plants absorb nitrogen-containing compounds from the soil; animals obtain nitrogen in the form of proteins in the foods they eat. These proteins are broken down by digestion to amino acids, which are then used in the synthesis of animal proteins.

The nitrogen cycle involves decomposers and other soil bacteria. Figure 8-10 shows the various components of the nitrogen cycle, which are described below.

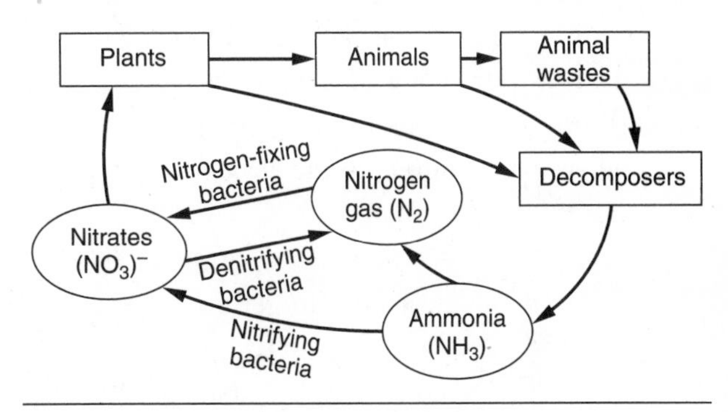

Figure 8-10. The nitrogen cycle.

Nitrogen-fixing bacteria, which live in nodules inside the roots of some plants (refer to Figure 8-2), convert free nitrogen gas (N_2) from the air into nitrogen-containing compounds called nitrates (NO_3). *Nitrates* are absorbed from the soil by plants and used in protein synthesis. Animals that eat plants convert the nitrogen-containing plant proteins into animal proteins. The nitrogenous wastes of living animals, and the nitrogen compounds in the remains of dead plants and animals, are broken down by decomposers and converted to ammonia (NH_3). *Nitrifying bacteria* in the soil convert ammonia into nitrates, which can be used again by plants. *Denitrifying bacteria* break down some nitrates into free nitrogen (N_2), which is released into the atmosphere as a gas.

QUESTIONS

MULTIPLE CHOICE

67. Carbon dioxide is added to the atmosphere by A. photosynthesis in plants B. evaporation of water C. respiration in animals only D. respiration in plants and animals

68. Oxygen (O_2) is added to the atmosphere by A. evaporation and photosynthesis B. respiration in plants C. photosynthesis only D. denitrifying bacteria

69. Which of the following processes is *not* involved in the water cycle? A. condensation B. nitrification C. evaporation D. transpiration

70. The processes involved in the recycling of carbon, hydrogen, and oxygen are
A. evaporation and condensation
B. photosynthesis and respiration
C. nitrification and denitrification
D. respiration and transpiration

71. Nitrogen is both removed from the atmosphere and returned to the atmosphere by the activities of A. plants only B. animals only C. plants and animals D. bacteria

72. Animals obtain their nitrogen from A. proteins in their food B. nitrates in the soil C. gas in the atmosphere D. bacteria in their intestines

Base your answers to questions 73 through 75 on the diagram below, which represents a cycle in nature, and on your knowledge of biology.

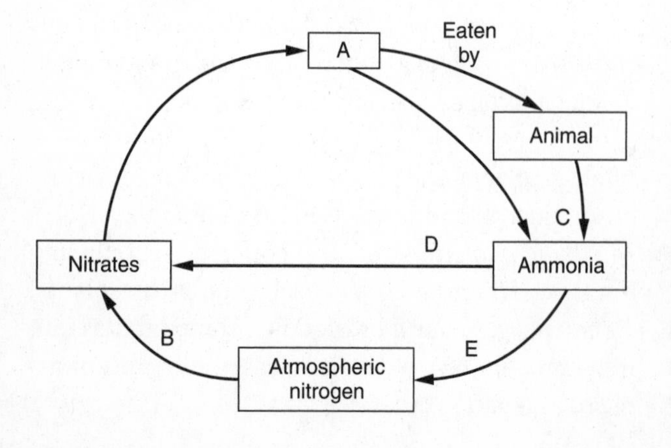

73. The cycle represented by the diagram is the A. nitrogen cycle B. carbon cycle C. water cycle D. oxygen cycle

74. Nitrifying bacteria in the soil are represented by the letter A. *A* B. *E* C. *C* D. *D*

75. The letter *B* most likely represents
A. bacteria of decay B. denitrifying bacteria C. a legume (peanut plant) D. nitrogen-fixing bacteria

OPEN RESPONSE

76. Describe how carbon dioxide and oxygen are recycled by the processes of respiration and photosynthesis.

77. Explain why both nitrogen-fixing bacteria and nitrifying bacteria are important for the survival of plants.

ECOSYSTEM FORMATION

Ecosystems tend to change over a long period of time until a stable one is formed. Both the living (biotic) and nonliving (abiotic) parts of the ecosystem change.

Succession

The replacement of one kind of community by another in an ecosystem is called ecological, or biological, **succession**. *Ecological succession* is usually a long-term process, happening over the course of many years (and many generations of different plants and animals). The kind of stable ecosystem that eventually develops in a particular geographical area depends on the region's climate.

Pioneer Organisms. Depending on climate and other abiotic environmental factors, succession on land can begin in an area that has no living things and end with a forest. Succession begins with *pioneer organisms*, which are the first plants, or plant-like organisms, to populate an area. Lichens and algae are typical pioneer organisms on bare rock, such as that found on a newly emerged volcanic island (Figure 8-11).

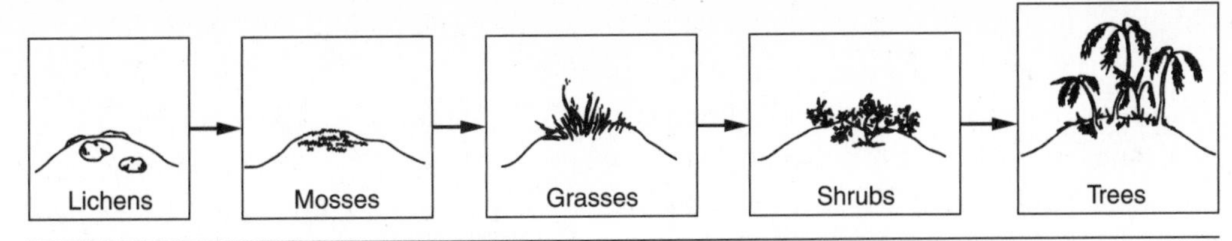

Figure 8-11. Ecological succession occurs, over time, on a new island.

Starting with pioneer plants, each community modifies the environment, often making it less favorable for itself and more favorable for other kinds of communities. (For example, some plants may change the soil composition over time.) One sequence of plant succession in New York State might be lichens → grasses → shrubs → conifers (pine trees) → deciduous (beech and maple) woodlands.

Since plants are the basic source of food for a community, the types of plants present in a community determine the types of animals that can live in the community. As the plant populations change, the animal populations also change.

Climax Communities. Succession ends with the development of a *climax community* in which populations of plants and animals exist in balance with each other and with the environment. In New York State, for example, the oak-hickory and hemlock-beech-maple associations represent two climax communities. In the Midwest, where there is less rain, grasslands are the typical climax community.

The climax community remains stable until a catastrophic change, such as a volcanic eruption or forest fire, alters or destroys it. Thereafter, succession begins again, leading to the development of a new climax community. This new community may be of the same type as the previous one or, if the catastrophe has changed the environment in some important way, it may be of another kind.

Biodiversity

In addition to the factors mentioned above, a stable community or stable ecosystem requires **biodiversity**. This term refers to the presence of a wide range of different species of organisms living and interacting with each other and with their nonliving environment. These organisms play a variety of roles that contribute to the overall stability of an ecosystem. For example, green plants and algae act as producers; fungi and bacteria act as decomposers, recycling vital materials; and animals act as consumers. Some roles are quite apparent while others may not be so obvious. Nevertheless, the removal of any one species from its natural environment may have profound negative effects on the overall health of the ecosystem. Climate change, human activity, and invasion by nonnative (exotic) species all threaten biodiversity by bringing some organisms to the brink of extinction or by outcompeting them in their natural environment. In addition, the introduction of some *invasive species* can destroy populations of native organisms because they have no natural defenses against them.

For example, the mongoose (a small Asian mammal) was introduced to Puerto Rico to control its rat population and protect its sugarcane crop. However, the rats learned to avoid the mongooses, which then began to eat other animals, such as lizards. The decrease in lizards led to unexpected negative result—an increase in June beetle larvae—a pest that ate much of the sugarcane (Figure 8-12).

Figure 8-12. The mongoose was introduced to Puerto Rico to control its rat population. However, this produced harmful results, such as an increase in the population of June beetle larvae.

Biodiversity also increases the probability that at least some organisms would be able to survive a catastrophic environmental event, such as climate change or a volcanic eruption. In time, the surviving organisms could reestablish a healthy community.

In addition, stable ecosystems that are rich in species, such as tropical rain forests, contain a wealth of genetic material that may have beneficial uses in medicine, agriculture, or other areas. Today, tropical forests, wetlands, coral reefs, and other ecosystems that are rich in biodiversity are being destroyed at an alarming rate, mainly due to human activities such as logging, mining, overhunting, habitat destruction, and polluting (some of which may increase global warming and its effects). Once species are lost to extinction, they can never be recovered. Careful protection of diverse habitats and their living resources is critical to preserving the biodiversity of Earth, not only for the needs of humans or individual ecosystems but also for the health and stability of the entire planet.

QUESTIONS

MULTIPLE CHOICE

78. The natural replacement of one community by another until a climax stage is reached is known as A. ecological balance B. organic evolution C. dynamic equilibrium D. ecological succession

79. In an ecological succession in New York State, lichens growing on bare rock are considered to be A. climax species B. pioneer organisms C. primary consumers D. decomposers

80. One of the first organisms to become established in an ecological succession leading to a pond community would be A. grasses B. algae C. minnows D. deciduous trees

81. Ecological succession ends with the development of a stable A. climax community B. pioneer community C. ecological niche D. abiotic community

82. Which two groups of organisms are most likely to be pioneer organisms? A. songbirds and squirrels B. lichens and algae C. deer and black bears D. oak and hickory trees

83. After a major forest fire occurs, an area that was once wooded is converted to barren soil. Which of the following sequences describes the most likely series of changes in vegetation after the fire?

A. shrubs → maples → pines → grasses

B. maples → pines → grasses → shrubs

C. pines → shrubs → maples → grasses

D. grasses → shrubs → pines → maples

84. Biodiversity in an ecosystem is important because it A. allows one species to dominate the others in its habitat B. slows down the pace at which species evolve C. provides stability to the ecosystem D. limits the amount of variation among organisms

85. Stable ecosystems are characterized by A. only two major species interacting with each other B. an infinite amount of available resources C. a variety of different species interacting with one another D. very little recycling of materials between the biotic and abiotic components

Base your answer to question 86 on the diagram below, which represents a process in nature, and on your knowledge of biology.

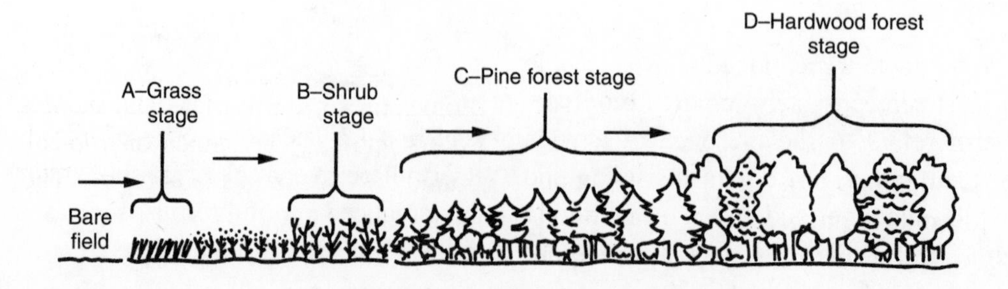

86. Stage *D* in the diagram is located on land that was once a bare field. The sequence of stages leading from the bare field to stage *D* best illustrates the process known as A. replication B. recycling C. feedback D. succession

87. A greater stability of the biosphere would most likely result from A. decreased finite resources B. increased deforestation C. increased biodiversity D. decreased consumer populations

OPEN RESPONSE

88. List the stages that precede a beech-maple forest in New York State. Identify the pioneer organism and the climax community in this succession.

89. Compare a natural meadow with a cornfield in terms of biodiversity. In your answer, be sure to address the:
   - number of species that live in each habitat;
   - number of interactions among species that occur in each habitat;
   - relative ability of each habitat to survive a natural disaster that might occur.

Base your answers to questions 90 through 92 on the diagram below, which represents the changes in an ecosystem over a period of 100 years, and on your knowledge of biology.

90. State one biological explanation for the changes in types of vegetation observed from *A* through *C*.

91. Identify one human activity that could be responsible for the change from *C* to *D*.

92. Predict what would happen to the soil *and* vegetation of this ecosystem after stage *F*, assuming no natural disaster or human interference.

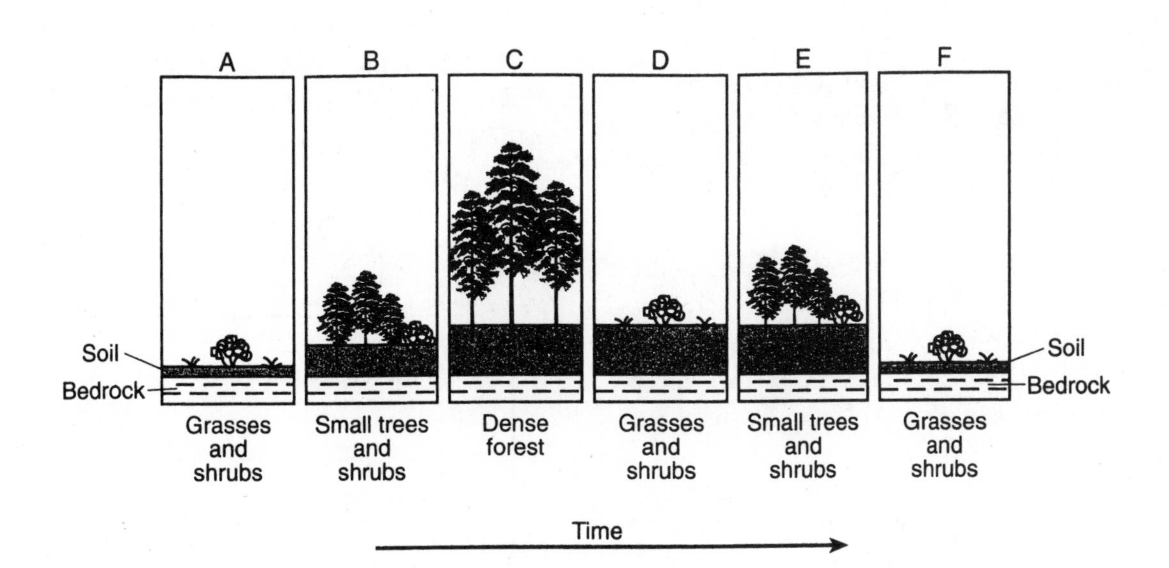

BIOMES

Earth can be divided into broad geographic regions by climate. The kind of climax ecosystem that develops in these large climatic areas is called a **biome**. Biomes may be terrestrial (land biomes) or aquatic (water biomes). The stretch of tropical rain forests around the equator is a land biome. The ocean is an aquatic biome.

Table 8-1. **The Major Terrestrial Biomes on Earth**

Biome	*Characteristics*	*Plants*	*Animals*
Tundra	Permanently frozen subsoil	Lichens, mosses, grasses	Snowy owl, caribou
Taiga	Long, severe winters; summers with thawing subsoil	Conifers	Moose, black bear
Temperate forest	Moderate precipitation; cold winters; warm summers	Deciduous trees (maple, oak, beech)	Fox, deer, gray squirrel
Tropical forest	Heavy rainfall; constant warmth	Many broad-leaved plant species	Snake, monkey, leopard
Grassland	Variability in rainfall and temperature; strong winds	Grasses	Antelope, bison, prairie dog
Desert	Sparse rainfall; extreme daily temperature fluctuations	Drought-resistant plants and succulents	Lizard, tortoise, kangaroo rat

Terrestrial Biomes

The major plant and animal associations (biomes) on land are determined by the large climate zones of Earth. These climate zones are, in turn, determined by geographic factors, including *latitude* (distance north or south of the equator) and *altitude* (distance above or below sea level). Other major geographic features, including large bodies of water, mountains, and deserts, modify the climate of nearby regions.

Climate includes the temperature range and the amounts of precipitation and solar radiation received by a region. The presence or absence of water is a major limiting factor for terrestrial biomes and determines the kinds of plant and animal communities that can be established.

Kinds of Terrestrial Biomes. Land biomes are described in terms of, and sometimes named for, the dominant kind of climax vegetation found there. Table 8-1 lists the major land biomes, their characteristics, dominant plant life, and some representative animals.

Effects of Latitude and Altitude. At the equator, the temperature and amount of rainfall remain relatively constant throughout the year. With increasing distance from the equator, temperature and rainfall show more variation during the year.

Increasing altitude may have the same effect on climate as increasing latitude. Thus, the temperature and kind of climax vegetation found at the top of a high mountain near the equator may be very much like that of a sea-level region far north of the equator. This relationship is shown in Figure 8-13.

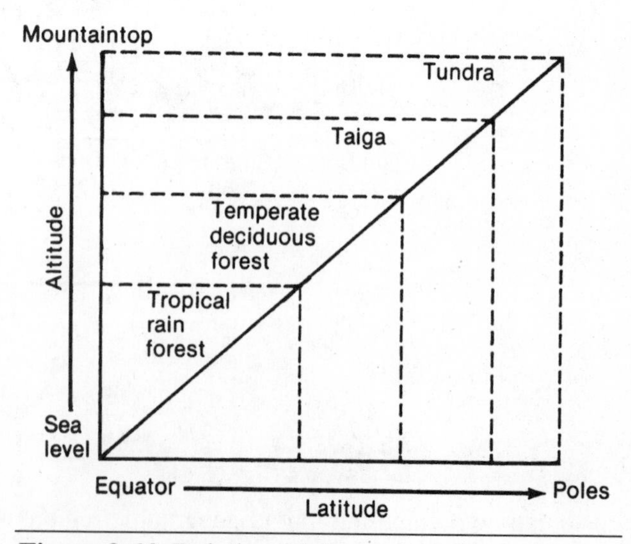

Figure 8-13. Relationship between latitude and altitude and terrestrial biomes.

Aquatic Biomes

Aquatic biomes make up the largest ecosystem on Earth. More than 70 percent of Earth's surface is covered by water; the majority of living things on Earth are water-dwellers.

Aquatic biomes are more stable than terrestrial biomes; they show less variation in temperature because water has a greater capacity to absorb and hold heat. The kinds and numbers of organisms present in an aquatic biome are affected by various abiotic factors, such as the water temperature, amounts of dissolved oxygen and carbon dioxide, intensity of light, and the kinds and amounts of dissolved minerals and suspended particles in the water (Figure 8-14).

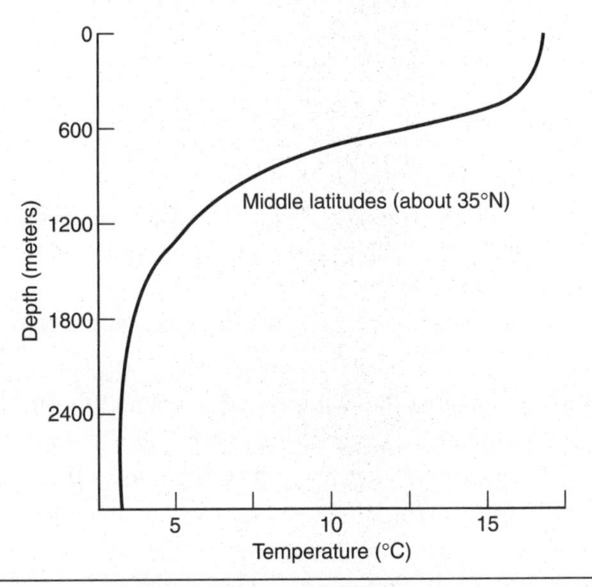

Figure 8-14. The relationship between ocean depth and water temperature: These factors, among others, have an effect on the types of organisms found in an aquatic biome.

Aquatic organisms are well adapted for the removal of dissolved oxygen from water. They also have adaptations for maintaining a proper water balance in their cells. (Water balance is affected by the concentration of salts in the water.)

In aquatic biomes, most photosynthesis takes place near the surface of the water, since light intensity is strongest there. At greater depths, where sunlight does not penetrate, there is no photosynthesis. However, as discussed earlier in this chapter, another type of food-making reaction takes place on parts of the ocean floor; chemosynthesis supports entire communities of organisms very different from those found elsewhere in the ocean or on land.

Marine Biome. The marine, or saltwater, biome includes all the oceans of Earth, which actually make up one continuous body of water. Most of the water on Earth is contained within the saltwater biome (Figure 8-15). The most important characteristics of the marine biome are that it: (a) is the most stable environment on Earth; (b) absorbs and holds large quantities of solar heat, thereby stabilizing Earth's temperature; (c) contains a relatively constant supply of nutrients and dissolved salts; (d) serves as a habitat for a large number and wide variety of organisms; and (e) includes the area in which most of the photosynthesis on Earth occurs (in coastal waters, along the edges of landmasses).

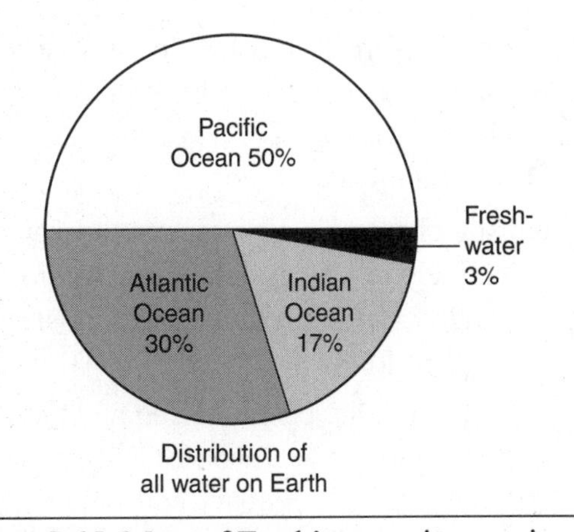

Figure 8-15. Most of Earth's water is contained within the marine biome.

Freshwater Biomes. The freshwater biome includes ponds, lakes, and rivers. Because these are separate bodies of water, they vary widely in size, temperature, oxygen and carbon dioxide concentrations, amounts of suspended particles, current velocity, and rate of succession.

Ponds and lakes tend to fill in over time. Dead plant material and sediment accumulate on the bottom and around the banks, gradually making the body of water shallower and smaller (Figure 8-16 on page 174). Thus, in all but the largest lakes, there is a gradual succession from a freshwater to a terrestrial climax community.

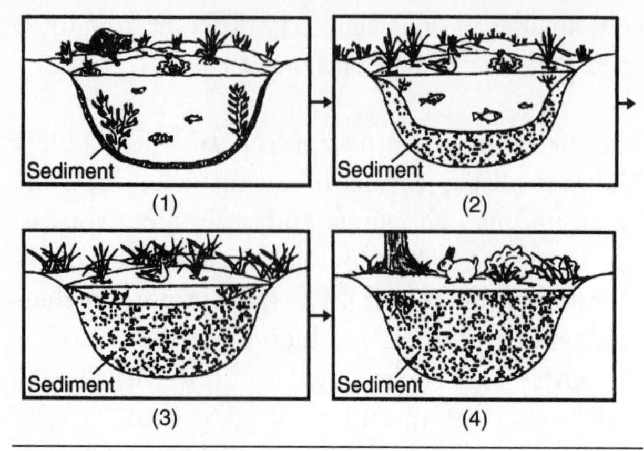

Figure 8-16. Over time, ponds tend to fill in as natural materials accumulate on the sides and bottom.

QUESTIONS

MULTIPLE CHOICE

93. In which of the following biomes does most of the photosynthesis on Earth occur? A. forests B. oceans C. deserts D. grasslands

94. Drastic changes in air temperature would be *least* likely to affect which of these biomes? A. tundra B. temperate forest C. marine D. tropical forest

95. Land biomes are characterized and named according to the A. secondary consumers in the food webs B. primary consumers in the food webs C. climax vegetation in the region D. pioneer vegetation in the region

96. The largest and most stable ecosystems are the A. aquatic biomes B. terrestrial biomes C. high-altitude biomes D. high-latitude biomes

97. Which is the most common sequence of major land biomes encountered when going from the equator to the polar region? A. tundra, taiga, temperate forest, tropical forest B. tropical forest, temperate forest, taiga, tundra C. temperate forest, tropical forest, taiga, tundra D. tropical forest, temperate forest, tundra, taiga

98. Which biome is characterized by its ability to absorb and hold large quantities of solar heat, which helps to regulate Earth's temperature? A. desert B. marine C. grassland D. taiga

99. Generally, an increase in altitude has the same effect on the habitat of organisms as an increase in A. latitude B. moisture C. available light D. longitude

For each description given in questions 100 through 103, select the biome from the list below that most closely matches that description.

A. Desert
B. Grassland
C. Taiga
D. Temperate deciduous forest
E. Tundra

100. This area has a short growing season and low precipitation, mostly in the form of snow. The soil is permanently frozen and the vegetation includes lichens and mosses. A. *A* B. *B* C. *C* D. *D* E. *E*

101. This area has 25 to 50 centimeters of rainfall annually. The growing season does not produce trees, but the soil is rich and well suited for growing crops such as wheat and corn. Grazing animals are found here. A. *A* B. *B* C. *C* D. *D* E. *E*

102. There are many lakes in this area and the vegetation is coniferous forest composed mainly of spruce and fir. There are many large animals, such as bear and deer. A. *A* B. *B* C. *C* D. *D* E. *E*

103. This area has broad-leaved trees, which shed their leaves in the fall. Winters are fairly cold, and the summers are warm with well-distributed rainfall. A. *A* B. *B* C. *C* D. *D* E. *E*

OPEN RESPONSE

104. How are latitude and altitude similar in terms of how they affect the types of organisms that can live in a biome?

105. Describe the two main types of aquatic biomes; list four important abiotic factors that affect the kinds of organisms that live in them.

106. Explain why a coastal city may experience less fluctuation in temperatures during the winter and summer than a city farther inland, even though both cities may be at the same latitude.

READING COMPREHENSION

Base your answers to questions 107 through 110 on the information below and on your knowledge of biology. Source: *Science News* (May 28, 2005): Vol. 167, No. 22, p. 350.

Pesticide Makes Bees Bumble

A naturally derived pesticide previously considered safe for insect pollinators may hamper the foraging of wild bees, researchers report.

Bumblebee larvae raised on pollen spiked with spinosad, an insecticide mixture of chemicals made by bacteria, grow up to be slow, clumsy foragers, say Lora Morandin of Simon Fraser University in Burnaby, British Columbia, and her colleagues in the July *Pest Management Sciences*. As adults, the bees suffer from muscle tremors and take longer to penetrate complex flower structures than do bees nourished as larvae with untainted pollen, the researchers found.

Previous studies on bees hadn't focused on sublethal effects of pesticides or on larvae, says Morandin. Moreover, researchers had looked mostly at domesticated honeybee colonies, which farmers can move before spraying a field. Wild bee colonies don't relocate during spraying, and so these bees probably suffer higher exposures to pesticides.

At least one-third of the food produced in developed countries relies on bees, birds, and other pollinators. Wild bees probably make a significant contribution, though their role hasn't been quantified. In a forthcoming *Ecological Applications*, Morandin shows that canola plants in fields with large wild bee populations produce more seeds than do plants in fields with fewer wild bees.

Mark Winston, also of Simon Fraser University and a coauthor of the pesticide study, notes that bees were affected by spinosad concentrations that insects might reasonably encounter in a crop field. Even so, he adds, the pesticide, which farmers use on many crops against a wide range of insects, shouldn't necessarily be scrapped. Rather, he says, "timing, dose, and formulation may need to be managed in order to use this pesticide properly."

107. What effect does spinosad appear to have on wild bees' foraging (food-getting) ability?

108. Why are wild bee populations more susceptible to this insecticide than domestic bee colonies are?

109. Why is the fact that "at least one-third of the food produced in developed countries relies on bees, birds, and other pollinators" important to this article?

110. According to one of the researchers, what are some ways in which this pesticide problem can be handled?

CHAPTER 9 Scientific Inquiry and Laboratory Skills

Standards SIS1, SIS2, SIS3, SIS4 (Scientific Inquiry Skills)
BROAD CONCEPT: All students need to achieve a sufficient level of scientific literacy to enable them to succeed. To achieve this, students need to develop skills that allow them to search out, describe, and explain natural phenomena and designed artifacts. Scientific inquiry, experimentation, and design involve practice (skills) in direct relationship to knowledge; content knowledge *and* skills are necessary to inquire about the natural and human-made worlds.

As part of any high school biology course, students are expected to master a number of specific scientific inquiry, experimental, and design skills. Some skills involve the use of the **scientific method**, while others involve the use of actual laboratory techniques and procedures. Both sets of skills will be covered in this chapter.

Skills Using the Scientific Method

- Formulate a question or define a problem for investigation and develop a hypothesis to be tested in an investigation.
- Distinguish between controls and variables in an experiment.
- Collect, organize, graph, and analyze data.
- Make predictions based on experimental data.
- Formulate generalizations or conclusions based on the investigation.
- Apply the conclusion to other experimental situations.

Skills Involving Laboratory Procedures

- Given a laboratory problem, select suitable lab materials, safety equipment, and appropriate observation methods.
- Demonstrate safety skills in heating materials in test tubes or beakers, use of chemicals, and handling dissection equipment.
- Identify the parts of a compound light microscope and their functions. Use the microscope correctly under low power and high power.
- Determine the size of microscopic specimens in micrometers.
- Prepare wet mounts of plant and animal cells and apply stains, including iodine and methylene blue.
- With the use of a compound light microscope, identify cell parts, including the nucleus, cytoplasm, chloroplasts, and cell walls.
- Use indicators, such as pH paper, Benedict's solution (or Fehling's solution), iodine solution (or Lugol's solution), and bromthymol blue. Interpret changes shown by the indicators, or *reagents*.
- Use measurement instruments, such as metric rulers, Celsius thermometers, graduated cylinders, pipettes, and scales.
- Dissect plant or animal specimens, exposing major structures for examination.

THE SCIENTIFIC METHOD

Defining a Problem and Developing a Hypothesis. Scientists conduct research to answer a question or to solve a problem. The first step in planning a research project is to define the problem to be solved; this is usually stated in the form of a question, ideally one that can be answered with a "Yes" or a "No." For example, a researcher can ask: "Are enzymes affected by temperature?" The next step is to develop a possible solution to the problem. This proposed explanation, or *hypothesis*, is the statement that identifies the factor to be tested in the experiment. The hypothesis is an "educated guess" because it is based on some previous scientific knowledge. A scientist studying the enzyme amylase might want to measure the rate of enzyme action at various temperatures. The basic hypothesis would be that the rate at which amylase breaks down, or *hydrolyzes*, starch is affected by temperature. It could be stated: "If temperature is changed, then the rate of amylase activity will also change."

Designing and Conducting an Experiment. Biologists use controlled experiments when doing research. In a controlled experiment, there are actually two setups: an *experimental* setup and a *control* setup. The experimental and control setups are identical except for the single factor, or **variable**, that is being tested. The experimental group is exposed to the variable, but the control group is not; all other conditions are kept the same. Any changes observed during the experiment can then be explained in terms of the variable. Thus, in an experiment to determine the effect of temperature on the rate of enzyme action, temperature would be the variable.

The experimental setup in this case would involve testing the enzyme amylase's activity at several different temperatures; the control setup would remain at only one temperature. The control is used for comparison, that is, to determine whether or not the variable has any effect on the experimental group. The researcher can measure the rate of enzyme activity at the different temperatures and then compare the results with that of the control group. If the rates of activity change, then it would be reasonable to conclude that temperature does have an effect on the enzyme's activity.

Collecting, Organizing, and Graphing Data. During an experiment, the scientist makes careful observations and measurements. These data are the results of the experiment. The data may be recorded in a notebook in the form of a chart or data table. Often, the results are plotted on a graph. Line graphs are a particularly useful way to display data because they provide a visual representation of how two factors are related.

In a line graph that depicts the effect of temperature on the rate of enzyme action, you can clearly see the relationship (Figure 9-1). For example, as temperature increases from 0°C to 40°C, enzyme action increases and more starch is hydrolyzed. Above 40°C, enzyme action decreases and less starch is hydrolyzed. The information on the *x*-axis is called the *independent variable* because it is independently determined by the scientist; the measurements on the *y*-axis are called the *dependent variable* because each value depends on changes made to the independent variable. Scientists also use computers to record and organize experimental results. The computer can analyze the data and print reports, which scientists can then use to form their conclusions.

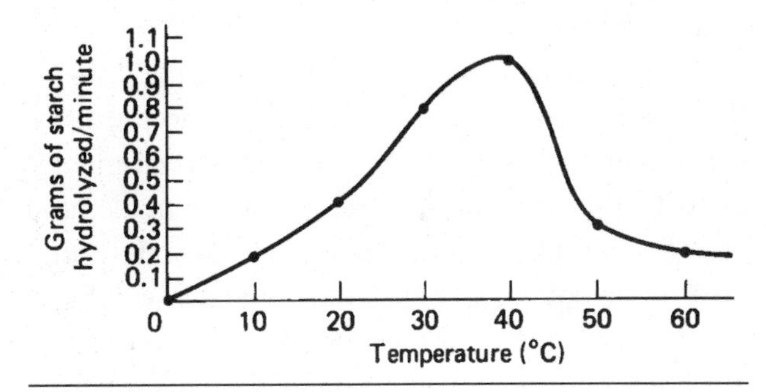

Figure 9-1. A line graph.

The relationship between two varying factors can also be shown clearly in a data table. The data table in Figure 9-2 shows the same information as the line graph in Figure 9-1.

Temperature (°C)	Grams of starch hydrolyzed per minute
0	0.0
10	0.2
20	0.4
30	0.8
40	1.0
50	0.3
60	0.2

Figure 9-2. A data table.

Making Predictions Based on Experimental Data. Scientists may make predictions based on experimental data. The validity of these predictions can then be tested by further experimentation.

For example, on the basis of the data shown in Figures 9-1 and 9-2, a scientist might predict that the number of grams of starch hydrolyzed at normal body temperature (37°C) would be between 0.8 and 1.0 gram/minute. Further measurements might show that the prediction was correct, or they might show that at 37°C the rate was higher than 1.0 gram/minute. Scientists must be extremely careful not to make any assumptions that are not supported by the data.

Making Generalizations and Drawing Conclusions. The results of an experiment are collected and analyzed. For a conclusion to be valid, the experiment must be repeated many times, produce similar results, and all the results must be included in the analysis. The conclusion is based solely on the experimental data. The more times an experiment is repeated and produces the same or similar results, the more valid the conclusion will be.

In the experiment on the effect of temperature on the rate of action of amylase, the data in the table show that the enzyme functions most efficiently at 40°C. However, if measurements were made only at 10° intervals, you could not say definitely that 40°C is the optimum temperature for amylase without making measurements at other intermediate temperatures. Still, it is probably safe to conclude that the optimum temperature is close to 40°C.

Student-Designed Lab Experiments

A key skill in a science course is the ability to develop one's own procedure for investigating a problem. Investigations generally begin with an observation that prompts a question in one's mind. The researcher then develops a procedure to run a controlled experiment that follows the scientific method to investigate the problem. An experiment conducted according to the scientific method would include the following: a research problem, a hypothesis, experimental materials, a procedure, a means to gather and record data, an analysis of the data, and then a conclusion that supports or refutes the hypothesis. *Note:* A hypothesis is never *proven*; rather, it is said to be *supported*, or not, by the data.

Here is an example of a student-designed lab that begins with an observation or a statement.

Scenario: A garden supply company has developed a new product called Bunny Hop-Away. The product is supposed to keep rabbits away from your garden plants. The researcher is skeptical about the product and want to know if it really works.

Problem: Does Bunny Hop-Away keep rabbits away from garden plants?

Hypothesis: Bunny Hop-Away will keep rabbits away from lettuce plants.

Materials: 200 lettuce plants of the same variety; Bunny Hop-Away product; 20–30 rabbits

Procedures:

1. Divide the lettuce plants into two groups of 100 each and plant in the garden in two spots, at least 15 meters (about 50 feet) away from each other.
2. Spray one group of lettuce plants (the experimental) with the product Bunny Hop-Away. Do not spray the other group of lettuce plants (the control).
3. Release the rabbits into the garden near the lettuce plants.
4. Count the number of times rabbits approach the treated and untreated lettuce plants for a total of 30 minutes.
5. Repeat the experiment two more times after allowing the rabbits some time to rest and digest. For example, repeat the procedure once the next day and once the day after that.

Data:

Time (minutes)	***Number of rabbits that approach treated plants***	***Number of rabbits that approach untreated plants***
5	0	3
10	0	7
15	1	12
20	0	14
25	1	21
30	0	23

Analysis: If the researcher observes that rabbits approach mostly, or only, the untreated plants and stay away from the treated plants, then the hypothesis has been supported.

Conclusion: Bunny Hop-Away is effective in keeping rabbits away from lettuce (garden) plants.

QUESTIONS

MULTIPLE CHOICE

1. The diagram below represents a setup at the beginning of a laboratory investigation.

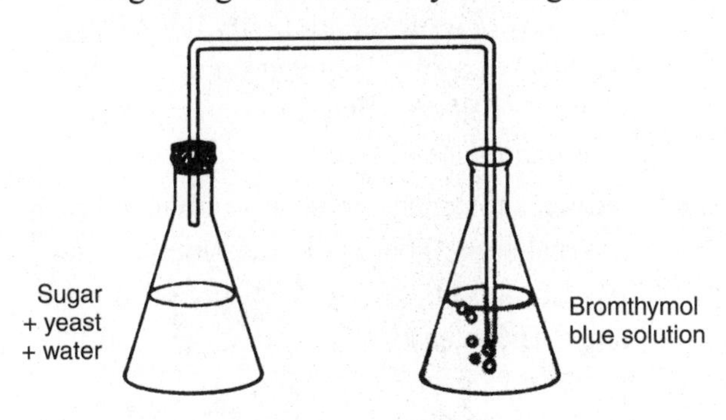

Which hypothesis would most likely be supported by observing and collecting data from this investigation? A. The fermentation of a yeast-sugar solution results in the production of carbon dioxide. B. Yeast cells contain simple sugars. C. Oxygen is released when a yeast-sugar solution is illuminated with green light. D. Yeast cells contain starches.

Base your answers to questions 2 and 3 on the information and data table below and on your knowledge of biology.

A green plant was placed in a test tube, and a light was placed at varying distances from the plant. The bubbles of oxygen given off by the plant were counted. The table shows the data collected during this experiment.

Distance of light from plant (cm)	Number of bubbles per minute
10	60
20	25
30	10
40	5

2. The variable in this investigation is the A. color of the light used B. distance between the light and the plant C. size of the test tube D. type of plant used

3. Which conclusion can be drawn from this investigation? A. As the distance from the light increases, the number of bubbles produced decreases. B. As the distance from the light increases, the number of bubbles produced increases. C. As the distance from the light decreases, the number of bubbles produced decreases. D. There is no relationship between the number of bubbles produced and the distance of the plant from the light.

Base your answers to questions 4 through 6 on the following information, diagram, and data table, and on your knowledge of biology.

A student is studying the effect of temperature on the digestive action of the enzyme gastric protease, which is contained in gastric fluid. An investigation is set up using five identical test tubes, each containing 40 milliliters of gastric fluid and 20 millimeters of glass tubing filled with cooked, chopped up egg white, as shown below. After 48 hours, the amount of egg white digested in each tube was measured. The data collected are shown in the data table below.

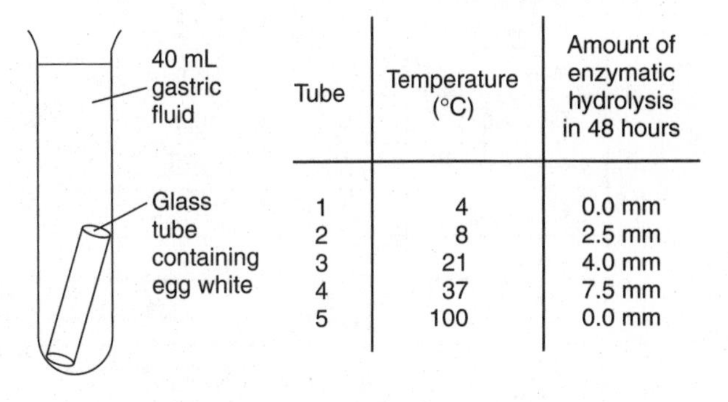

Tube	Temperature (°C)	Amount of enzymatic hydrolysis in 48 hours
1	4	0.0 mm
2	8	2.5 mm
3	21	4.0 mm
4	37	7.5 mm
5	100	0.0 mm

4. Which is the independent variable in this investigation? A. gastric fluid B. length of glass tubing C. temperature D. time

5. If an additional test tube were set up identical to the other test tubes and placed at a temperature of 15°C for 48 hours, what amount of hydrolysis might be expected? A. less than 2.5 mm B. between 2.5 mm and 4.0 mm C. between 4.0 mm and 7.5 mm D. more than 7.5 mm

6. Which set of axes shown below would produce the best graph for plotting the data from the results of this investigation?
A. 1 B. 2 C. 3 D. 4

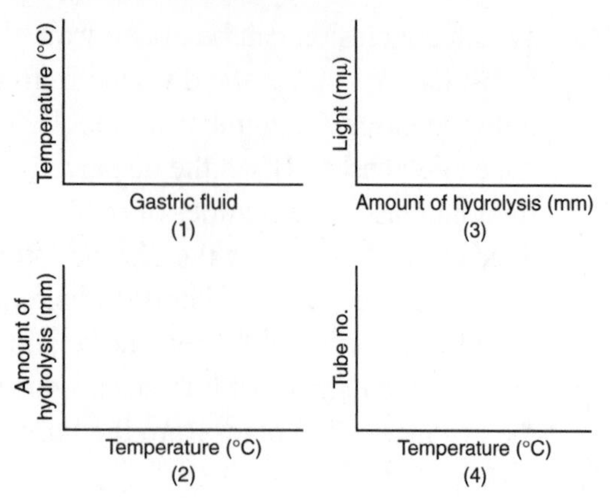

Base your answers to questions 7 through 11 on the information and two charts below and on your knowledge of biology.

Chart I shows the percentages of certain materials in the blood entering the kidney and the percentages of the same materials in the urine leaving the body. Chart II shows the number of molecules in the beginning and at the end of the kidney tubule for every 100 molecules of each substance entering the glomerulus.

Chart I

Substance	% of blood	% of urine
Protein	7.0	7.0
Water	91.5	96.0
Glucose	0.1	0.0
Sodium	0.33	0.29
Potassium	0.02	0.24
Urea	0.03	2.7

Chart II

	Number of Molecules		
Substance	In blood entering glomerulus	Beginning of tubule	End of tubule
Protein	100	0	0
Water	100	30	1
Glucose	100	20	0
Sodium	100	30	1
Potassium	100	23	12
Urea	100	50	90

7. According to Chart I, which of these substances is more highly concentrated in the urine than in the blood? A. water B. sodium C. protein D. glucose

8. According to Charts I and II, which substance enters the tubules but does *not* appear in the urine leaving the body? A. protein B. water C. glucose D. potassium

9. According to the data, which substance did *not* pass out of the blood into the tubule? A. water B. urea C. glucose D. protein

10. The data in the two charts would best aid a biologist in understanding the function of the A. heart of a frog B. nephron of a human C. nerve cell of a fish D. contractile vacuole of a paramecium

11. Which substances enter the tubule and then are reabsorbed back into the blood as they pass through the tubule? A. urea and potassium B. water and sodium C. urea and protein D. protein and glucose

Base your answers to questions 12 through 14 on the information provided by the graph below. The graph shows the average growth rate in grams for 38 pairs of newborn rats. One member of each pair was injected with anterior pituitary extract. The other member of each pair served as a control.

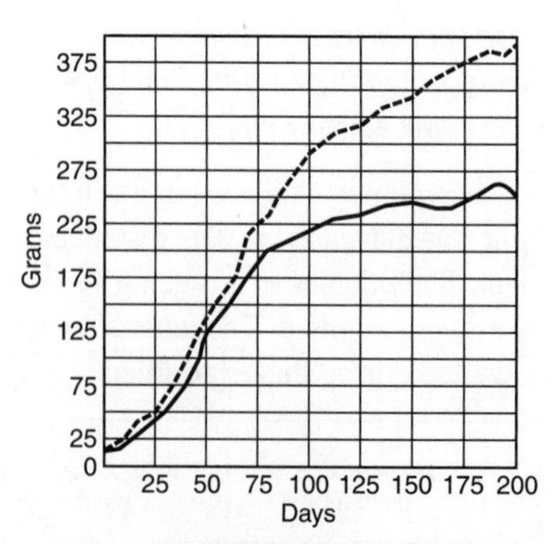

12. At 75 days, what was the average weight of the rats injected with pituitary extract? A. 65 grams B. 125 grams C. 200 grams D. 225 grams

13. Based on the graph, it can be correctly concluded that the pituitary extract A. is essential for life B. determines when a rat will be born C. affects the growth of rats D. affects the growth of all animals

14. The graph shows the relationship between the weight of treated and untreated rats and the A. age of the rats B. sex of the rats C. size of the rats' pituitary glands D. type of food fed to the rats

Base your answers to questions 15 and 16 on the following information, diagrams, and data table, and on your knowledge of laboratory procedures used in biology.

Diagrams *A* through *E* show the general appearance of five tree fruits that were used by a science class in an experiment to determine the length of time necessary for each type of fruit to fall from a second-floor balcony to the lobby floor of their school. One hundred fruits of each type were selected by the students, and the average time of fall for each type of fruit is shown in the table that follows.

Fruits (not drawn to scale)

Silver maple (A)

Norway maple (B)

White ash (C)

Red oak (D)

Shagbark hickory (E)

Tree type	*Average fall time of 100 fruits*
Silver maple	3.2 sec
Norway maple	4.9 sec
White ash	1.5 sec
Red oak	0.8 sec
Shagbark hickory	0.8 sec

15. Based on this experimental evidence, what inference seems most likely to be true concerning the distribution of these fruits during windstorms in nature? A. Silver maple fruits would land closer to the base of their parent tree than would shagbark hickory fruits. B. White ash fruits would land farther from the base of their parent tree than would silver maple fruits. C. White ash fruits would land closer to the base of their parent tree than would shagbark hickory fruits. D. Norway maple fruits would land farther from the base of their parent tree than would silver maple fruits.

16. Which graph best shows the average fall time for each fruit type tested during this experiment? A. 1 B. 2 C. 3 D. 4

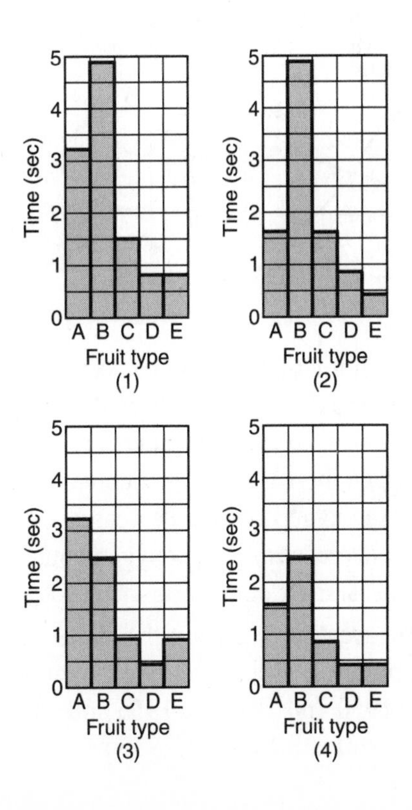

17. The graph below was developed as a result of an investigation of bacterial counts of three identical cultures grown at different temperatures. Which conclusion might be correctly drawn from this graph?

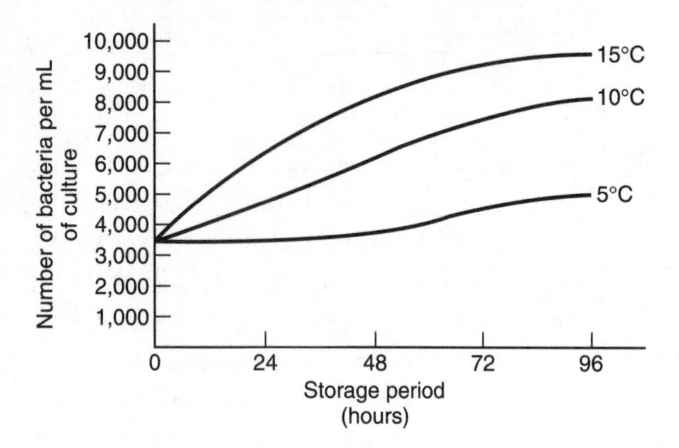

A. The culture contains no bacteria. B. Refrigeration retards bacterial reproduction. C. Temperature is unrelated to the bacteria reproduction rate. D. Bacteria cannot grow at a temperature of 5°C.

Base your answers to questions 18 through 20 on the following graphs, which show data on some environmental factors affecting a large Northeastern lake.

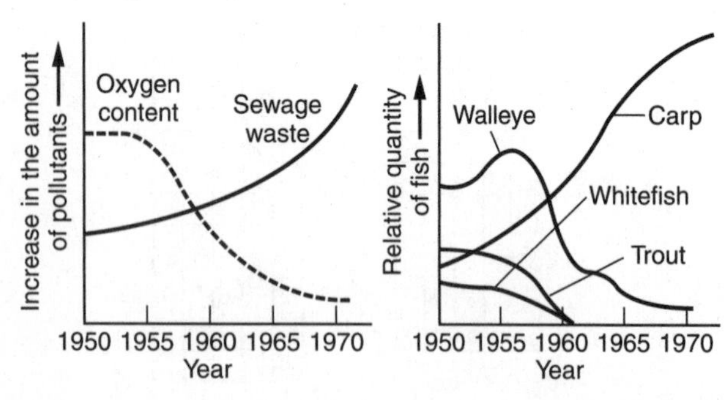

18. Which relationship can be correctly inferred from the data presented? A. As sewage waste increases, oxygen content decreases. B. As sewage increases, oxygen content increases. C. As oxygen content decreases, carp population decreases. D. As oxygen content decreases, trout population increases.

19. The greatest change in the lake's whitefish population occurred between which years? A. 1950 and 1955 B. 1955 and 1960 C. 1960 and 1965 D. 1965 and 1970

20. Which of the fish species appears able to withstand the greatest degree of oxygen depletion? A. trout B. carp C. walleye D. whitefish

LABORATORY SKILLS AND PROCEDURES

Selecting Suitable Lab Equipment. Knowledge of the correct lab equipment is essential for planning and carrying out an experiment. Figure 9-3 illustrates some basic laboratory equipment that you should know.

Safety in the Laboratory. Following are some safety precautions that you should practice in the laboratory.

- Do not handle chemicals or equipment unless you are told by your teacher to do so.
- If any of your lab equipment appears to be broken or unusual, do not use it. Report it to your teacher.
- Report any personal injury or damage to clothing to your teacher immediately.
- Report any unsafe activities to your teacher immediately.
- Wear appropriate safety equipment, such as safety goggles and lab apron. Tie back long hair; secure dangling jewelry and loose sleeves.
- Never taste or directly inhale unknown chemicals. Never eat or drink in the lab.
- Never pour chemicals back into stock bottles; never exchange bottle stoppers (between bottles); and never place the stopper on the lab table.
- When heating a liquid in a test tube, always wear safety goggles and point the opening away from yourself and all others.
- Handle all sharp instruments with care, moving slowly and deliberately.

Using a Compound Light Microscope. Review the parts of the compound light microscope and their functions by studying Figure 1-2 and Table 1-1 (on pages 18–19).

In using the compound microscope, the observer should begin by viewing the specimen with the low-power objective, focusing first with the coarse adjust-

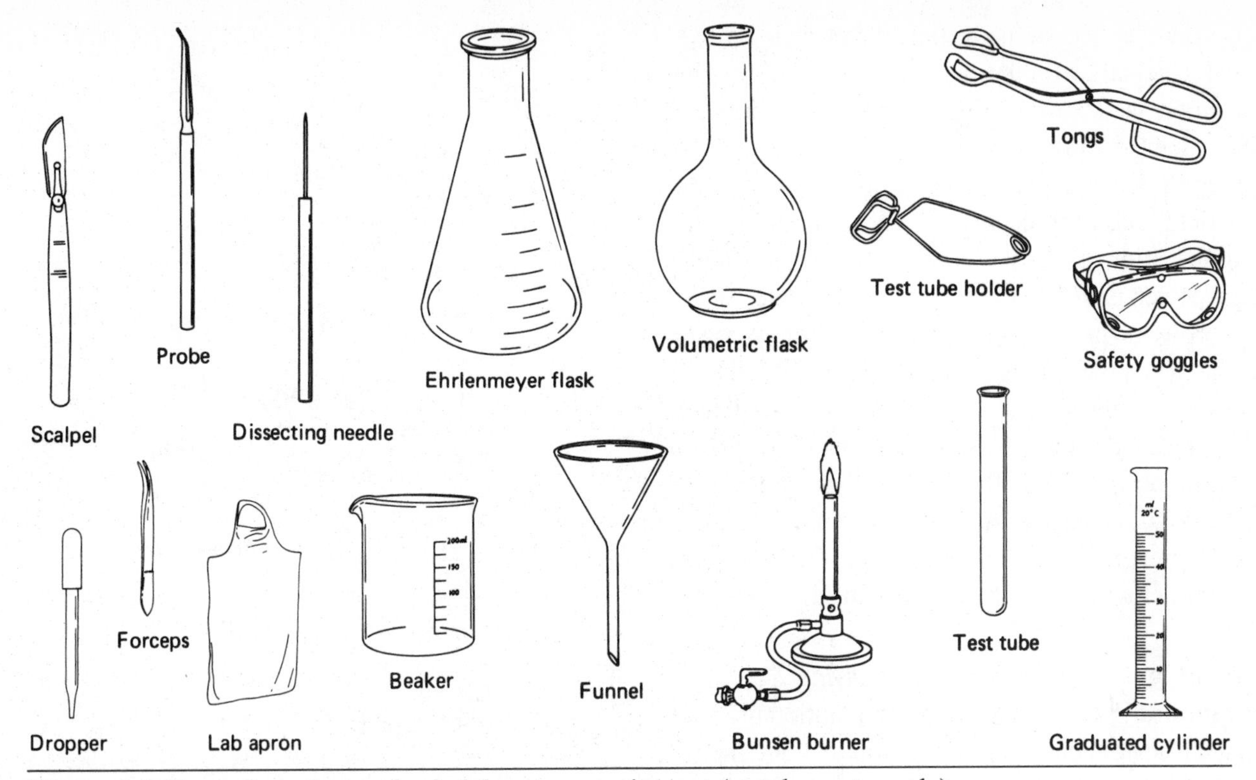

Figure 9-3. Examples of some basic laboratory equipment (not drawn to scale).

ment, then with the fine adjustment. The objectives can then be switched from low power to high power. All focusing under high power should be done with the fine adjustment. The field appears dimmer under high power than under low power. Open the diaphragm to allow more light to reach the specimen.

Due to the nature of lenses, the image of an object seen under a microscope is enlarged, reversed (backward), and inverted (upside down). When viewed through the microscope, an organism that appears to be moving to the right is actually moving to the left. An organism that appears to be moving toward the observer, or up, is actually moving away from the observer, or down.

Determining the Size of Microscopic Specimens. To determine the size of a specimen being examined under a microscope, you must know the diameter of the microscope field. You can actually measure the field diameter with a clear plastic metric ruler. Place the ruler over the opening in the stage of the microscope, as shown in Figure 9-4. Focus on the ruler markings and adjust the position of the ruler so that a millimeter marking is at the left.

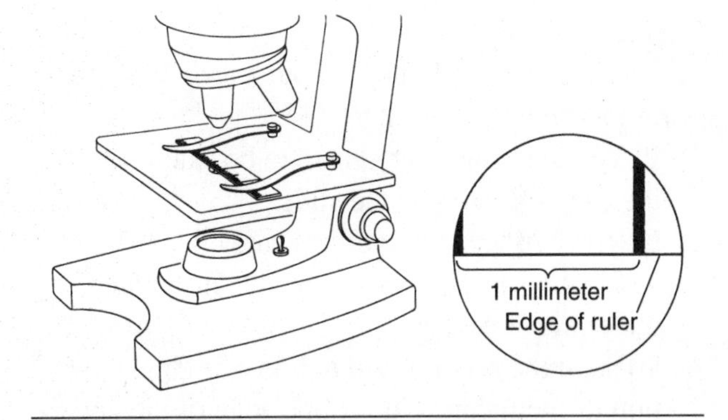

Figure 9-4. Measuring with a microscope.

Once you have estimated the field diameter under low power, you can estimate the size of specimens observed under low power by how much of the field they cover. For example, if the diameter of the field is 1.5 mm and a specimen is about one-third the diameter of the field, the specimen is about 0.5 mm in length.

The unit most commonly used in measuring microscopic specimens is the *micrometer*, symbol μm, which is one-thousandth of a millimeter. *Note:* The micrometer is also referred to as a *micron*.

1 mm = 1000 μm   1 μm = 0.001 mm

In the example given, the field diameter is 1.5 mm, which is equal to 1500 μm ($1.5 \times 1000 = 1500$). The specimen is 0.5 mm long, which equals 500 μm ($0.5 \times 1000 = 500$).

When you switch from low power to high power, the field diameter decreases. For example, if the magnification under low power is 100× and under high power is 400×, then the field diameter under high power will be one-fourth that under low power. If the low-power magnification is 100× and the high-power magnification is 500×, then the diameter of the high-power field will be one-fifth that of the low-power field. Remember that under high power you are seeing *less* of the specimen, but with greater magnification. The size of the specimen does not change, only the diameter of the field.

Preparing a Wet Mount and Applying Stains. A **wet mount** is a temporary slide preparation used for viewing specimens with a compound light microscope. Any specimen to be examined must be thin enough for light to pass through it.

The preparation of a wet mount involves the following steps:

1. Use a medicine dropper to put a drop of water in the center of the slide.
2. Place the tissue or organism to be examined in the drop of water on the slide.
3. Cover the specimen with a coverslip, as shown in Figure 9-5. Try not to trap any air bubbles under the coverslip.
4. To stain the section, add a drop of iodine solution or methylene blue at one edge of the coverslip. Touch a small piece of paper towel to the opposite side of the coverslip to draw the stain across the slide and through the specimen.

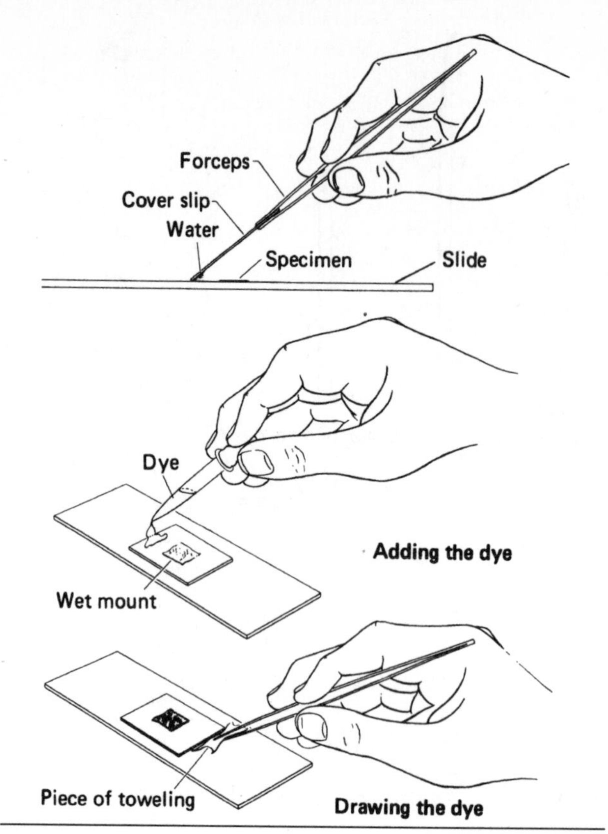

Figure 9-5. Making a wet mount (above) and staining a specimen.

Identifying Cell Parts with a Compound Light Microscope. Review the structure of plant cells and animal cells (covered in Chapter 1).

Unstained cells viewed with a compound light microscope show relatively little detail. The use of stains, such as iodine or methylene blue, enhances contrast. With such stains, the nucleus becomes clearly visible, and in plant and algal cells the cell wall becomes visible, too. Chloroplasts are visible as small oval green structures. Most other cell organelles, including mitochondria and the endoplasmic reticulum, are not usually visible with the compound light microscope; they are too small to be magnified by it.

Using Indicators and Interpreting Changes. Indicators are used to test for the presence of specific substances or chemical characteristics.

Litmus paper is an indicator used to determine whether a solution is an acid or a base. Litmus papers are small strips of filter paper that have been soaked in a solution of litmus (a plant dye) and then dried. An acid turns blue litmus paper red (or pink). A base turns red litmus paper blue. Litmus paper will get darker when it is dipped in a test solution, but that is because it is wet. If the paper gets darker, but does not change color, it is a negative test (that is, the solution was neither an acid nor a base).

pH paper is an indicator used to determine the actual pH (concentration of hydrogen ions) of a solution. When a piece of pH paper is dipped into a test solution, it changes color. The color of the pH paper is then matched against a color chart, which shows the pH value. The pH scale ranges from 0 to 14. According to this scale, pH values under 7 are acidic; above 7 are basic; and exactly 7 are neutral (neither acid nor base).

Bromthymol blue is an indicator used to detect carbon dioxide. In the presence of carbon dioxide, bromthymol blue turns to bromthymol yellow. If the carbon dioxide is removed, the indicator changes back to bromthymol blue.

Benedict's solution is an indicator used to test for simple sugars. When heated in the presence of simple sugars, Benedict's solution turns from blue to yellow, green, or brick red, depending on the sugar concentration. *Fehling's solution* also may be used to test for simple sugars.

Lugol's, or *iodine, solution* is an indicator used to test for starch. In the presence of starch, Lugol solution turns from red-orange to blue-black.

Biuret's solution is an indicator used to test for protein. In the presence of protein, Biuret's solution turns from light blue to purple.

Using Measurement Instruments. The following tools are used for making scientific measurements.

- *Metric ruler*. The basic unit of length in the metric system is the meter, abbreviated m. One meter contains 100 centimeters (cm). As shown in Figure 9-6, metric rulers are generally calibrated in centimeters and millimeters (mm). Each centimeter contains 10 millimeters, thus each meter is equal to 1000 mm.

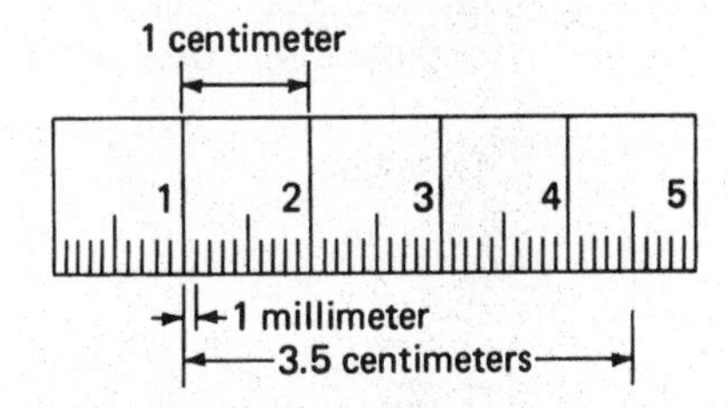

Figure 9-6. A metric ruler.

- *Celsius thermometer*. In the metric system, temperature is commonly measured in degrees Celsius. On the Celsius scale, 0°C is the freezing point of water, 21°C is room temperature, and 100°C is the boiling point of water. Figure 9-7 shows a thermometer calibrated in degrees Celsius. Note that each degree is marked by a short line (such as 37°C), and every tenth degree is labeled with the number (such as 30°C and 40°C).

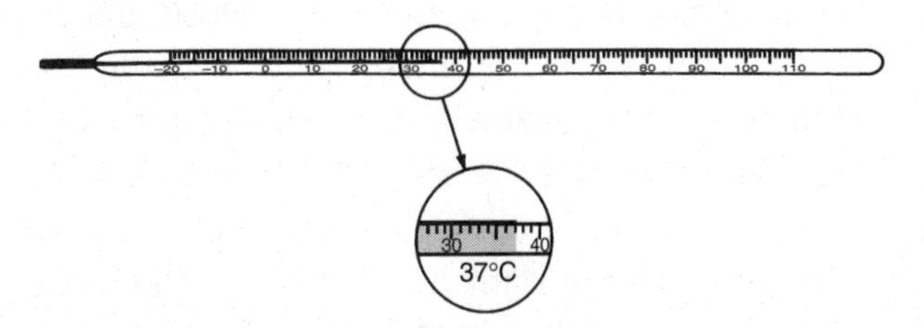

Figure 9-7. A Celsius thermometer.

- *Graduated cylinder*. The basic unit used for measuring the volume of a liquid in the metric system is the liter, abbreviated L. A liter contains 1000 milliliters (mL). Most laboratory measurements involve milliliters rather than liters.

The volume of a liquid is frequently measured in graduated cylinders, which come in many sizes. When you need an accurately measured amount of liquid, use a graduated cylinder of appropriate size—that is, to measure 5 mL of liquid, use a 10-mL graduated cylinder, not a 1000-mL graduated cylinder.

The surface of water and similar liquids curves upward along the sides of a cylinder. This curved surface, or *meniscus*, is caused by the strong attraction of liquid molecules to the glass surface (Figure 9-8). For an accurate measurement, the reading should be done at eye level, and the measurement should be read at the bottom of the meniscus. With other types of liquids—for example, mercury—the meniscus curves the other way (that is, downward); in such cases, the measurement should be read across the top of the meniscus.

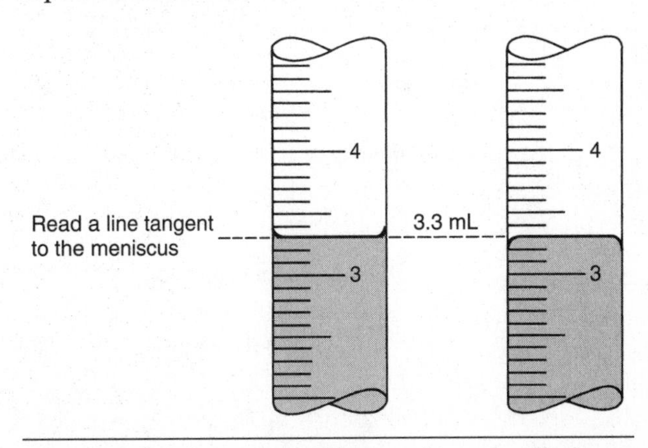

Figure 9-8. Measuring with a graduated cylinder.

Dissecting Plant and Animal Specimens. Dissections are done to expose major structures for examination. The specimen is generally placed in a dissection pan and fastened down with pins or strings. While doing a dissection, you should be very careful with the dissection instruments, which are sharp. Scalpels, forceps, scissors, and stereomicroscopes are used. You should also be careful in cutting into and handling the specimen so that you do not damage important structures. Follow all instructions and record your observations by making labeled diagrams as you proceed with the dissection. Since the specimens are preserved in a chemical, it is also very important to wear rubber gloves, a lab apron, and safety goggles to protect your eyes.

QUESTIONS

MULTIPLE CHOICE

Base your answers to questions 21 through 23 on the four sets of lab materials listed below and on your knowledge of biology.

21. Which set should a student select to test for the presence of a carbohydrate in food? A. Set *A* B. Set *B* C. Set *C* D. Set *D*

22. Which set should a student select to determine the location of the ovules in the ovary of a flower? A. Set *A* B. Set *B* C. Set *C* D. Set *D*

23. Which set should a student use to observe chloroplasts in elodea (a water plant)? A. Set *A* B. Set *B* C. Set *C* D. Set *D*

24. To view cells under the high power of a compound light microscope, a student places a slide of the cells on the stage and moves the stage clips over to secure the slide. She then moves the high-power objective into place and focuses on the slide with the coarse adjustment. Two steps in this procedure are incorrect. For this procedure to be correct, she should have focused under A. low power using coarse and fine adjustments and then under high power using only the fine adjustment B. high power first, then low power using only the fine adjustment C. low power using the coarse and fine adjustments and then under high power using coarse and fine adjustments D. low power using the fine adjustment and then under high power using only the fine adjustment

Set A	Set B	Set C	Set D
Light source	Droppers	Scalpel	Compound microscope
Colored filters	Benedict solution	Forceps	Glass slides
Test tubes	Iodine	Scissors	Water
Test-tube stand	Test tubes	Pan with wax bottom	Forceps
	Test-tube rack	Stereo-microscope	
	Heat source	Pins	
	Safety goggles	Safety goggles	

Base your answers to questions 25 and 26 on the following diagram of a compound light microscope.

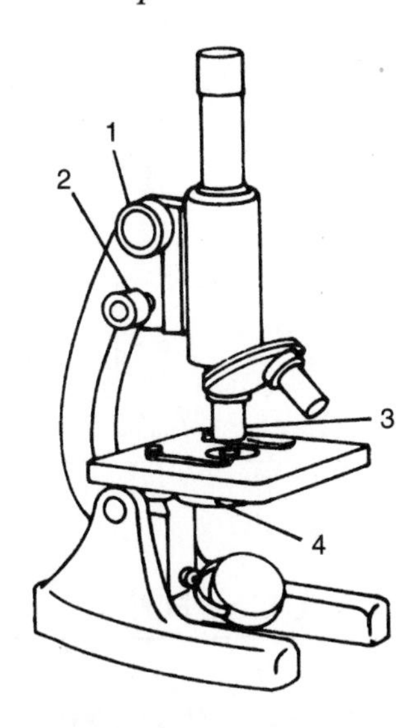

25. The part labeled 1 is used to A. increase the amount of light reaching the specimen B. focus with the high-power objective C. hold the lenses in place D. focus with the low-power objective

26. To adjust the amount of light reaching the specimen, you would use the part labeled A. 1 B. 2 C. 3 D. 4

Base your answers to questions 27 through 29 on the information below and on your knowledge of biology.

A student prepares a wet mount of onion epidermis and observes it under three powers of magnification with a compound light microscope (40×, 100×, and 400×).

27. An adjustment should be made to allow more light to pass through the specimen when the student changes the magnification from A. 100× to 400× B. 400× to 100× C. 400× to 40× D. 100× to 40×

28. Iodine stain is added to the slide. Under 400× magnification, the student should be able to observe a A. mitochondrion B. nucleus C. ribosome D. centriole

29. A specimen that is suitable for observation under this microscope should be A. stained with Benedict's solution B. moving and respiring C. alive and reproducing D. thin and transparent

30. A microscope is supplied with 10× and 15× eyepieces, and with 10× and 44× objectives. What is the maximum magnification that can be obtained from this microscope? A. 59× B. 150× C. 440× D. 660×

31. Under low power (100×), a row of eight cells can fit across the field of a certain microscope. How many of these cells could be viewed in the high power (400×) visual field of this microscope? A. 1 B. 2 C. 8 D. 32

32. A compound light microscope has a 10× ocular, a 10× low-power objective, and a 40× high-power objective. A student noted that under high power, four cells end to end extended across the diameter of the field. If the microscope were switched to low power, approximately how many cells would fit across the field? A. 1 B. 8 C. 16 D. 4

33. The diagram below shows a section of a metric ruler scale as seen through a compound light microscope. If each division represents 1 millimeter, what is the approximate width of the microscope's field of view in micrometers (μm)? A. 3700 B. 4200 C. 4500 D. 5000

Base your answers to questions 34 through 37 on your knowledge of biology and on the diagrams below, which represent fields of view under the low power of the same compound microscope (100×). Diagram A *shows the millimeter divisions of a plastic ruler, and diagram* B *shows a sample of stained onion epidermal cells.*

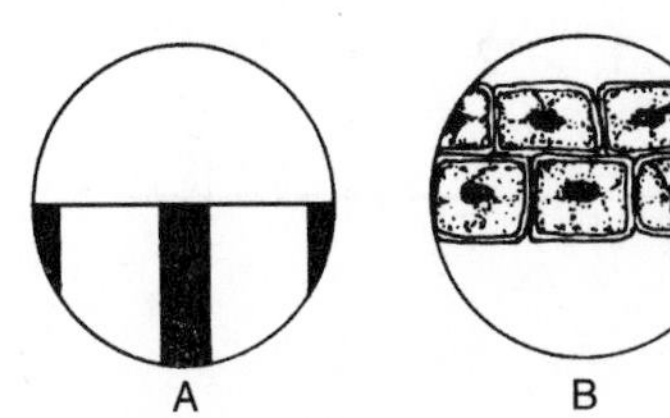

34. Structure *X* in diagram *B* was most likely stained by adding A. water B. iodine solution C. Benedict's solution D. bromthymol blue

35. Structure *X* in diagram *B* indicates A. a nucleus B. a mitochondrion C. the cell wall D. the cytoplasm

36. The diameter of the field of view in diagram *A* is approximately A. 500 μm B. 1000 μm C. 1500 μm D. 2000 μm

37. What is the approximate length of each onion epidermal cell in field *B*? A. 200 μm B. 660 μm C. 1000 μm D. 2500 μm

38. Iodine solution is used to test for the presence of A. proteins B. simple sugars C. oxygen D. starch

39. In the presence of carbon dioxide, bromthymol blue A. shows no color change B. turns yellow C. turns blue-black D. turns red-orange

40. Benedict's solution is used to test for A. disaccharides B. oxygen C. starch D. simple sugars

41. Which piece of equipment should be used to transfer a protist onto a microscope slide? A. scissors B. dissecting needles C. medicine dropper D. forceps

42. While a student is heating a liquid in a test tube, the mouth of the tube should always be A. corked with a rubber stopper B. pointed toward the student C. allowed to cool off D. aimed away from everybody

Base your answer to question 43 on the diagram of a peppered moth and the metric ruler below.

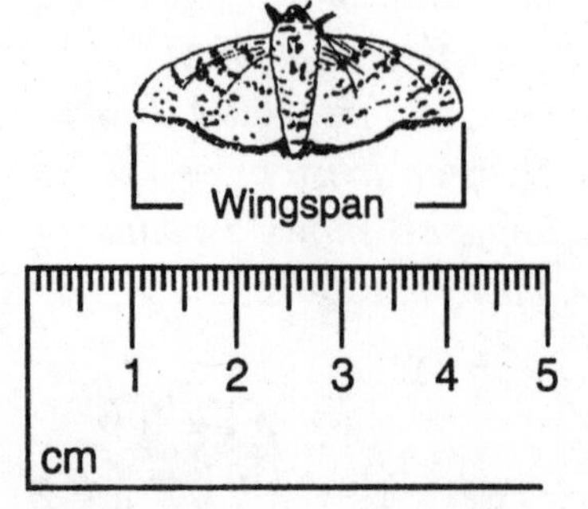

43. Which row in the table below best represents the ratio of body length to wingspan of the peppered moth? A. Row 1 B. Row 2 C. Row 3 D. Row 4

Row	Body Length:Wingspan
(1)	1:1
(2)	2:1
(3)	1:2
(4)	2:2

QUESTIONS

MULTIPLE CHOICE AND OPEN RESPONSE

Base your answer to question 44 on the following information and data table.

To determine which colors of light are best used by plants for photosynthesis, three types of underwater green plants of similar mass were subjected to the same intensity of light of different colors for the same amount of time. All other environmental conditions were kept the same. After 15 minutes, a video camera was used to record the number of bubbles of gas (oxygen) each plant gave off in a 30-second time period. Each type of plant was tested six times. The average of the data for each plant type is shown in the table below.

Average Number of Bubbles Given Off in 30 Seconds

Plant Type	*Red Light*	*Yellow Light*	*Green Light*	*Blue Light*
Plant #1	35	11	5	47
Plant #2	48	8	2	63
Plant #3	28	9	6	39

44. Which statement is a valid inference based on the data? A. Each plant carried on photosynthesis best in a different color of light. B. Red light is better for photosynthesis than blue light. C. These types of plants make food at the fastest rates with red and blue light. D. Water must filter out red and green light.

Base your answers to questions 45 through 48 on the following information and diagram and on your knowledge of biology.

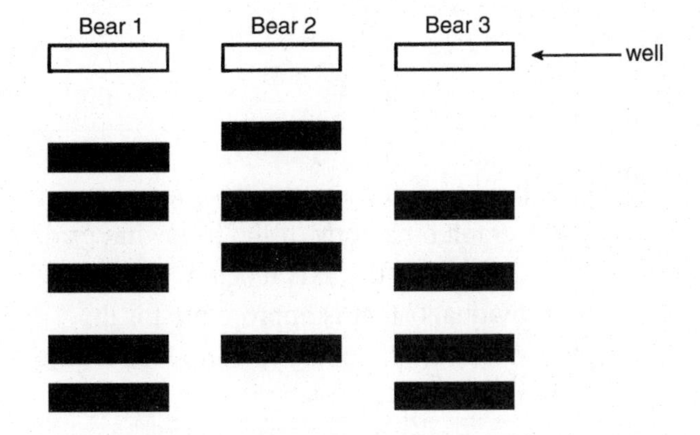

The diagram shows the results of a test that was done using DNA samples from three bears of different species. Each DNA sample was cut into fragments using a specific enzyme and placed in the wells as indicated below. The DNA fragments were then separated using gel electrophoresis.

45. Which *two* bears are most closely related? Support your answer with data from the test results.

46. Identify one additional way to determine the evolutionary relationship of these bears.

47. Gel electrophoresis was used to separate the bears' DNA fragments on the basis of their A. size B. color C. functions D. chromosomes

48. Identify one procedure, other than electrophoresis, that is used in the laboratory to separate the different types of molecules in a liquid mixture.

Base your answer to question 49 on the following information.

A student measures his pulse rate while he is watching television and records it. Next, he walks to a friend's house nearby and, when he arrives, measures and records his pulse rate again. He and his friend then decide to run to the mall a few blocks away. On arriving at the mall, the student measures and records his pulse rate once again. Finally, after sitting and talking for a half hour, the student measures and records his pulse rate one last time.

49. Which of the following graphs best illustrates the expected changes in the student's pulse rate according to the activities described above? A. 1 B. 2 C. 3 D. 4

(1)

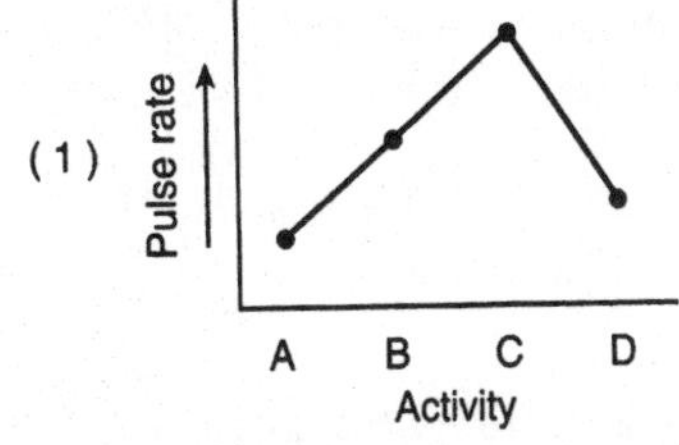

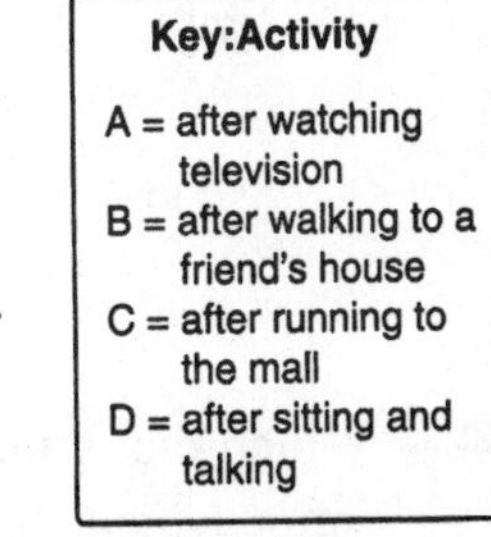

(2)

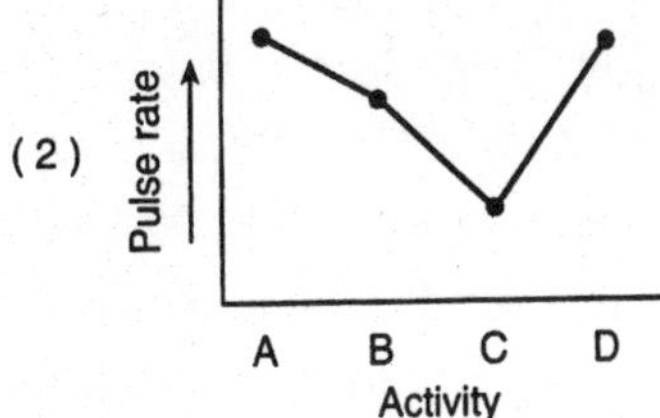

(3)

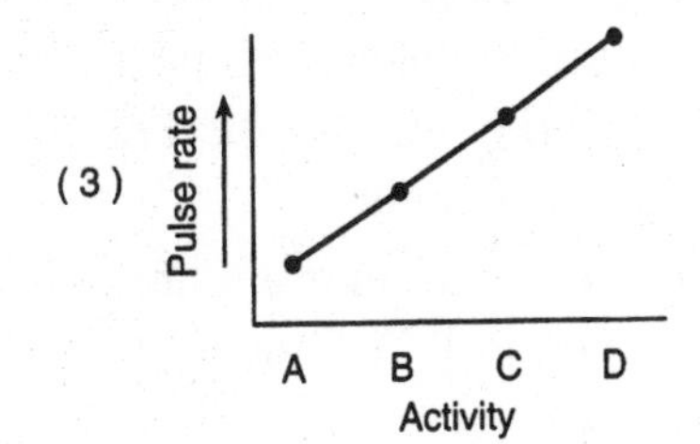

(4)

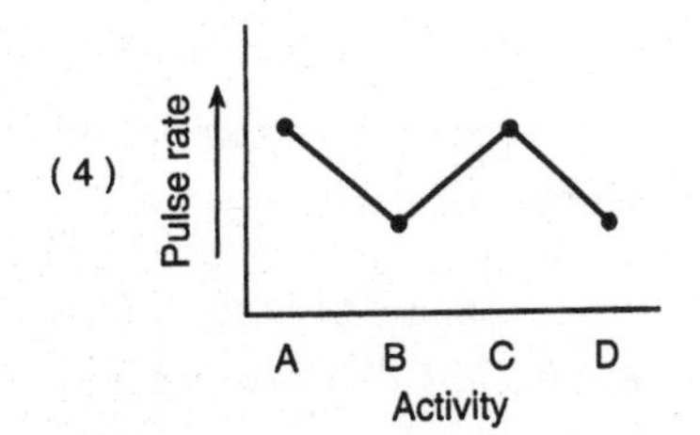

Base your answers to questions 50 through 52 on the following information and data table and on your knowledge of biology.

It has been hypothesized that a chemical known as BW prevents colds. To test this hypothesis, 20,000 volunteers were divided into four groups (of 5000 each). Each volunteer took a small pill every morning for one year. The contents of the pill taken by the members of each group are shown in the table below.

Group	Number of Volunteers	Contents of Pill	Percent Developing Colds
1	5000	5 grams of sugar	20
2	5000	5 grams of sugar 1 gram of BW	19
3	5000	5 grams of sugar 3 grams of BW	21
4	5000	5 grams of sugar 9 grams of BW	15

50. Which factor most likely had the greatest influence on these experimental results? A. color of the pills B. amount of sugar added C. number of volunteers in each group D. health history of the volunteers

51. Which statement is a valid inference based on the results? A. Sugar reduced the number of colds. B. Sugar increased the number of colds. C. BW is always effective in the prevention of colds. D. BW may not be effective in the prevention of colds.

52. Which group served as the control in this investigation? A. Group 1 B. Group 2 C. Group 3 D. Group 4

Base your answers to questions 53 through 56 on the information and data table below and on your knowledge of biology.

A student added two species of single-celled organisms, *Paramecium caudatum* and *Didinium nasutum*, to the same culture medium. Each day, the number of individuals of each species was determined and recorded. The results are shown in the table below.

Culture Population

Day	*Number of Paramecium*	*Number of Didinium*
0	25	2
1	60	5
2	150	10
3	50	30
4	25	20
5	0	2
6	0	0

53. Use the information in the table to construct a line graph on a copy of the following grid. Mark a scale on the axis labeled "Number of Individuals" that is appropriate for the plotted *Didinium* population and for plotting the *Paramecium* population.

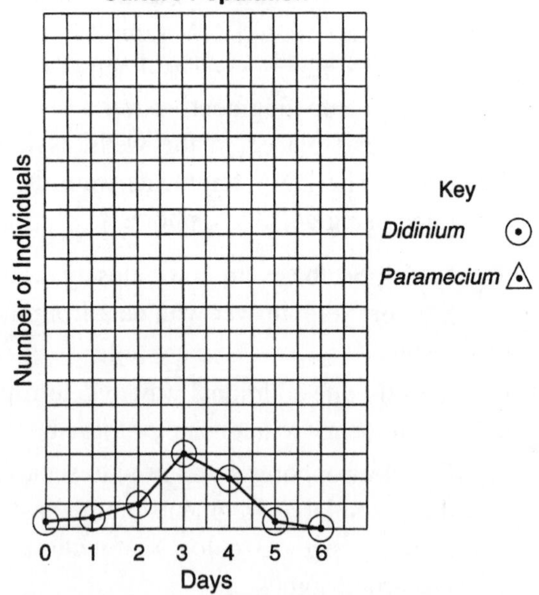

54. Plot the data for *Paramecium* on the grid. Surround each data point with a small triangle and connect the points.

55. What evidence in the data indicates that *Didinium* could be a predator of the *Paramecium*?

56. State *two* possible reasons why the two populations died off between days 4 and 6.

57. Molecules *A* and *B* are both organic molecules found in many cells. When tested, it is found that molecule *A* cannot pass through a cell (plasma) membrane, but molecule *B* easily passes through. State one way the two molecules could differ that would account for the differences in their ability to pass through the plasma membrane.

58. If vegetables become wilted, they can often be made crisp again by soaking them in water. However, they may lose a few nutrients during this process. Using the concept of diffusion and concentration, state why some nutrients would leave the plant cells.

Base your answer to question 59 on the information and diagram below.

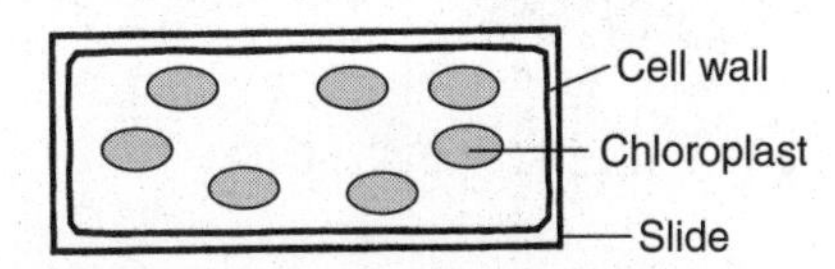

Elodea cell in freshwater

Elodea is a plant that lives in freshwater. The diagram represents one Elodea leaf cell in its normal freshwater environment.

59. Show how the contents of the Elodea cell would change if the cell were placed in saltwater for several minutes by completing, in your notebook, a copy of the diagram "Elodea cell in saltwater," shown below. Draw and label the location of the cell membrane (not shown in either diagram).

Elodea cell in saltwater

60. A scientist conducted an experiment in which he fed mice large amounts of the amino acid cysteine. He observed that this amino acid protected mouse chromosomes from damage by toxic chemicals. The scientist then claimed that cysteine, added to the diet of all animals, would protect their chromosomes from damage. State whether or not this is a valid claim. Support your answer.

Practice Test 1

SESSION 1

DIRECTIONS

This session contains twenty-one multiple-choice questions and two open-response questions. You may work out solutions to multiple-choice questions in your notebook.

1 Which of the following would be the complementary **mRNA** sequence for the DNA sequence GAACCT?

A. GAACCU
B. CTTGGA
C. CUUGGA
D. GAACCT

2 What is the first step that occurs during protein synthesis?

A. The mRNA attaches to a ribosome.
B. Amino acids are linked together.
C. The tRNA attaches to a codon.
D. DNA serves as a template for mRNA.

3 Which of these would be the **most** important limiting factor for organisms in a desert?

A. predators
B. temperature
C. water
D. wind

4 Which parts of the DNA molecule would be held together by weak hydrogen bonds?

A. phosphate and sugar
B. phosphate and adenine
C. cytosine and sugar
D. cytosine and guanine

Base your answers to questions 5 and 6 on the figure below.

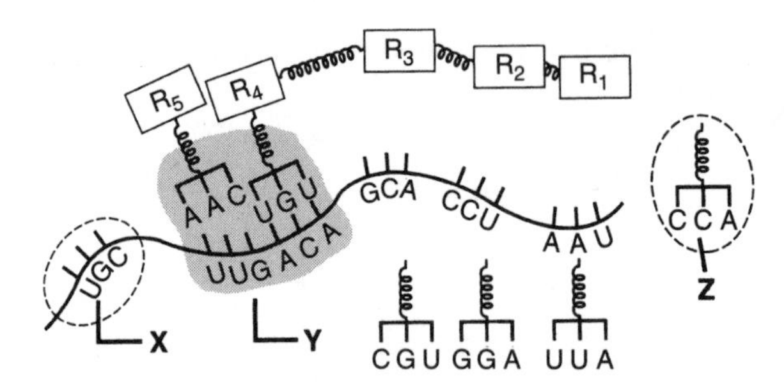

5 What structure is represented by the letter *Y*?

A. mitochondria
B. chloroplast
C. nucleus
D. ribosome

6 The original template for this process would be

A. ribosomal RNA
B. a DNA strand
C. transfer RNA
D. messenger RNA

7 There is a loss of energy at each trophic level. Which process brings new energy into an ecosystem?

A. decomposition
B. photosynthesis
C. nitrogen fixation
D. ecological succession

You may work out solutions to multiple-choice questions 8 through 11 in your notebook.

The following section focuses on a procedure that is used in genetic engineering. Read the information and study the diagram below and use the data to answer the four multiple-choice questions and one open-response question that follow.

Genetic engineering involves the transfer of genetic material from one organism to another. This recombination of genes results in the formation of recombinant DNA. Using gene-splicing techniques, donor genes from one organism can be inserted into the DNA of another organism.

Human genes that control the synthesis of important proteins have been introduced into bacterial cells, where they function as part of the bacterial DNA. In this way, bacterial cells are being used to synthesize some substances needed by humans.

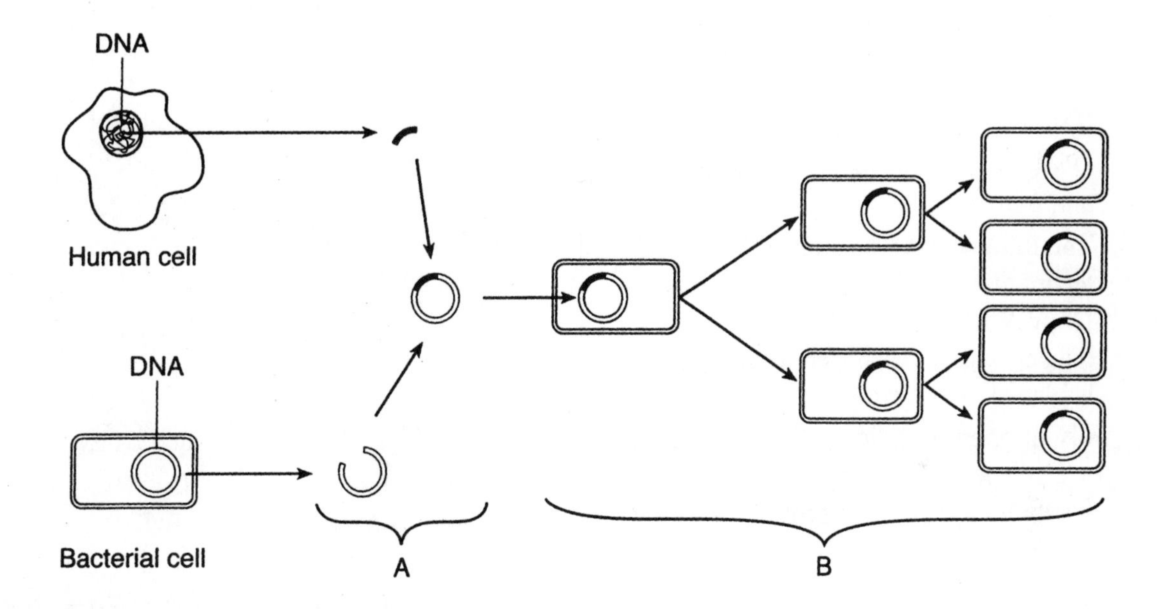

8 In the procedure indicated by letter *A*, the DNA segments from human and bacterial DNA are joined by the action of which of the following?

A. starch molecules
B. simple sugars
C. enzymes
D. hormones

9 Which process is indicated by letter *B*?

A. natural selection
B. asexual reproduction
C. sexual reproduction
D. gene deletion

10 In step *A*, a small piece of human DNA is inserted into a piece of bacterial DNA. What is this naturally occurring circular piece of DNA called?

A. a plasmid
B. a hormone
C. a donor
D. an enzyme

11 The segment of bacterial DNA that receives the donor DNA is said to function as a (an)

A. hormone
B. vector
C. protein
D. antigen

Question 12 is an open-response question.

- **BE SURE TO ANSWER AND LABEL ALL PARTS OF THE QUESTION.**
- **Show all your work (diagrams, tables, or computations).**
- **If you do the work in your head, explain in writing how you did the work.**

12 The techniques of genetic engineering make it possible to treat some diseases caused by human genetic disorders.

a. Provide an example of a disease treated by a product of genetic engineering. Then explain the role of each of the following items in making recombinant DNA: *restriction enzymes, plasmids, and ligase.*

b. The following is a scrambled list of the techniques used in making recombinant DNA. Write these steps in the correct sequence and, for each step, explain why it is placed in that order.

- Cut open plasmid with restriction enzyme.
- Obtain synthesized protein from the bacteria.
- Clone bacterial cells with rDNA plasmids.
- Insert donor DNA into the open plasmid.
- Cut out donor DNA with restriction enzyme.
- Add ligase to bond donor DNA and plasmid.

You may work out solutions to multiple-choice questions 13 through 22 in your notebook.

13 A change in a species' genetic code that may affect its survival can result from

A. a struggle for food
B. competition
C. a lower birth rate
D. a mutation

14 Which pyramid below would be considered **most** energy efficient?

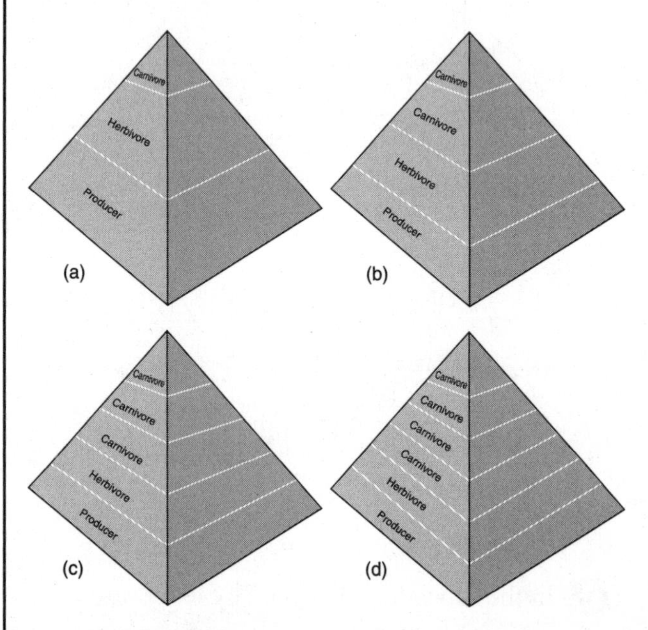

A. pyramid *a*
B. pyramid *b*
C. pyramid *c*
D. pyramid *d*

15 The Principle of Independent Assortment states that genes for different traits

A. exchange alleles during gamete formation
B. separate during gamete formation
C. stay together during gamete formation
D. separate after gamete formation

16 Two black mice mated and produced 24 black and 8 white offspring. Black hair is dominant to white hair. What can you determine about the genes of both parents?

A. Both parents are homozygous dominant.
B. Both parents are homozygous recessive.
C. Both parents are heterozygous.
D. One parent is heterozygous; the other is homozygous dominant.

17 Which is the **most** correct statement concerning amino acids?

A. They are the waste products of protein synthesis.
B. They are the molecular building blocks of starch.
C. They are stored as fat molecules in the human liver.
D. They are the molecular building blocks of proteins.

18 The diagram below represents the evolutionary relationships of several species. Which statement about these species is correct?

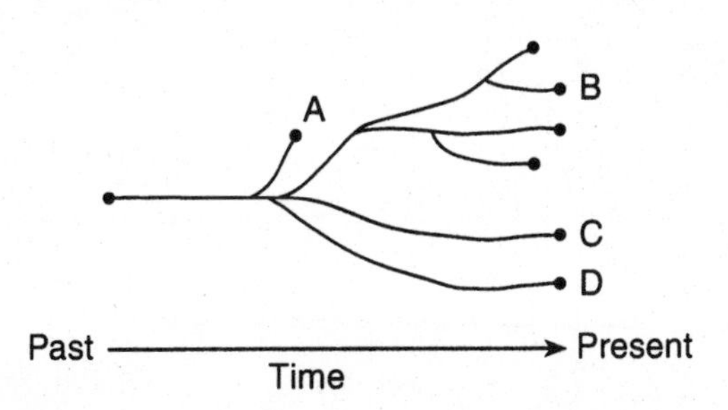

A. Species *A*, *B*, *C*, and *D* came from totally different ancestors.
B. Species *B* evolved directly from ancient species *A*.
C. Species *A*, *B*, and *C* can now interbreed successfully.
D. Species *C* is more closely related to species *D* than to species *B*.

19 A brown mink and a silver mink mated and produced brown offspring. When the brown offspring were crossed with each other, the phenotypic ratio of their offspring was 3 brown to 1 silver. These results are **best** explained by

A. independent assortment and crossing-over
B. dominance, segregation, and recombination
C. codominance, segregation, and recombination
D. incomplete dominance inheritance

20 Haploid gametes are produced by which of the following processes?

A. mitosis
B. meiosis
C. fertilization
D. fission

21 A certain organic compound contains only the elements carbon, hydrogen, and oxygen in a ration of 1:2:1. This compound is probably a

A. nucleic acid
B. monosaccharide
C. protein
D. fatty acid

22 ATP is the compound that is synthesized when

A. chemical bonds are formed between carbon atoms during photosynthesis
B. energy stored in chemical bonds is released during cellular respiration
C. energy stored in nitrogen molecules is released, forming proteins
D. digestive enzymes break down DNA into smaller parts

Question 23 is an open-response question.

- **BE SURE TO ANSWER AND LABEL ALL PARTS OF THE QUESTION.**
- **Show all your work (diagrams, tables, or computations).**
- **If you do the work in your head, explain in writing how you did the work.**

 The diagram below shows some interactions between several organisms located in a meadow environment.

a. A rapid **decrease** in the frog population results in a change in the hawk population. State how the hawk population may change. Support your answer.

b. Identify **one** cell structure found in the producers in this meadow ecosystem that is **not** found in the consumers. State the importance of this cell structure.

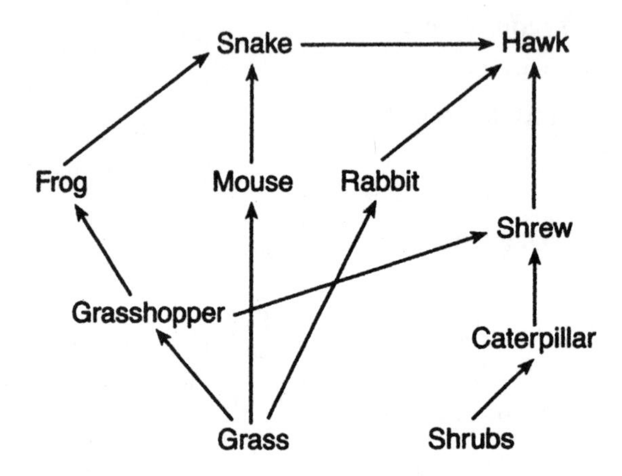

SESSION 2

DIRECTIONS

This session contains nineteen multiple-choice questions and three open-response questions. You may work out solutions to multiple-choice questions in your notebook.

24 Biologists have determined that the cell membrane is composed **mostly** of

A. proteins and starch
B. proteins and cellulose
C. lipids and starch
D. lipids and proteins

25 Which end product of respiration is of the greatest benefit to an organism?

A. glucose molecules
B. carbon dioxide
C. ATP molecules
D. water molecules

Base your answer to question 26 on the information and graph below.

The graph shows the relative amounts of product formed by the action of an enzyme in a solution with a pH of 6 at different temperatures.

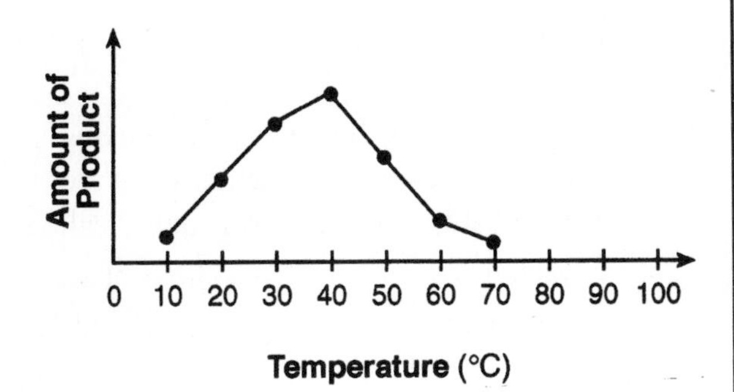

26 Based on the data given, you could predict that if the experiment is repeated at a pH of 4, the amount of product formed at each temperature will be

A. equal to the amount produced at pH 6
B. greater than the amount produced at pH 6
C. less than the amount produced at pH 6
D. impossible to predict based on this graph

27 Which feature do chloroplasts and mitochondria have in common?

A. They produce carbohydrates.
B. They contain their own DNA.
C. They recycle cellular materials.
D. They give structure to the cell.

28 What type of energy conversion occurs during photosynthesis?

A. light energy into heat energy
B. chemical energy into light energy
C. light energy into chemical energy
D. heat energy into light energy

29 Which change would **increase** the rate of photosynthesis in a plant?

A. a decrease in the amount of CO_2 available
B. an increase in the amount of CO_2 available
C. a decrease in the amount of H_2O available
D. an increase in the amount of O_2 available

30 To which group of organic compounds do starch and glycogen belong?

A. peptides
B. polypeptides
C. disaccharides
D. polysaccharides

31 A cell that is in the process of aerobic respiration

A. uses less oxygen than in anaerobic respiration
B. produces more ATP than in anaerobic respiration
C. uses less carbon dioxide than in anaerobic respiration
D. produces more alcohol than in anaerobic respiration

Question 32 is an open-response question.

- **BE SURE TO ANSWER AND LABEL ALL PARTS OF THE QUESTION.**
- **Show all your work (diagrams, tables, or computations).**
- **If you do the work in your head, explain in writing how you did the work.**

32 Smallpox is a disease caused by a specific virus, while the common cold can be caused by over 100 different viruses. It has been possible to develop a vaccine to prevent smallpox, but it is difficult to develop a vaccine to prevent the common cold.

a. Identify the substance in a vaccine that makes the particular vaccine effective.

b. Explain the relationship between a vaccine and white blood cell activity; state why the response of the immune system to a vaccine is specific.

c. State **one** reason why it is difficult to develop a vaccine against the common cold.

You may work out solutions to multiple-choice questions 33 through 43 in your notebook.

33 What type of organic molecule is the **most** preferred energy source for a cell?

A. lipids

B. proteins

C. carbohydrates

D. nucleic acids

34 The figure below could be used to illustrate all of the following types of transport except

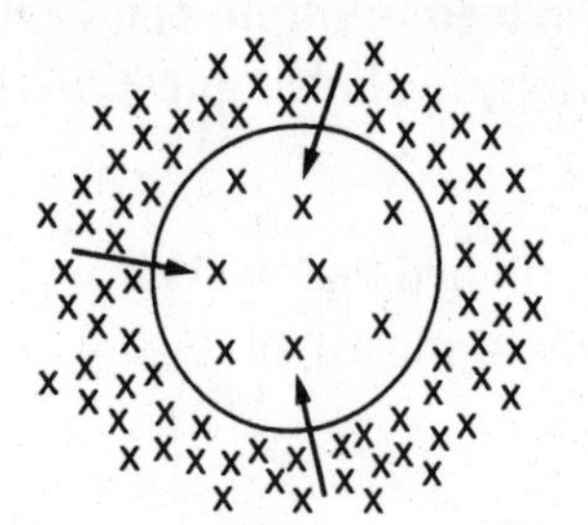

Figure A Substance moving into the cell

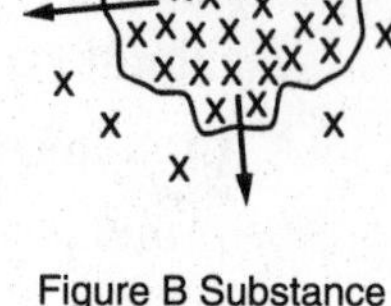

Figure B Substance moving out of the cell

A. diffusion

B. osmosis

C. active transport

D. passive transport

35 Which of the following is the complementary nitrogenous base for thymine on a strand of DNA?

A. adenine

B. cytosine

C. guanine

D. uracil

36 All the organelles in a cell work together to carry out which of these activities?

A. passive transport

B. active transport

C. information storage

D. metabolic processes

37 While viewing a slide of rapidly moving sperm cells, a student concludes that these cells probably have many organelles called

A. vacuoles

B. ribosomes

C. chloroplasts

D. mitochondria

38 Using a microscope, a botanist observed a plant cell in a drop of water. She added a 20 percent salt solution to the slide and then observed it again. She drew a diagram of the plant cell before (*A*) and after (*B*) the salt solution was added. Why does the plant cell look different in diagrams *A* and *B*?

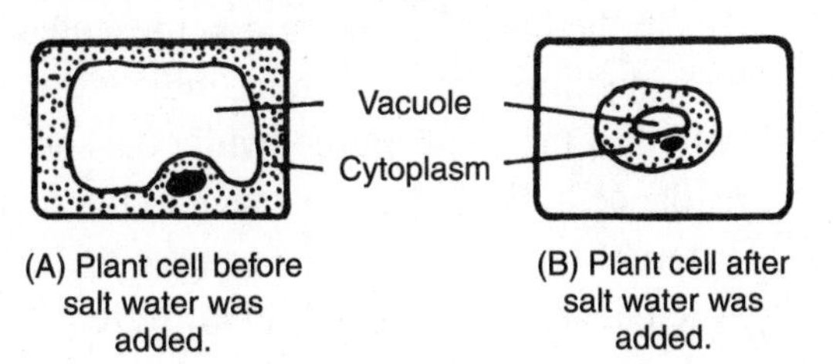

(A) Plant cell before salt water was added. (B) Plant cell after salt water was added.

A. More salt flowed out of the cell than into the cell.
B. More salt flowed into the cell than out of the cell.
C. More water flowed into the cell than out of the cell.
D. More water flowed out of the cell than into the cell.

39 Which of the following cells contains a membrane-bound nucleus?

A. prokaryotic cell
B. bacterial cell
C. viral cell
D. eukaryotic cell

40 Molecules will move across a cell membrane until their concentration

A. is hypotonic inside
B. is hypertonic inside
C. reaches equilibrium
D. causes plasmolysis

41 Which word equation represents the process of photosynthesis?

A. carbon dioxide + water → glucose + oxygen + water
B. glucose → alcohol + carbon dioxide
C. maltose + water → + glucose + glucose
D. glucose + oxygen → carbon dioxide + water

42 Which structures in the figure below enable an observer to identify it as a plant cell?

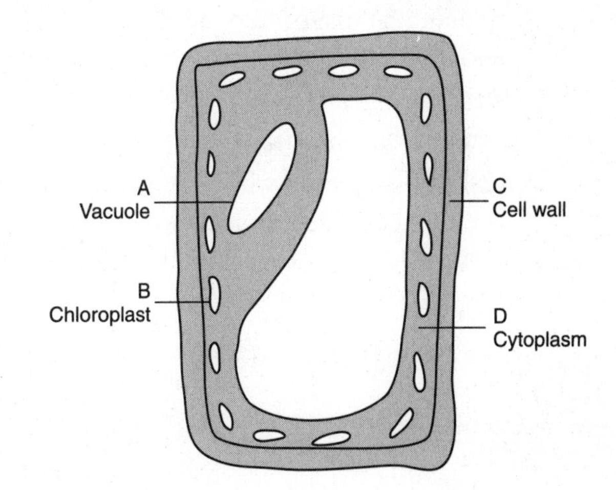

A. structures *A* and *B*
B. structures *B* and *C*
C. structures *A* and *C*
D. structures *B* and *D*

43 During protein synthesis, the amino acid sequence is determined by the

A. speed at which translation occurs
B. number of mitochondria in the cell
C. genetic code in the DNA molecule
D. number of ribosomes in the cell

Questions 44 and 45 are open-response questions.

- **BE SURE TO ANSWER AND LABEL ALL PARTS OF EACH QUESTION.**
- **Show all your work (diagrams, tables, or computations).**
- **If you do the work in your head, explain in writing how you did the work.**

44 Enzyme molecules are affected by changes in conditions within organisms. A prolonged, excessively high body temperature during an illness can be fatal to a person.

a. Explain the role of enzymes in the human body.

b. Describe the effect of a high body temperature on enzyme activity.

c. Identify the reason why a high body temperature can result in death.

Base your answers to question 45 on the diagram and information below.

45 Honeybees have a very cooperative way of living. Scout bees find food, return to the hive, and do the "waggle dance" to communicate the location of the food source to other bees in the hive. The waggle, represented by the wavy line in the diagram, indicates the direction of the food source, while the speed of the dance indicates the distance to the food. Different species of honeybees use the same basic dance pattern in slightly different ways, as shown in the data table.

a. What is the relationship between the distance to the food source and the number of waggle runs in 15 seconds?

b. Explain how waggle-dance behavior might **increase** the reproductive success of the bees.

c. The number of waggle runs for each distance (in 15 seconds) is very similar for both bee species. Explain why this similar behavior is **most likely** the result of natural selection.

Number of Waggle Runs in 15 Seconds		Distance to Food (feet)
Giant Honeybee	Indian Honeybee	
10.6	10.5	50
9.6	8.3	200
6.7	4.4	1000
4.8	2.8	2000

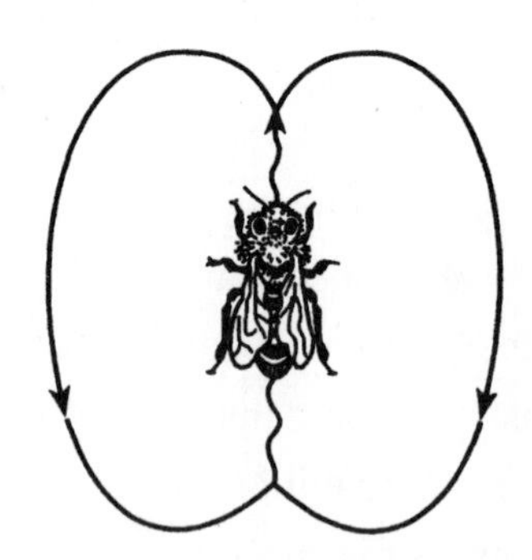

Practice Test 2

SESSION 1

DIRECTIONS

This session contains twenty-one multiple-choice questions and two open-response questions. You may work out solutions to multiple-choice questions in your notebook.

1 A plant cell is placed on a slide with salt water (the cell's "environment"). After several minutes, the cell shrinks due to the diffusion of

A. salt from the plant cell to the environment

B. salt from the environment into the plant cell

C. water from the plant cell to the environment

D. water from the environment into the plant cell

2 According to the graph below, about how many cells per milliliter (mL) of culture would be present after 90 minutes?

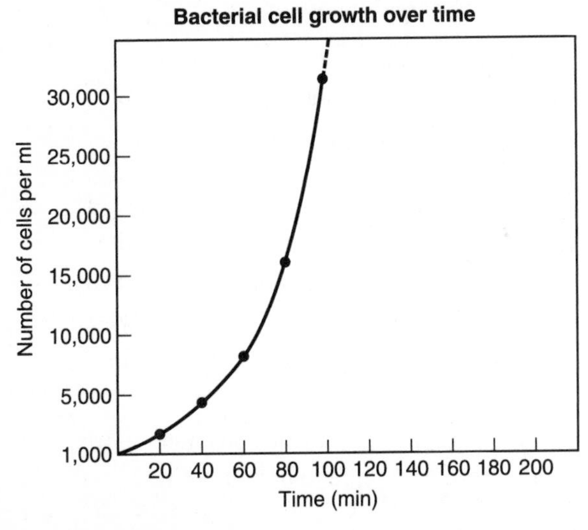

A. 9,000

B. 14,000

C. 22,000

D. 29,000

3 The movement of molecules into a cell is **most** dependent on which of the following?

A. selectivity of the cell membrane

B. selectivity of the cell wall

C. number of vacuoles in the cell

D. number of mitochondria in the cell

4 The graph below shows the world's fish catch in millions of metric tons (mmt) from 1950–1990. According to the graph, over that period of time the world's fish catch increased approximately how much?

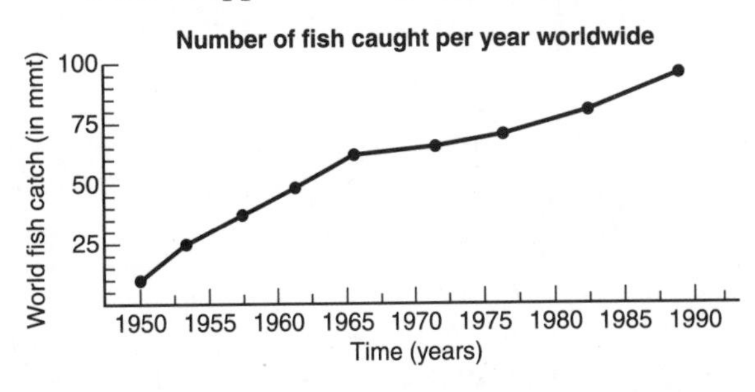

A. three times

B. five times

C. seven times

D. nine times

5 Andy conducted a science experiment and was upset that the results from the experiment did not support his original hypothesis. Andy should take which of the following actions?

A. Perform a different experiment to support his hypothesis.

B. Ignore the results and redo the same experiment to support it.

C. Repeat the experiment and change his hypothesis if necessary.
D. Publish the results of the experiment without repeating it.

6 Rachel knows that her cat and her friend's cat are both afraid of lightning. She concludes that all cats are afraid of lightning. Rachel's conclusion is based on

A. a controlled experiment
B. observation and reasoning
C. scientific research
D. emotional intuition

7 In the following diagram, what substance is represented by the letter *X*?

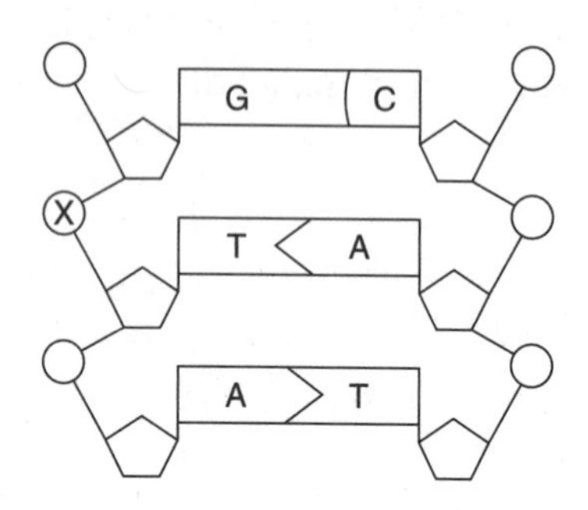

A. hydrogen
B. deoxyribose
C. phosphate
D. thymine

You may work out solutions to multiple-choice questions 8 through 11 in your notebook.

The following section focuses on the role of the pituitary gland in regulating the function of other endocrine glands. Read the information and study the diagram below and use the data to answer the four multiple-choice questions and one open-response question that follow.

Each arrow in the diagram represents a different hormone released by the pituitary gland that then stimulates the gland indicated in the diagram. All structures are present in the same organism.

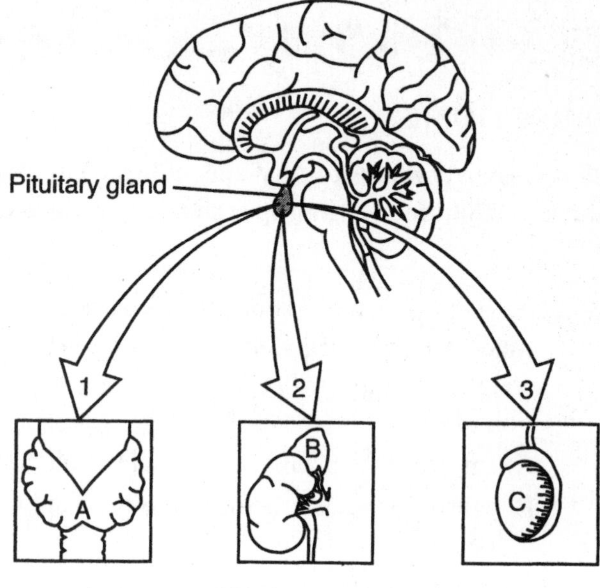

(Not drawn to scale)

8 The pituitary gland may release hormone 2 when blood pressure drops. Hormone 2 causes gland B to release a different hormone that raises blood pressure which, in turn, stops the secretion of hormone 2. The interaction of these hormones is an example of

A. a DNA base substitution
B. manipulation of genetic instructions
C. a negative feedback mechanism
D. an antigen-antibody reaction

9 What would **most likely** occur if the interaction is blocked between the pituitary and gland C, the site of meiosis in males?

A. The level of progesterone would start to increase.
B. The pituitary would produce another hormone to replace hormone 3.

C. Gland A would begin to interact with hormone 3 to maintain homeostasis.

D. The level of testosterone in the body may start to decrease.

10 Why does hormone 1 influence the action of gland A but not gland B or C?

A. Every activity in gland A is different from the activities in glands B and C.

B. The cells of glands B and C contain different receptors than the cells of gland A.

C. Each gland contains cells that have different base sequences in their DNA.

D. The distance a chemical can travel is influenced by both pH and temperature.

11 Which of these features is **not** associated with the endocrine system?

A. regulation of calcium levels in the blood

B. development of a goiter

C. secondary sex characteristics

D. reflex reactions of muscles

Question 12 is an open-response question.

- **BE SURE TO ANSWER AND LABEL ALL PARTS OF THE QUESTION.**
- **Show all your work (diagrams, tables, or computations).**
- **If you do the work in your head, explain in writing how you did the work.**

12 The pituitary gland is often called the "master gland" of the body.

a. Explain **why** the pituitary is thought of as a master gland. Give **one** example to justify your answer.

b. Identify **two** ways in which the functioning of the endocrine system is very different from that of the nervous system. In what way are the two systems similar?

You may work out solutions to multiple-choice questions 13 through 22 in your notebook.

13 A nerve cell needs to maintain a higher concentration of sodium ions inside the cell than outside of it. The sodium ions move across the cell membrane into the nerve cell by a process called

A. facilitated diffusion

B. passive transport

C. active transport

D. diffusion

14 Which substance would be **least** likely to pass through a cell membrane by diffusion?

A. carbon dioxide

B. glucose

C. oxygen

D. water

15 The pie chart below shows the proportion of elements in the human body. Which conclusion is best supported by this chart?

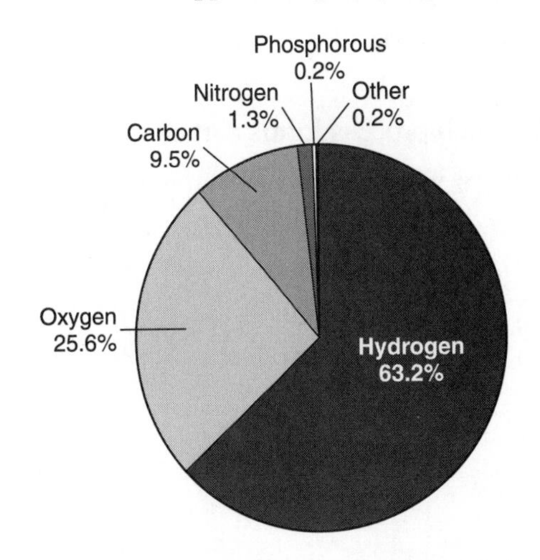

A. The human body is composed of six different elements.

B. Oxygen and carbon are the most abundant elements in the human body.

C. There is more nitrogen than phosphorus in the human body.
D. Hydrogen has a greater mass than carbon in the human body.

16 Which of these cells do **not** have cell walls?
A. plant cells
B. fungal cells
C. bacterial cells
D. animal cells

17 The cellular organelle that functions in the intracellular transport of cellular molecules is the
A. endoplasmic reticulum
B. cell membrane
C. ribosome
D. cell wall

18 The substances that most directly control the rate of reaction during cellular respiration are known as
A. enzymes
B. phosphates
C. monosaccharides
D. disaccharides

19 Which process forms carbon dioxide and water as waste products?
A. digestion
B. protein synthesis
C. cellular respiration
D. photosynthesis

20 The type of organism that is **not** included in the pyramid below is the

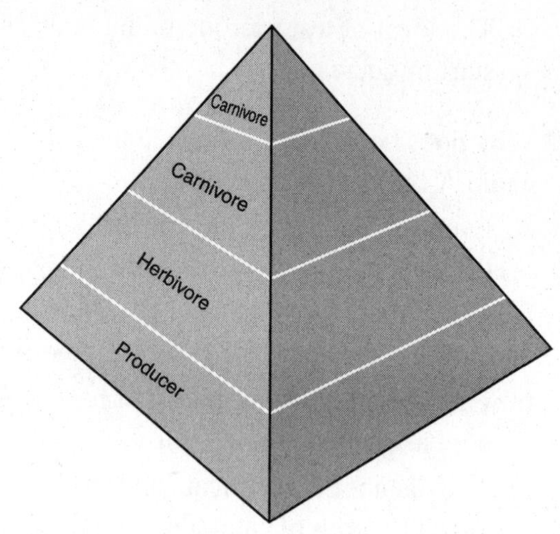

A. plant or alga
B. plant eater
C. meat eater
D. decomposer

21 In animal cells, the energy used to convert ADP to ATP comes directly from
A. hormones
B. sunlight
C. organic molecules
D. inorganic molecules

22 What determines the specific shape of a protein molecule?
A. whether it is organic or inorganic
B. the sequence of its amino acids
C. the number of lipids inside the protein
D. the number of chromosomes in the cell

Question 23 is an open-response question.

- **BE SURE TO ANSWER AND LABEL ALL PARTS OF THE QUESTION.**
- **Show all your work (diagrams, tables, or computations).**
- **If you do the work in your head, explain in writing how you did the work.**

23. The diagram below illustrates some steps in genetic engineering.

a. State **one** way that enzymes are being used in step 2.

b. What happens to the genetic information in step 3? Why is this process useful?

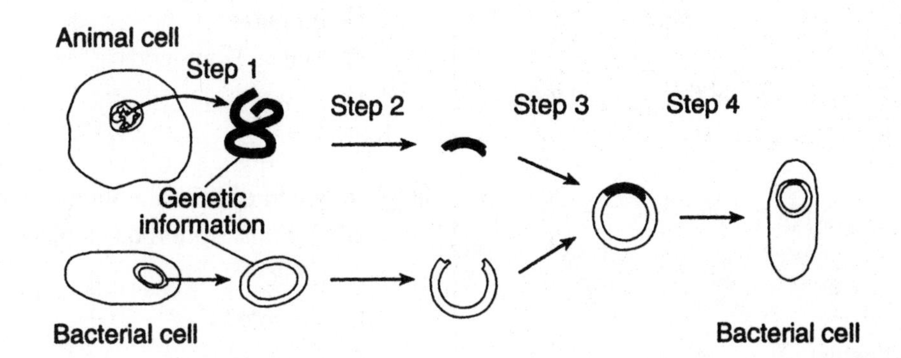

SESSION 2

DIRECTIONS

This session contains nineteen multiple-choice questions and three open-response questions. You may work out solutions to multiple-choice questions in your notebook.

24 Which substance stores glucose in the human liver?

A. chitin
B. starch
C. glycogen
D. cellulose

25 The process shown in the following diagram results in the formation of

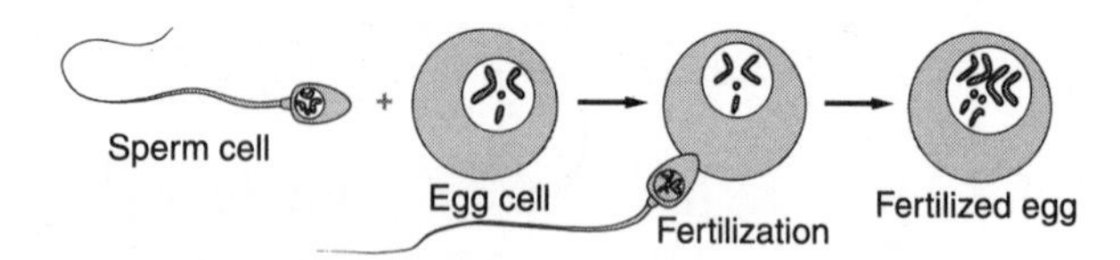

A. diploid cells
B. haploid cells
C. sex cells
D. gametes

26 Before a cell divides, its DNA and associated proteins condense to form the

A. chromosomes
B. genes
C. centrioles
D. centromeres

27 Of the events listed below, which mitotic event happens last?

A. the replication of chromosomes
B. disintegration of the nuclear membrane
C. appearance of the spindle fibers
D. separation of the chromatids by the spindles

28 Which process involves the pairing of a codon with an anticodon on a tRNA?

A. transcription
B. synthesis
C. replication
D. translation

29 Protists such as the amoeba reproduce by means of

A. spores
B. budding
C. binary fission
D. conjugation

30 In which of the following organelles does DNA transcription takes place?

A. nucleus
B. vacuole
C. ribosome
D. lysosome

31 RNA receives information from DNA by

A. binding with the DNA and forming a triple helix
B. matching with the complementary bases of a DNA strand
C. making an exact copy of the DNA molecule
D. accepting proteins through the nuclear membrane pores

Question 32 is an open-response question.

- **BE SURE TO ANSWER AND LABEL ALL PARTS OF THE QUESTION.**
- **Show all your work (diagrams, tables, or computations).**
- **If you do the work in your head, explain in writing how you did the work.**

Characteristics of Four Plant Species

Plant Species	*Seeds*	*Leaves*	*Pattern of Vascular Bundles (structures in stem)*	*Type of Chlorophyll Present*
A	round/small	needlelike	scattered bundles	chlorophyll a and b
B	long/pointed	needlelike	circular bundles	chlorophyll a and c
C	round/small	needlelike	scattered bundles	chlorophyll a and b
D	round/small	needlelike	scattered bundles	chlorophyll b

32 A series of investigations was performed on four different plant species. The results of these investigations are recorded in the data table above.

a. Based on these data, which **two** plant species appear to be the **most** closely related? Support your answer.

b. What additional information could be gathered to support your answer?

c. State **one** reason why scientists might want to know if two plant species are closely related.

You may work out solutions to multiple-choice questions 33 through 43 in your notebook.

33 Some concepts included in Darwin's theory of natural selection are represented in the chart below. Which concept would be correctly placed in box *X*?

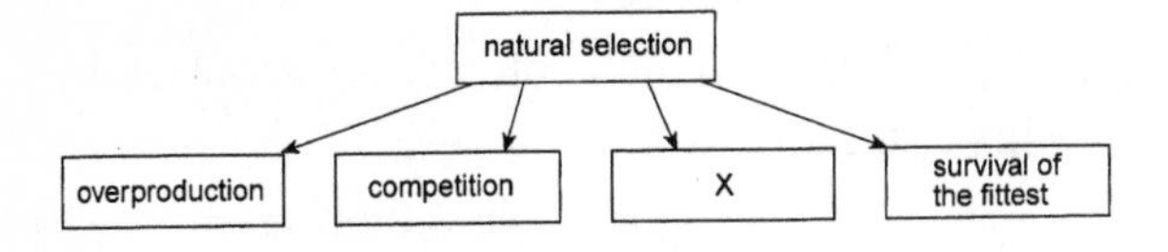

A. use and disuse
B. variation in traits
C. changes in nucleic acids
D. inheriting acquired traits

34 A DNA nucleotide unit consists of which of the following structures?

A. phosphate group, a six-carbon sugar, and a hydrogen base
B. phosphate group, a five-carbon sugar, and a nitrogenous base
C. phosphate group, a six-carbon sugar, and a nitrogenous base
D. hydrogen group, a five-carbon sugar, and a nitrogenous base

35 One way in which RNA differs from DNA is that

A. RNA is double stranded, while DNA is single stranded
B. RNA contains thymine, while DNA contains uracil
C. RNA is found only in the nucleus of eukaryotes
D. RNA contains ribose, while DNA contains deoxyribose

Base your answers to questions 36 and 37 on the graph below.

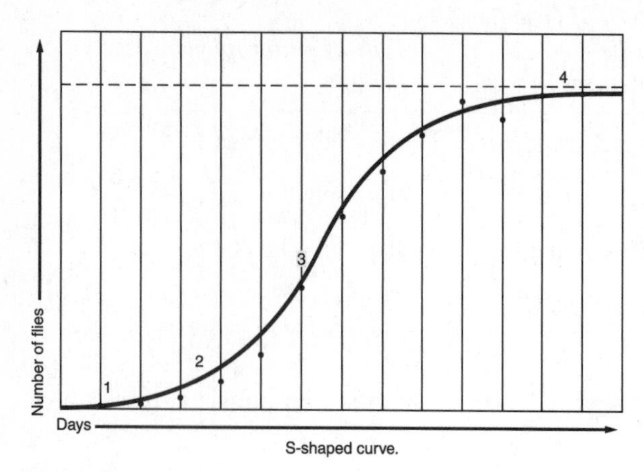

36 Which number represents the population in its most rapid phase of growth?

A. 1 C. 3
B. 2 D. 4

37 At which point has the population reached the carrying capacity?

A. 1 C. 3
B. 2 D. 4

38 The changes that occur in an ecosystem over a long period of time are called

A. natural selections
B. geological succession
C. ecological succession
D. biological diversity

39 Which set of body parts represents homologous structures?

A. wings of an insect and wings of a hummingbird
B. tentacles of an octopus and flippers of a whale
C. front legs of a butterfly and leg bones of a human
D. front leg bones of a dog and wing bones of a bat

40 Where do the final steps of aerobic cellular respiration occur?

A. along the endoplasmic reticulum
B. throughout the cytoplasm
C. on the surface of the ribosomes
D. inside the mitochondria

41 Which organisms in the food web below occupy similar trophic levels?

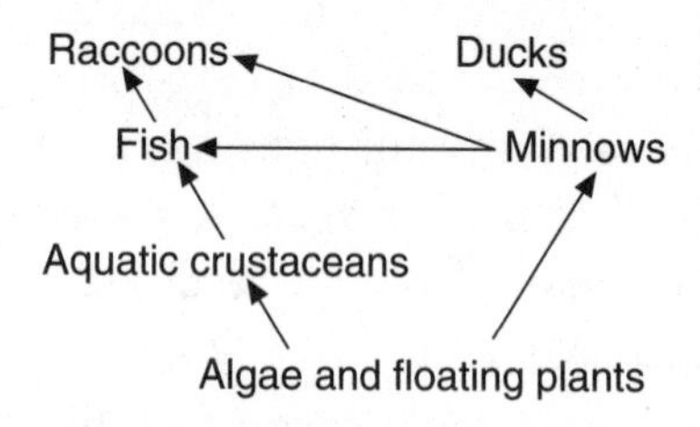

A. algae and minnows
B. ducks and crustaceans
C. raccoons and minnows
D. minnows and crustaceans

42 A population of squirrels is divided in two by a large valley. Over time, they become so different that they can no longer interbreed. This change in the two groups is due to

A. hybridization
B. fossilization
C. geographic isolation
D. primary succession

43 A current idea in the field of classification divides life into three broad categories, as shown below. Which concept is supported by this diagram?

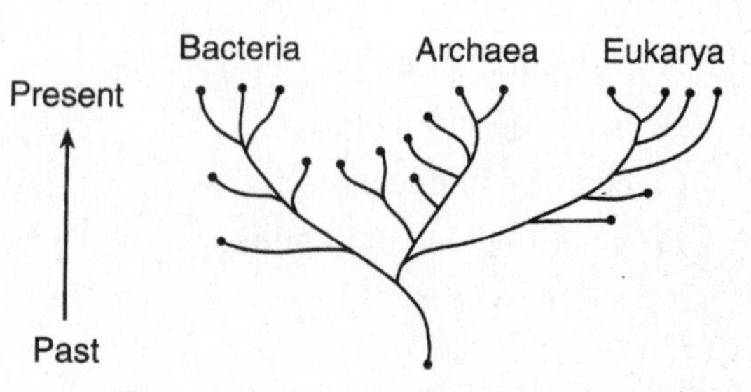

A. All evolutionary pathways eventually lead to present-day organisms.
B. Evolutionary pathways proceed in only one set direction over time.
C. All evolutionary pathways continue for the same total length of time.
D. Evolutionary pathways diverge and proceed in many different directions.

Questions 44 and 45 are open-response questions.

- **BE SURE TO ANSWER AND LABEL ALL PARTS OF EACH QUESTION.**
- **Show all your work (diagrams, tables, or computations).**
- **If you do the work in your head, explain in writing how you did the work.**

Base your answer to question 44 on the diagram and information below.

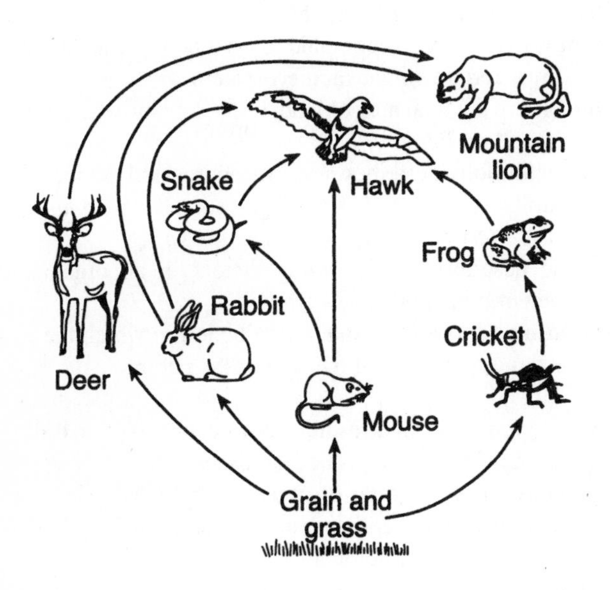

44 The organisms in the food web live near large cattle ranches. Over many years, mountain lions occasionally killed a few cattle. One year, a few of the ranchers hunted and killed many mountain lions to prevent future loss of their cattle. Later, many ranchers noticed that animals from this food web were eating large amounts of grain from their fields.

a. Identify **two** specific populations that most likely increased in number after the mountain lion population decreased (from being hunted). Support your answer.
b. What kind of consumers are the **two** organisms you identified? What role does the mountain lion fill in this food web?
c. Explain how the killing of mountain lions affected other ranchers in the community.

45 Just like complex organisms, cells are able to survive by coordinating various activities. Complex organisms have a variety of systems, and cells have a variety of organelles, that work together for survival.

a. State the names of **two** organelles and describe the function of each one.
b. Provide an explanation of how these **two** organelles work together in the cell.
c. Identify **one** organelle and **one** system in the human body that have similar functions.

Glossary

abiotic describes all the nonliving factors in an organism's environment, e.g., soil and water

acquired immunity type of immunity people get when they develop antibodies (i.e., active immunity) or receive antibodies (i.e., passive immunity) to a particular antigen or pathogen

active immunity type of immunity people get when exposed to an antigen or pathogen directly, or in a vaccine, and then develop antibodies to it

active transport movement of substances across a membrane from an area of lower concentration to an area of higher concentration; requires energy

adaptations special characteristics that make an organism well suited for a particular environment

AIDS (acquired immunodeficiency syndrome) an immunodeficiency disease, caused by HIV in humans

algae plantlike organisms, often single-celled, that carry out photosynthesis

alleles the two different versions of a gene for a particular trait

allergic reactions conditions caused by an overreaction of the immune system to antigens

alternation of generations the life cycle of plants that includes sexually reproducing and asexually reproducing generations

amino acids organic compounds that are the building blocks (subunits) of proteins

antibiotics chemicals that kill specific microorganisms (usually bacteria); frequently used to combat infectious diseases in people and animals

antibodies protein molecules that are produced to defend against foreign objects in the body; antibodies bind to specific antigens

antigens protein molecules on the surface of a foreign object that are detected by the immune system, causing it to produce antibodies

artificial selection see *selective breeding*

asexual reproduction the process that requires only one parent to pass on genetic information to offspring, e.g., budding

atmosphere the blanket of gases that covers Earth; usually called "air"

atoms the smallest units of an element that can combine with other elements; make up all matter

ATP (adenosine triphosphate) the substance (derived from energy stored in glucose) used by cells as an immediate source of chemical energy

autoimmune diseases occur when an overactive immune system starts to attack its own normal body tissues

autotrophic describes a self-feeding (producer) organism such as a plant that obtains its energy from inorganic sources

bacteria single-celled organisms that have no nuclear membrane to contain their DNA molecule

bacteriophages refers to viruses that infect, and destroy, bacterial cells

B cells special kinds of white blood cells that respond to specific antigens by producing antibody proteins that bind to it; include memory B cells

behavior every action that an animal takes, either learned or instinctive; usually to aid survival

biochemistry the chemistry of living things; provides evidence of common ancestry (evolution)

biodiversity the variety of different species in a community or ecosystem

biology the study of living things

biomes very large geographic areas that are characterized by a certain climate and ecosystem

biosphere the total area of land, water, and air where life is found on Earth

biotechnology describes new procedures and devices that utilize discoveries in biology; see also *recombinant DNA technology* and *genetic engineering*

biotic describes all the living factors in an organism's environment or ecosystem

budding a form of asexual reproduction in which the offspring grows out of the side of the parent

cancer a disease that results from uncontrolled cell division, which damages normal tissues

carbohydrates a group of organic compounds (includes simple sugars and starch), made up of hydrogen, oxygen, and carbon

carbon one of the six most important chemical elements for living things; carbon atoms form the backbone of nearly all organic compounds

carbon dioxide the inorganic molecule from which plants get carbon for photosynthesis; waste product of cellular respiration; a greenhouse gas

carnivores animals that obtain their energy by eating other animals; see also *consumers* and *heterotrophic*

carrying capacity the size of a population that can be supported by an ecosystem

catalysts substances that increase the rate of a chemical reaction, but are not changed during the reaction

cell the smallest living unit of an organism; all organisms are made up of at least one cell

cell cycle the complete sequence of events in a cell's life, from formation to cell division or death

cell membrane a selectively permeable membrane that separates the inside and outside of a cell and regulates the substances that pass through it; see also *plasma membrane*

cell theory the idea that all organisms are made up of one or more cells, which are the basic unit of structure and function, and which come only from other cells

cellular respiration the (aerobic) process that uses oxygen to create ATP for energy use by cells; see also *respiration*

central nervous system the brain and nerve (spinal) cord

centrifuge equipment used to separate materials of different densities from one another

chloroplasts the organelles in plants and algae that contain chlorophyll pigment and carry out photosynthesis

chromosomes structures composed of DNA that contain the genetic (hereditary) material

circulation the movement of blood throughout the body of an animal

classification the grouping and naming of organisms according to their evolutionary relationships and shared characteristics

cloning the production of identical genetic copies of an organism by asexual reproduction (or from the cell of another individual)

codon a sequence of three nucleotide bases that makes up the "word" coding for an amino acid

community all the populations of different species that interact within a particular area

competition the struggle between organisms for limited resources such as food and space

compound light microscope tool that uses two lens types to magnify the image of a specimen

consumers organisms that obtain their energy by feeding on other organisms; heterotrophs

coordination the means by which body systems work together to maintain homeostasis

cytokinesis the division of the cytoplasm, during and after mitosis, to form two new cells

cytoplasm the watery fluid (gel) that fills a cell, surrounding its organelles

decomposers heterotrophic organisms that obtain their energy by feeding on decaying organisms

density the number of individuals in a population in a given area; affects population growth

dependent variable in an experiment, the change that occurs because of the independent variable

development the changes that occur in an organism, from fertilization (the zygote) until death

deviations changes in the body's normal functions that are detected by control mechanisms, which maintain a balanced internal environment

differentiation the formation of specialized cells and tissues from less specialized parent cells through controlled gene expression

diffusion the movement of molecules from an area of higher concentration to an area of lower concentration

digestion the process of breaking down food particles into molecules small enough to be absorbed by cells

disease a disruption of homeostasis; a condition in a living body that impairs normal functioning

diversity the variety of different traits in a species or different species in an ecosystem

DNA (deoxyribonucleic acid) the nucleic acid molecules that contain the hereditary material and instructions for cellular activities in all organisms

dominant the allele that is expressed, i.e., that shows its trait even if only one copy is present

dynamic equilibrium in the body, a state of homeostasis in which conditions fluctuate yet always stay within certain limits

ecology the biological study of the interactions of living things and their environment

ecosystem includes all the living and nonliving factors that interact in a specific area

egg the female gamete that supplies half the genetic information to the zygote

electrophoresis technique by which DNA fragments are separated in a charged gel (to compare DNA from one person, or organism, to another)

element a substance that cannot be broken down into any simpler substances

embryo an organism in an early stage of development before it is born or germinated; the fertilized egg cell

energy what is required continuously by all living things to stay organized and remain alive

energy pyramid describes the flow of energy through an ecosystem; most energy is at the base (producers) and decreases at each higher level (consumers)

environment the physical surroundings of an organism, with which it interacts

enzymes proteins that act as catalysts to increase the rate of chemical reactions in living things

equilibrium in organisms, a balance in substances inside and outside the cell; in ecosystems, an overall stability in spite of cyclic changes

estrogen in females, along with progesterone, a major sex hormone that influences secondary sex characteristics and reproduction

eukaryotic cells refers to cells that contain a nucleus and membrane-bound organelles

evolution the change in organisms over time due to natural selection acting on genetic variations that enable them to adapt to changing environments

excretion the process by which metabolic wastes are removed from the body

expression the use of information in a gene to produce a particular trait, which can be modified by interactions with the environment

external development occurs when an embryo develops in the outside environment, not in the female's body

external fertilization occurs when the sperm and egg nuclei fuse in the outside environment, not in the female's body

extinct describes when a species or group of organisms is no longer alive, e.g., all dinosaurs

extinction the complete disappearance of all living members of a species (or group of species)

feedback mechanism a system that reverses an original response to a stimulus when the desired condition is reached; important for homeostasis; also called *feedback loops*

fertilization in sexual reproduction, fusion of the nuclei of an egg cell and a sperm cell to form a zygote

fetus in humans, a developing embryo after the first three months of development

food chain the direct transfer of energy from one organism to the next

food web the complex, interconnecting food chains in a community

fossils the traces or remains of dead organisms that have been preserved by natural processes

fungi (singular, **fungus**) kingdom of heterotrophic organisms such as mushrooms that obtain their energy by feeding on decaying organisms

gametes the male and female sex cells that combine to form a zygote during fertilization

gene expression see *expression*

genes the segments of DNA that contain the genetic information for a given trait or protein

genetic code the triplet code, i.e., the sequences of three nucleotide bases (codons), which code for specific amino acids

genetic engineering recombinant DNA technology, i.e., the insertion of genes from one organism into the genetic material of another; see also *biotechnology*

genetics the branch of biology that deals with patterns of inheritance

genetic variation the differences among offspring in their genetic makeup, due to recombination

genome all the genes that are present in an organism

genotype the genetic combination that determines a trait

genus a group that has one or more closely related, different species classified within it

geographic isolation the physical separation of related populations, by a natural barrier such as a river or mountain, that prevents interbreeding and leads to speciation

geologic time earth's history divided into vast units of time by which scientists mark important changes in Earth's climate, surface, and life-forms

global warming an increase in the average atmospheric temperature of Earth due to more heat-trapping CO_2 in the air, which causes the "greenhouse effect"

glucose a simple sugar that has six carbon atoms bonded together; a subunit of complex carbohydrates

habit a common type of behavior that results from repeating an action over and over again

habitat the place in which an organism lives; a specific environment and its living community

herbivores animals that obtain their energy by eating plants; see also *consumers* and *heterotrophic*

hereditary describes the genetic information that is passed from parents to offspring

heterotrophic describes an organism (animal) that obtains its energy by feeding on other living things; see also *consumers*

heterozygous describes when the two alleles for a trait are different, i.e., hybrid

homeostasis the ability of living things to detect external changes and to maintain a constant (i.e., stable) internal environment

homozygous describes when the two alleles for a trait are the same

hormones chemical messengers released into the bloodstream, which bind with receptor proteins to cause long-lasting changes in the body

host the organism that a parasite uses for food and shelter; the cell in which a virus reproduces

Human Genome Project scientific effort that mapped all three billion base pairs of human DNA

hydrogen one of the six most important chemical elements for living things

immune system recognizes and attacks specific invaders, e.g., bacteria, to protect the body against infection and disease

immunity the ability to resist or prevent infection by a particular microbe

immunodeficiency disease occurs when the body's immune system is underactive because it has been weakened, e.g., by HIV

incomplete dominance when the heterozygous offspring shows an intermediate phenotype for a trait

independent variable in an experiment, the one factor that might explain the observation

infectious diseases refers to diseases caused by pathogenic microorganisms that may be passed from one person to another; see also *disease*

inheritance the process by which traits are passed from one generation to the next

innate immunity type of immunity that is present at birth and has nothing to do with exposure to pathogens or antigens

inorganic compounds that do not contain both carbon and hydrogen atoms; in cells, substances that allow chemical reactions to take place; in ecosystems, substances that are cycled between living things and the environment

instincts complex, inborn behaviors that aid survival, e.g., reflexes

insulin a hormone secreted by the pancreas that maintains normal blood sugar levels

intercellular fluid (ICF) the fluid that surrounds all cells and contains dissolved substances

internal development occurs when the embryo develops within the female's body

internal fertilization occurs when the sperm fertilizes the egg cell within the female's body

karyotype a photograph showing all 23 chromosome pairs from a human embryonic or fetal cell

kingdom the largest taxonomic group into which scientists categorize related living things

life functions the basic processes carried out by all living things; also called *life processes*

life span the length of time between the birth and death of an organism

limiting factors the specific environmental requirements that can limit where an organism lives

linked occurs when the genes for one trait are inherited along with genes for another trait, e.g., red hair and freckles, because they are on the same pair of chromosomes

lipids the group of organic compounds that includes fats and oils; store energy for long-term use

locomotion a property of living things; the ability to move from place to place; see also *movement*

malfunction occurs when an organ or body system stops functioning properly, which may lead to disease or death

meiosis the division of one parent cell into four daughter cells; reduces the number of chromosomes to one-half the normal number, i.e., produces gametes

membrane see *cell membrane*

metabolism the chemical reactions (building up; breaking down) that take place in an organism

microbes microscopic organisms that may cause disease when they invade another organism's body; microorganisms, e.g., bacteria and viruses

mitochondria the organelles in which the cell's energy is released

mitosis the division of one cell's nucleus (i.e., the chromosomes) into two identical daughter cell nuclei

molecules the smallest unit of a compound; made up of atoms held together by a chemical bond

movement the flow of materials between the cell and its environment, i.e., transport; a property of living things, i.e., locomotion

multicellular describes organisms that are made up of more than one cell

mutation an error in the linear sequence (gene) of a DNA molecule; a change in genetic material

natural selection the process by which organisms having the most adaptive traits for an environment are more likely to survive and reproduce

nerve cells in animals, the cells that receive and send nerve impulses to other nerve cells and to other types of cells; also called *neurons*

niche an organism's specific role in, or interaction with, its ecosystem; how an organism survives

nitrogen one of the six most important chemical elements for living things

nucleic acids the organic compounds that include DNA and RNA

nucleotides the building blocks, or subunits, of DNA and RNA; they include four types of nitrogen bases, which occur in two pairs

nucleus the dense region of a (eukaryotic) cell that contains the genetic material

nutrients important molecules in food, such as lipids, proteins, and vitamins

nutrition the life processes by which organisms take in and utilize nutrients

observation what is made when you notice a natural event; part of the scientific method

omnivores animals that consume both plant and animal matter; see also *consumers* and *heterotrophic*

organ a structure made up of similar tissues that work together to perform the same task, e.g., the liver; describes a level of organization in living things

organelles structures within a cell that perform a particular task, e.g., the vacuole

organic describes compounds that contain both carbon and hydrogen

organic compounds contain carbon and hydrogen; found in living things

organisms living things; life-forms

organ system a group of organs that work together to perform a major task, e.g., the digestive system

osmosis the diffusion of water molecules across a membrane

ovaries the female reproductive organs that produce the mature egg cells and secrete the female sex hormones

oxygen one of the six most important chemical elements for living things; released as a result of photosynthesis; essential to cellular (aerobic) respiration

pancreas gland that secretes pancreatic juice (contains enzymes that aid digestion) and insulin (maintains normal blood sugar levels)

parasites organisms that live in or on another organism (the host), deriving nutrients and energy from it, causing it harm

passive immunity type of immunity people get when antibodies are passed from one person to another, e.g., through the placenta to a fetus or by breast-feeding

passive transport movement of substances across a membrane; requires no use of energy

pathogens microscopic organisms that cause diseases, such as certain bacteria and viruses; see also *microbes*

pedigree a chart that shows the inheritance patterns of certain genetic traits in a family for several generations

pesticides chemicals used to kill agricultural pests, mainly insects, some of which have evolved resistance to the chemicals

pH a measurement (on a scale of 0–14) of the hydrogen ion concentration of a solution, i.e., how acidic or basic it is

phenotype the physical expression of a trait in an organism

pheromones chemicals produced by animals that function in communication

phloem in plants, the vascular tissue that carries dissolved food and other substances from the leaves to the rest of the plant

photosynthesis the process that, in the presence of light energy, produces chemical energy (glucose) and water

placenta the structure that forms in the uterus of mammals to nourish a developing embryo and remove its waste products

plasma membrane separates the inside and outside of a cell; see also *cell membrane*

pollination in angiosperms, the transfer of pollen grains from the anther to the stigma

pollutants harmful substances that are not normally found in the environment

polygenic expression describes a trait that is determined by more than one gene, such as height

polymer a large molecule, such as blood protein hemoglobin, made up of smaller molecules

population all the individuals of the same species that live in the same area at one time

predator an organism that feeds on another living organism (the prey); a consumer

predator-prey a relationship in which the prey is usually killed right away by the predator

prey a living organism that is eaten by another organism (the predator)

producers the autotrophic organisms on the first trophic level, which obtain their energy from inorganic sources (and make organic compounds), usually by photosynthesis

progesterone in females, along with estrogen, a major sex hormone; see also *estrogen*

prokaryotic cells refers to cells that do not contain a nucleus or any membrane-bound organelles

proteins a group of organic compounds that are made up of chains of amino acids

protists the kingdom that includes eukaryotic unicellular and multicellular organisms that live in aquatic or moist environments; algae and protozoa

Punnett square a box-shaped chart that is used to show all possible combinations of gametes and their probabilities when two organisms are crossed

radiation a form of energy that can cause genetic mutations in sex cells and body cells, e.g., x-rays

receptors molecules that play an important role in the interactions between cells, e.g., molecules that bind with hormones

recessive the allele that is hidden, i.e., that does not show its trait, unless two copies are present

recombinant DNA technology describes the methods used to remove and join (i.e., recombine) genes from one cell to another; see also *genetic engineering*

recombination the formation of new combinations of genetic material, from crossing-over between chromosomes during meiosis or from genetic engineering; produces greater genetic variability

reduction division during meiosis, a diploid cell divides to produce two new cells with the haploid number of chromosomes

reflex a type of innate behavior that happens very quickly; aids survival

replicate the process by which DNA in the nucleus of a cell makes a copy of itself during mitosis and protein synthesis

reproduction the production of offspring (i.e., passing on of hereditary information), either by sexual or asexual means

reproductive isolation situation in which a population is physically separated from others of its kind and changes so much that it can no longer interbreed with them; leads to speciation

residue the remains of dead organisms, which are recycled in ecosystems by the activities of bacteria and fungi

resistance describes natural (genetic) ability of some bacteria to survive exposure to antibiotics

response an organism's reaction to an internal or external stimulus; can be inborn or learned

respiration in the lungs, the process of exchanging gases; in cells, the process that releases the chemical energy stored in food; see also *cellular respiration*

restriction enzyme recognizes a specific sequence of four to six base pairs and cuts the DNA

ribosomes the organelles at which protein synthesis occurs, and which contain RNA

RNA (ribonucleic acid) the nucleic acid molecules that help translate genetic information and carry out protein synthesis; the genetic material of some viruses

saprophytes organisms that obtain nutrients from dead or decaying organisms, e.g., the fungi

scavengers animals that feed on the remains of a dead animal, rather than hunt and kill animals

science a body of knowledge about our world; the process of acquiring that knowledge

scientific inquiry the nature of scientific thinking and learning

scientific method an organized approach to problem solving

selective breeding the method by which humans encourage the development of preferred traits by allowing only those plants or animals that have those traits to breed; artificial selection

selective permeability the ability of a cell membrane to determine which molecules can pass through it

sex cells the male and female gametes, which fuse to form a zygote; each has one-half the normal chromosome number as a result of meiosis

sex-linked genes describes traits controlled by genes from the female parent because the genes on the X chromosome have no corresponding alleles on the Y chromosome

sexual reproduction the process that requires two parents (a male and a female) to pass on genetic information to offspring

simple sugars single sugars that have six carbon atoms, e.g., glucose

solar energy radiant energy from the sun, i.e., sunlight, which is a renewable resource

species a group of similar, related organisms that can breed and produce fertile offspring

sperm the male gamete that supplies half the genetic information to the zygote

stability in ecosystems, the tendency to resist change (usually, the greater the species diversity, the more stable the ecosystem); in organisms, the tendency to maintain constant internal conditions; see also *homeostasis*

starch a complex carbohydrate (polysaccharide) made up of many glucose molecules; used for energy storage in plants

stem cells human cells that have not yet developed into mature cells; have the ability to become almost any type of tissue

stimulus (plural, **stimuli)** any event, change, or condition in the environment that causes an organism to make a response, i.e., to react

subunits the four types of nucleotide bases that make up the DNA molecule

succession the gradual replacement of one ecological community by another until reaching a point of stability; usually a series of slow changes

symbiosis a close relationship between two or more different organisms that live together; each partner may either help, harm, or have no effect on the other partner

synthesis the building of compounds that are essential to life, e.g., protein synthesis

system an organized group of structures that works together to perform a task; see also *organ system*

tapeworms parasitic flatworms that absorb nutrients from the hosts directly through their bodies

taxonomy the branch of science that deals with the classification of organisms

T cells special kinds of white blood cells that detect and destroy cells that have been infected by microbes (i.e., killer T cells) or assist other B cells and T cells (i.e., helper T cells)

technology the process of using scientific knowledge and other resources to develop new products and processes

template in DNA replication, the original molecule that serves as the pattern to make a copy

territory the area in which an animal lives, and which it usually defends from other animals

testcross test conducted (cross with homozygous recessive) to determine the true genotype of a plant or animal that shows the dominant phenotype (physical trait)

testes (singular, **testis)** the pair of male reproductive organs that produces the sperm cells

testosterone in males, the main sex hormone that influences secondary sex characteristics and reproduction

theory a general statement supported by many scientific observations; represents the most logical explanation of the evidence

theory of evolution explains how all organisms have developed and changed over time; see also *evolution*

tissues a group of similar cells that work together to perform the same function; describes a level of organization in living things

toxins chemicals that can harm a developing fetus if taken during pregnancy; also, chemicals that may get passed, and increase, from one trophic level to the next as they move up the food chain

trait a genetic characteristic, determined by the combination of alleles

transcription the process by which a DNA strand is copied into an RNA strand

translation the process by which RNA functions to assemble amino acid sequences at the ribosome

transport involves the absorption and movement of materials throughout an organism's body

trophic levels the feeding levels through which energy flows in a food chain

unicellular describes single-celled organisms, i.e., organisms made up of only one cell

uterus in mammals, the reproductive organ that holds the developing embryo; also called *womb*

vaccination method used (usually injection) to prepare the immune system to fight off infection by a specific pathogen

vacuoles in plant and animal cells, the organelles that store materials, such as food and wastes

variability see *genetic variation*

vascular refers to specialized conducting tissues that transport materials throughout the bodies of plants and animals

vectors special molecules that can move pieces of DNA from the cell of one organism to another

virus a microscopic, nonliving particle of genetic material that can replicate only within a host cell, where it usually causes harm

wet mount describes technique used for staining and viewing a microscopic specimen

white blood cells several types of (immune system) cells that work to protect the body from disease-causing microbes and foreign substances

xylem in plants, the vascular tissue that carries water and dissolved minerals upward from the roots

zygote the fertilized egg cell that is formed when the nuclei of two gametes (a male and a female) fuse; contains the diploid number of chromosomes; develops into the embryo

Index